# METHODS IN MOLECULAR BIOLOGY™



*Series Editor*
**John M. Walker**
**School of Life Sciences**
**University of Hertfordshire**
**Hatfield, Hertfordshire, AL10 9AB, UK**

For other titles published in this series, go to
www.springer.com/series/7651

# Cell Cycle Checkpoints

## Methods and Protocols

Edited by

**Willis X. Li**

*Department of Medicine, University of California, San Diego, La Jolla, CA, USA*

Humana Press

*Editor*
Willis X. Li
Department of Medicine
University of California, San Diego
La Jolla, CA 92093-0063, USA
willisli@ucsd.edu

ISSN 1064-3745 e-ISSN 1940-6029
ISBN 978-1-4939-6181-8 ISBN 978-1-61779-273-1 (eBook)
DOI 10.1007/978-1-61779-273-1
Springer New York Dordrecht Heidelberg London



Printed on acid-free paper

Humana Press is part of Springer Science+Business Media (www.springer.com)

# Preface

Cell cycle checkpoints control the fidelity and orderly progression of eukaryotic cell division. These checkpoints prevent progression into the next phase of the cell cycle until the processes at the current phase have been properly completed. The most studied cell cycle checkpoints act at transition points, such as from G1 to S, G2 to M, or the metaphase to anaphase transition. Genotoxic stresses from byproducts of cellular metabolism or environmental sources can also activate cell cycle checkpoints, halting cell cycle progression until the damages are repaired.

By controlling the orderly progression of critical cell cycle events such as DNA replication and chromosome segregation, and ensuring proper repair of damaged DNA, cell cycle checkpoints function to ensure genome integrity. Cell cycle checkpoint pathways consist of sensors that detect DNA damage or incomplete cell cycle processes, signal transducers that halt cell cycle progression, and effectors that execute damage repair or a commitment to apoptosis, if the damage cannot be repaired.

Research in the past decade or so has elucidated many of the molecular mechanisms involved in various aspects of cell cycle checkpoints. Mechanisms of checkpoint control are not only the focus of investigators interested in cell cycle regulation, but are also of interest to researchers studying cancer development. It is becoming increasingly clear that loss of cell cycle checkpoints, which leads to genomic instability, is a hallmark of tumorigenesis.

The aim of *Cell Cycle Checkpoints* is to provide detailed descriptions of a wide variety of methodologies currently employed by researchers in the field of cell cycle checkpoints. The methodologies include those commonly used in the mammalian, yeast, *Caenorhabditis elegans*, *Drosophila*, and *Xenopus* model systems. Each chapter describes a specific technique or protocol, such as a method for inducing cell cycle checkpoints in a particular model system, for synchronizing a population of cells to allow observations of cell cycle progression, for identifying genes involved in checkpoint regulation, or for studying particular protein components of cell cycle checkpoint pathways. Most of the chapters are aimed at researchers new to the field of cell cycle checkpoints and describe every step of the methodology in sufficient detail to allow for successful application.

The editor wishes to thank the authors of all the chapters whose contributions made this volume possible, and Dr. John Walker for his valuable guidance and encouragement during the whole process of putting this book together.

*Willis X. Li*

# Contents

## Contributors

BYEONG J. CHA • *Lisa Muma Weitz Advanced Microscopy and Cell Imaging Laboratory, Department of Pathology and Cell Biology, College of Medicine, University of South Florida, Tampa, FL, USA*

MARILA CORDEIRO-STONE • *Department of Pathology, University of North Carolina, Chapel Hill, NC, USA*

YUN DAI • *Hematology/Oncology, Virginia Commonwealth University, Richmond, VA, USA*

JEAN M. DAVIDSON • *Department of Biology, University of North Carolina, Chapel Hill, NC, USA*

TOVAH A. DAY • *Dana Farber Cancer Institute, Boston, MA, USA*

MICHAEL DERAN • *Department of Biomedical Genetics, Department of Biochemistry and Biophysics, University of Rochester Medical Center, Rochester, NY, USA*

ROBERT J. DURONIO • *Department of Biology, Program in Molecular Biology and Biotechnology, Lineberger Comprehensive Cancer Center, University of North Carolina, Chapel Hill, NC, USA*

JEAN GAUTIER • *Institute for Cancer Genetics, Columbia University, New York, NY, USA*

MARA GLADSTONE • *Department of Molecular, Cellular and Developmental Biology, University of Colorado, Boulder, CO, USA*

STEVEN GRANT • *Hematology/Oncology, Virginia Commonwealth University, Richmond, VA, USA*

HAYDEN HOMER • *Mammalian Oocyte and Embryo Research Laboratory, Division of Biosciences and Institute for Women's Health, University College London, London, UK; Reproductive Medicine Unit, EGA/University College London, London, UK*

MINGXIA HUANG • *Department of Biochemistry and Molecular Genetics, University of Colorado School of Medicine, Aurora, CO, USA*

JANNIE DE JONG • *Department of Cell Biology, University Medical Center Groningen, University of Groningen, Groningen, The Netherlands*

JACOB W. P. KEELING • *Department of Biochemistry and Molecular Biology, Oklahoma State University, Stillwater, OK, USA*

LEI LI • *Department of Experimental Radiation Oncology, Department of Genetics, M. D. Anderson Cancer Center, The University of Texas, Houston, TX, USA*

WILLIS X. LI • *Department of Medicine, University of California, San Diego, La Jolla, CA, USA*

LILI LIU • *Department of Biochemistry and Molecular Genetics, University of Colorado School of Medicine, Aurora, CO, USA*

JAMES J. MANFREDI • *Department of Oncological Sciences, Mount Sinai School of Medicine, New York, NY, USA*

RITA K. MILLER • *Department of Biochemistry and Molecular Biology, Oklahoma State University, Stillwater, OK, USA*

MATTHEW J. O'CONNELL • *Department of Oncological Sciences, Mount Sinai School of Medicine, New York, NY, USA*
MARY PULVINO • *Department of Biomedical Genetics, University of Rochester Medical Center, Rochester, NY, USA*
LOIS RESNICK-SILVERMAN • *Department of Oncological Sciences, Mount Sinai School of Medicine, New York, NY, USA*
NICHOLAS RHIND • *Biochemistry and Molecular Pharmacology, University of Massachusetts Medical School, Worcester, MA, USA*
BUNSYO SHIOTANI • *Harvard Medical School, Massachusetts General Hospital Cancer Center, Charlestown, MA, USA*
ODY C.M. SIBON • *Department of Cell Biology, University Medical Center Groningen, University of Groningen, Groningen, The Netherlands*
KATARZYNA SIUDEJA • *Department of Cell Biology, University Medical Center Groningen, University of Groningen, Groningen, The Netherlands*
CHRIS SPROUL • *Department of Pathology, University of North Carolina, Chapel Hill, NC, USA*
SEETHA V. SRINIVASAN • *Institute for Cancer Genetics, Columbia University, New York, NY, USA*
TIN TIN SU • *Department of Molecular, Cellular and Developmental Biology, University of Colorado, Boulder, CO, USA*
SAEKO TAKADA • *Department of Biochemistry, Molecular Biology, and Biophysics, University of Minnesota, Minneapolis, MN, USA*
CLAUDIA TAPIA-ALVEAL • *Department of Oncological Sciences, Mount Sinai School of Medicine, New York, NY, USA*
CYRUS VAZIRI • *Department of Pathology, University of North Carolina, Chapel Hill, NC, USA*
YUCAI WANG • *Department of Experimental Radiation Oncology, Department of Genetics, M. D. Anderson Cancer Center, The University of Texas, Houston, TX, USA*
NICHOLAS WILLIS • *Biochemistry and Molecular Pharmacology, University of Massachusetts Medical School, Worcester, MA, USA*
XIAORONG WU • *Department of Biochemistry and Molecular Genetics, University of Colorado School of Medicine, Aurora, CO, USA*
SHIAN-JANG YAN • *Department of Medicine, University of California, San Diego, La Jolla, CA, USA*
JIYONG ZHAO • *Department of Biomedical Genetics, University of Rochester Medical Center, Rochester, NY, USA*
LEE ZOU • *Massachusetts General Hospital Cancer Center, Boston, MA, USA; Department of Pathology, Harvard Medical School, Charlestown, MA, USA*

# Chapter 1

# Studying G2 DNA Damage Checkpoints Using the Fission Yeast *Schizosaccharomyces pombe*

Nicholas Willis and Nicholas Rhind

## Abstract

Using synchronized cells, one can directly measure delay in mitosis brought about by the G2 DNA damage checkpoint in response to exposure to exogenous DNA damaging agents. Scoring mitosis in the fission yeast *Schizosaccharomyces pombe* is relatively simple. Many techniques exist for synchronizing cells for such assays. We present a detailed explanation of the setup and use of centrifugal elutriation to synchronize cells in G2, exposure of cells to DNA damage, and measurement of mitotic progression and delay.

**Key words:** G2 DNA damage checkpoint, mitosis, septation index, elutriation, bleomycin, ionizing radiation.

## 1. Introduction

The G2 DNA damage checkpoint responds to DNA damage during G2, after replication is completed. This checkpoint prevents the G2 to M phase transition in the presence of DNA damage, allowing cells sufficient time to repair DNA damage before undertaking nuclear and cellular division (1). Much of what is known about this checkpoint has been discovered in the budding yeast *Saccharomyces cerevisiae* and the fission yeast *Schizosaccharomyces pombe* (2, 3). This chapter presents a protocol for studying the response of fission yeast to G2 DNA damage.

Fission yeast is easily manipulated and offers a short cell cycle, allowing for easy determination of the G2 DNA damage cell cycle response. In particular, it is straightforward to follow the passage from G2 though mitosis by microscopy. Fission yeast is a rod-shaped organism that divides by medial fission. After

W.X. Li (ed.), *Cell Cycle Checkpoints*, Methods in Molecular Biology 782,
DOI 10.1007/978-1-61779-273-1_1, © Springer Science+Business Media, LLC 2011

mitosis, cytokinesis is accomplished by the construction of a septum through the middle of the cell. Septum formation may be scored in live yeast since the septum is easily recognized under phase microscopy (*see* **Fig. 2b**). Timing of mitosis and delay due to exposure to DNA damage may be measured by scoring synchronous cells as the population progresses through M phase and septates.

Several techniques allow for the synchronization of yeast in G2. They can be divided into two broad categories: block-and-release and selection synchronization. Block-and-release approaches include nutrient starvation, drug-induced arrest, and use of cell division cycle (*cdc*) mutants, all of which block cells at a particular point in the cell cycle and allow for the release of a synchronous culture. Selection synchronization techniques rely on separation of a synchronous subset of a culture. Centrifugal elutriation, lactose gradient centrifugation, and ficoll step gradient centrifugation all represent techniques allowing the selection of small and synchronized cells from an asynchronous culture (4–6). The selection techniques have two important advantages over block-and-release synchronization. First, they do not severely disrupt the cell cycle, making them more sensitive and less artifact prone. Second, they may be employed to separate healthy small cells from an otherwise sick population containing dead and permanently arrested cells. This advantage allows the analysis of strains, particularly repair defective strains that display a degree of growth heterogeneity that would otherwise confound cell cycle analysis.

Centrifugal elutriation allows cell synchronization without significant known artifacts and with minimal cell cycle delay (7). In particular, elutriation allows recovery of cells with minimum cell stress and without the use of *cdc* alleles or drug exposure. Although elutriation suffers from limitations, including that only single cultures may be synchronized at a time and that only a limited number of cells can be recovered, we find it an efficient and robust approach to synchronizing cells.

We present here how to assemble and operate a centrifugal elutriation system, methods permitting sterile harvesting of cells, and basic protocols to expose synchronized cells to increasing doses of ionizing and ultra-violet radiation. Additionally, we describe in detail how to score and plot septation and mitotic indices for elutriated cultures.

## 2. Materials

### 2.1. Elutriator and Pump Setup

1. Tygon R-3603, 1/8" ID, 1/4" ED, 1/16" wall (VWR cat # 63009-170).
2. Kontes 3-way Stopcock (VWR cat # KT420163-4503).

3. WPA biowave Cell Density Meter (cat # CO8000, www.biochrom.co.uk/product/20/co8000-cell-density-meter.html.)
4. Masterflex L/S Economy Digital Drive and Easy-Load II Pump Head (Cole-Parmer Instrument Company, cat # C-07524-40, and C-77200-60).
5. Beckman J-20 with a JE-5 series rotor and 4-mL elutriation chamber (Beckman Instruments).
6. Flow cell cat # 73.4/SOG, www.starna.com.
7. Silicone pump tubing.

### 2.2. G2 Elutriation, DNA Damage, and Time-course

1. Graduated flasks for cell loading, recirculation, and collection.
2. YES rich media (yeast extract with supplements): 5 g/L yeast extract, 30 g/L glucose, 75 mg/L leucine, 75 mg/L uracil, 75 mg/L adenine, 75 mg/L histidine, autoclaved.
3. Bleomycin (Sigma cat # B-2434).
4. Stratalinker UV source, or equivalent.
5. Faxitron RX-650 X-ray source, www.faxitron.com, or equivalent.
6. Levy hemacytometer counting chamber (VWR cat # 15170-208).
7. Pall Life Sciences Supor Disk Filter, 25 $mm^2$ 0.2 μn (VWR cat # 28147-956).
8. Pall Life Sciences Supor Disk Filter, 47 $mm^2$ 0.2 μn (VWR cat # 28148-551).
9. Pall Life Sciences Vacuum manifold compatible with 25- or 47-$mm^2$ filters.

### 2.3. Sterile Elutriation

1. Two, 2-l flasks containing 1.0 l water, autoclaved.
2. One 500-mL bottle containing 150 mL water, autoclaved.
3. 50 mL bleach.
4. 500 mL 70% ethanol.

## 3. Methods

Elutriation selects cells based on size. The procedure depends on a special centrifuge rotor that allows liquid to be pumped through a spinning centrifuge chamber (**Fig. 1**). The centrifugal force experienced by cells as they are pumped through the chamber forces them down to the bottom of the chamber. This force is countered by the flow of the medium, which forces the cells

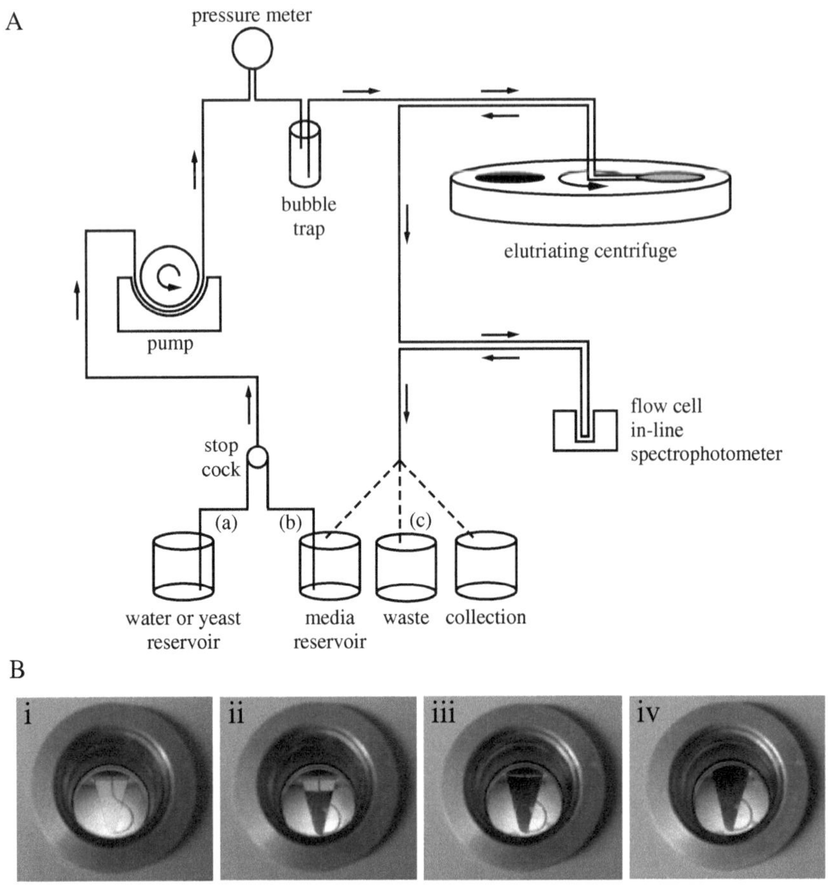


Fig. 1. The general setup for elutriation. (**a**) Two inlet tubes (*a*, *b*) are setup allowing quick switching between media, yeast, or water. Media from either reservoir are flowed through the system using a pump. The pressure meter allows determination of whether the system is blocked. The bubble trap smooths pressure pulses created by the pump and prevents air bubbles from entering and blocking the system with the centrifuge running. An in-line spectrophotometer is useful for monitoring cells escaping the centrifuge chamber when cells are recirculated and harvested. The outlet tubing (*c*) allows the user to recirculate media, remove waste, or collect cells. (**b**) The elutriator chamber, viewed during a run. (*i*) An empty chamber. (*ii*) A partially full chamber during cell loading. (*iii*) An almost full chamber; only a small gap remains between the cell front and the top of the chamber. At this point, no more cells need to be loaded, and the rotor speed can be reduced to allow cells to escape. Alternatively, the pump speed can be reduced, which will cause the cell front to retreat and make room for more cells. (*iv*) A full chamber with cells escaping.

to the top of the chamber. The hydrodynamics in the chamber sorts the cells by size, with smaller cells rising to the top. A synchronized strobe light and viewing window allows one to observe the cells in the chamber. By adjusting the opposing forces, it is possible to trap cells in the chamber and move them up to the point where only the smallest cells are washed out of the chamber. These small cells can be collected as a synchronous culture. Due to a particularity of the fission yeast cell cycle, the smallest cells in the culture are those that have just entered G2. These cells have just completed septation and replication and thus are early in the G2 phase of the cell cycle (*see* **Fig. 2a**).

### 3.1. Initial Elutriator Setup

1. Assemble the 4-mL centrifugation chamber according to manufacturer's instructions (*see* **Fig. 1**) (*see* **Note 1**).
2. Place a single inlet tube (a) in the water reservoir, the second inlet tube (b) in fresh media, the flow cell in the in-line spectrophotometer, and the outlet tube (c) in the waste (*see* **Note 2**).

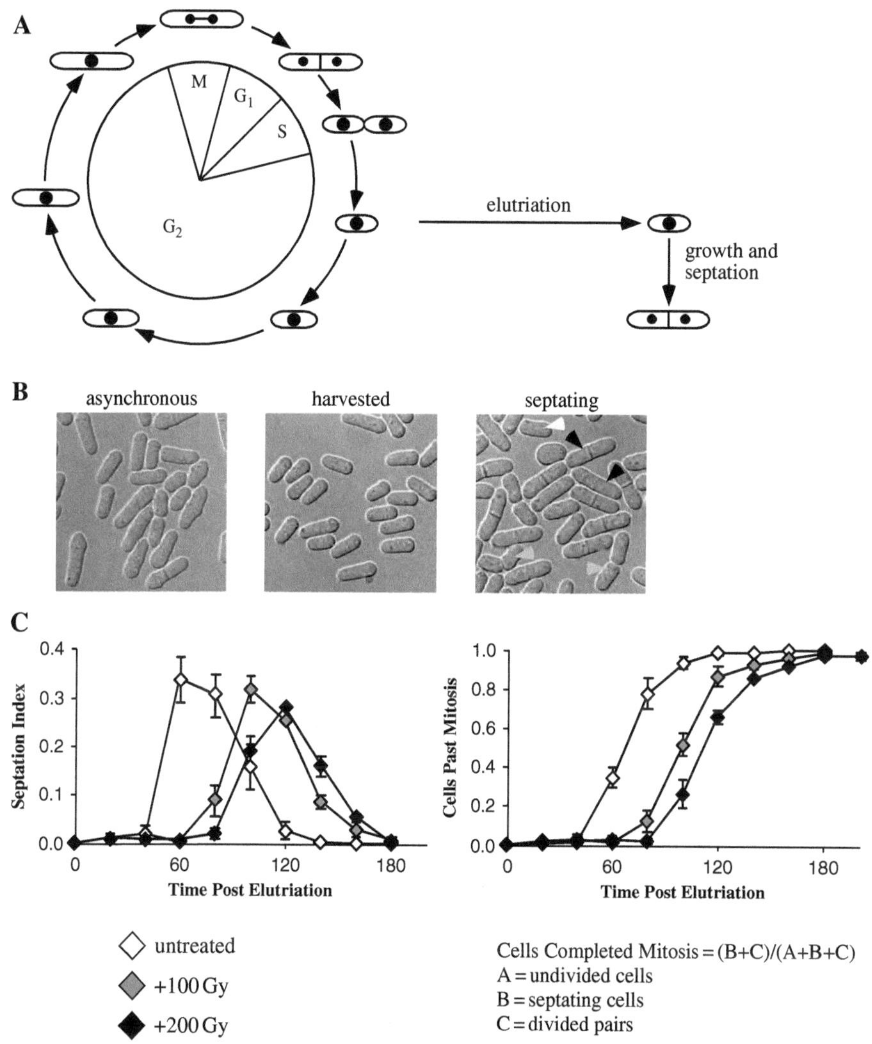


Fig. 2. G2 DNA damage response. (**a**) Cell cycle position is tied to cell morphology in fission yeast. Completion of mitosis and replication occur simultaneously, so the smallest cells in an asynchronous culture will be cells that have just entered G2 phase. Continued cell growth but no further septation is an indication of G2 DNA damage checkpoint-mediated arrest. (**b**) DIC images of an asynchronous culture, freshly elutriated population, and actively engaged in septation. *Black arrowheads* indicate septating cells, *gray arrowheads* indicate divided cell pairs, and the *white arrowhead* indicates half of a divided pair, a cell too short to be an undivided cell, but one that has detached from its division partner. (**c**) Examples of septation and mitotic progression plots and delays in mitosis brought about by exposure to increasing dose of ionizing radiation. Plots represent three independent experiments. Error bars represent the standard error of the mean.

3. Start the pump, set to 100 mL/min (*see* **Note 3**).
4. Fill the bubble trap and flood the entire system with water (*see* **Note 4**).
5. Start the centrifuge at 4100 rpm and set the centrifuge temperature to 30°C (*see* **Notes 5–7**).
6. Switch inlet source to (b) and flow fresh media into the system. Once the water is flushed out, transfer the outlet tubing (c) in the same reservoir of media permitting recirculation (*see* **Note 8**).
7. Adjust the strobe lamp to visualize the centrifugation chamber through the centrifuge port hole and reference the in-line spectrophotometer against the recirculating media.

### 3.2. Growth, Loading, and Elutriation of Cells

1. 500 mL cells should be grown to mid-log phase ($OD_{600}$ ~ 1.0) (*see* **Note 9**).
2. Transfer an inlet tube (a) to a reservoir containing the strain to be loaded.
3. Switch inlet source to (a) to load cells into the small 4-mL chamber at 30 mL/min flow rate at 4100 rpm until the cell front nears the top of chamber (*see* **Note 10**).
4. When chamber is full, switch stopcock to recirculate from the media reservoir.
5. Decrease centrifuge speed in 50-rpm decrements to eventually allow the smallest cells to escape the elutriation chamber. Monitor the escaping cell density with the in-line spectrophotometer. (*see* **Note 11**).
6. When the escaping cell density is between $OD_{600}$ 0.1 and 0.2, switch outlet tube (c) to an appropriate container to collect cells (*see* **Note 12**).
7. Check harvested cells under 40× power using hemacytometer to observe cell size (*see* **Note 13**).
8. Place the synchronous culture in an appropriate flask and incubate for the time-course at 30°C, 200 rpm.

### 3.3. Cell Counting and Septation Index Determination

1. For each time point, pipette 10 μL cells onto a hemacytometer and count cells (*see* **Note 14**).
2. Count cells every 20 min for 3–5 h (*see* **Note 15**).

### 3.4. Flushing out the Tubing, Centrifugation Chamber, and Housing Assembly

1. Stop the centrifuge and increase pump speed to 90 mL/min.
2. Flush the system with water and then purge with air.
3. Disconnect chamber and wash separately with mild detergent.

4. Connect the chamber inlet and outlet tubes to one another and flood the tubing and bubble trap with 70% ethanol followed by an excess of water (*see* **Note 16**).
5. Purge out all liquid from tubing.

#### 3.5. Sterile Elutriation Setup (See Note 17)

1. Flush the system with nonsterile water with the pump set to 100 mL/min.
2. Setup and start the centrifuge set to 2000 rpm.
3. Add 50 mL bleach to one flask containing 1 l of autoclaved water.
4. Flush out the water with the dilute bleach solution and allow recirculation for at least 20 min.
5. Flush out bleach with 500 mL 70% ethanol and allow recirculation for 20 min (*see* **Note 18**).
6. Flush out the alcohol using the remaining 1 l of sterile water (*see* **Note 19**).
7. Flood the system with sterile media to flush out the water.
8. Load samples as above.
9. Collect cells and sample time points using sterile technique (*see* **Note 20**).

#### 3.6. Treatment with Ionizing Radiation

1. Prepare 5-mL samples of G2 elutriated yeast for each dose to be tested.
2. Expose cells to appropriate doses of ionizing radiation (*see* **Note 21**).
3. Transfer irradiated cultures to culture tubes and incubate with shaking at 30°C for the time-course duration.
4. Count undivided cells, septating cells, and divided pairs every 20 min for 3–5 h (*see* **Notes 14** and **15**).

#### 3.7. Treatment with Ultraviolet Radiation

1. Setup a vacuum apparatus and prewet 25-mm$^2$ filters with yeast media (*see* **Note 22**).
2. Transfer up to 10 OD G2 elutriated culture and concentrate onto filters by vacuum.
3. Place filters yeast side up on 3 M blotting paper and expose in a Stratalinker (*see* **Note 23**).
4. Fold filters slightly and drop into culture tubes.
5. Add 5 mL media and vortex to release yeast into the medium.
6. Shake tubes at 30°C for the time-course.
7. Count undivided cells, septating cells, and divided pairs every 20 min for 3–5 h (*see* **Notes 14** and **15**).

## 4. Notes

***Initial Elutriator Setup***

1. Two elutriation chambers are available with 4- and 40-mL capacities. The procedure described above utilizes the smaller 4-mL chamber. For fission yeast, the 4-mL chamber holds approximately 400 ODs, allowing 40 ODs to be synchronously harvested (*see* **Note 9**). The 40-mL chamber capable of holding 4000 ODs may be used to harvest up to 400 ODs if more cells are required.
2. The stopcock is used to switch input between the two inlet sources (a, b), allowing seamless switching between various reservoirs of cells or media. Media recirculation (outlet (c) and inlet tube (b) are in the same vessel) allows the user increased preparation or lag time if needed.
3. Never shut off the pump while the centrifuge is running. If collecting cells in the centrifugation chamber, loss of pump pressure will result in cells being pushed "back" into the loading channel and clogging the system. If a clog of this nature occurs, first stop the centrifuge, then restart the pump and set the flow rate to 100 mL/min to attempt to force the cells out of the loading channel. If this method fails, try firmly tapping the side of the loading chamber to dislodge the clogging cells. If the clog cannot be cleared, the chamber must be dissembled and cells removed by hand.
4. Fill the bubble trap to about $^{3}/_{4}$ full; the air pocket in the bubble trap acts as a pressure buffer to dampen the flow pulses from the pump, which reduces turbulence in the elutriation chamber.
5. As the centrifuge speeds up, the remaining air bubbles will be forced from the rotor.
6. Monitor back pressure at the pressure meter. It should be a few psi when the chamber is empty of cells. When using the large, 40-mL chamber the psi may approach 10 psi when the chamber is almost full of cells. A back pressure greater than 10 psi indicates a block in the system, possibly due to bubbles trapped in the rotor. If the pressure remains above 10 psi, stop the centrifuge to allow bubbles to escape. Stopping the centrifuge will cause any loaded cells to be washed out of the chamber, but they can be collected and reloaded. If the pressure remains significantly above 10 psi, a tube joint will fail.
7. For temperature- or cold-sensitive strains, set the temperature of the elutriating centrifuge chamber to the appropriate temperature. The centrifuge is refrigerated but not heated, so to reach temperatures above the ambient temperature,

one must allow the centrifuge to run, to warm the chamber using heat generated from the centrifuge motor itself. Achieving 30°C from room temperature takes approximately 20 min centrifuge run time when set to 4100 rpm.

8. Always use fresh media and remove as much water as possible. Reuse of used media from the main culture or excessive recirculation with a chamber full of cells may cause cells to run out of sugar and begin respiration, producing $CO_2$ bubbles, which will reduce harvested cell homogeneity.

***Growth, Loading, and Elutriation of Cells***

9. Grow 500 ml culture in YES at 30°C, 200 rpm, to an optical density at 600 nm ($OD_{600}$) of about 1, starting from a mid-log 50-mL starter culture. $OD_{600}$ is used to follow yeast cell proliferation. Optical density is also used approximate cell number; 1 OD unit (abbreviated OD) is the number of cells required to give 1 mL of culture an optical density of 1. Therefore, a 100-mL culture at $OD_{600}$ of 0.1, a 10-mL culture at $OD_{600}$ of 1 and a 1-mL culture at $OD_{600}$ of 10 all contain 10 ODs of cells. 1 OD is approximately $2 \times 10^7$ cells. The 4- and 40-mL chambers hold approximately 400 and 4000 OD yeast cells, respectively. For the large chamber, several liters of cells may be grown to an optical density of 2.0. Sick strains may require more cells to fill the chamber, in excess of 800 or 8000 ODs of cells for the small and large chambers, respectively. For cultures that are extremely sick or flocculate at low OD, sonicate the culture using twenty 50-ms pulses at maximum power. Sonication also helps to break up divided pairs, thus increasing enrichment of small cells for any strain.
10. Begin loading the 40-mL chamber at 100 mL/min flow rate, 4100 rpm. To save time loading, after cells begin to accumulate in the chamber increase the pump speed to 130 mL/min until the chamber is close to full. When using the small chamber, it may be loaded at 100 mL/min for several minutes before the pump is reduced to 30 mL/min.
11. The pump and rotor speeds need to be balanced to bring the cell boundary to the top of the chamber and allow the smallest cells to escape. For course adjustment, to get the cell boundary to the top of the chamber, it is easiest to adjust the pump speed. Once cells begin to escape, adjusting the rotor speed gives one much finer control. Decreasing centrifuge speed subtly reduces forces on cells, allowing small changes in cell elutriation, whereas increasing pump speeds even by 1 mL/min causes dramatic changes and may force a heterogeneous population of cells out of the chamber. The flow cell and in-line spectrophotometer are used to follow the optical density of the media exiting the

centrifuge. Remember to tare the spec with the flow cell inserted and only after media is recirculating. Depending on how many cells are loaded in the chamber, they may escape between 3700 and 2900 rpm. The pump speed should be kept below 40 mL/min when harvesting cells. High pump flow causes turbulence within the chamber, reducing harvested cell synchrony. Once the collection of cells has begun, it is necessary to occasionally decrease the rotor speed to maintain the desired rate of cell elutriation, as the number of cells escaping will gradually decrease over time.

12. The optical density of the escaping cells should not exceed 0.4; to harvest the most synchronous population, aim to collect between an $OD_{600}$ of 0.1 and 0.2. Practically, it is best to limit recovery of cells to 10% of the loaded, asynchronous culture.

13. Strive for less than 1% of the collected cells to be septated in divided pairs, more than 1% indicates a poor synchronization.

***Cell Counting, Septation, and Mitotic Index Determination***

14. Count 100 cells. Typically, observed cells fall into three categories: undivided cells, septating cells, and divided pairs. Undivided cells appear small at the beginning of the timecourse and will grow and increase in cell length over time. Septating cells display a septum across the cell midsection and are typically longer. Divided pairs are those cells that have completed septation and show dimples around the septum (**Fig. 1b**). The septation index is simply the percentage of cells septating observed at any single time point. In a well-synchronized population, septation will peak with a value between 30 and 60%. For cultures incubated at 30°C, septation will peak around 80 min post elutriation, while cultures incubated at 25°C will peak around 120 min. Progression through mitosis may also be represented by plotting the total number of cells that have completed cytokinesis, which can be measured by the percentage of cells that are septated plus the percentage that have completed septation and formed a divided pair. Just after the septation index peaks, the mitotic index should approach 100%. This index may be calculated using the following equation (**Fig. 2b**).

$$\text{Cells Completed Mitosis} = (B + C) / (A + B + C)$$

$A$ = undivided single cells

$B$ = septating cells

$C$ = divided pairs

Undivided cells, septating cells, and divided pairs are each counted as a single cell. Divided pairs are counted as a single event because the pair of cells arose from a single mother cell. At later time points, undivided cells will be long and some divided pairs will have separated into single cells, thus count pairs of these small cells as single events (*see* **Fig. 2b**, right hand panel). For counting multiple rounds of mitosis, after most cells have completed septation and are divided pairs, sonicate the culture to separate divided pairs and continue counting assuming single cells are now undivided cells.

15. Septation takes approximately 20 min; thus, if counted every 20 min, each septation should be counted once. Therefore, the cumulative septation count across every time point should be around 100%; a greater or lesser value suggests over- or undercounting of septa.

***Flushing out the Elutriator Tubing, Chamber, and Housing Assembly***

16. To eliminate cell debris buildup inside the bubble trap and tubing, occasionally recirculate dilute bleach for 30 min, followed by flushing the system with 2 l of water.

***Sterile Elutriation Setup***

17. The elutriator is a nonsterile system harboring bacteria and fungi. If plating cells after synchronization to compare viability between treated and untreated samples, sterilize the system prior to loading cells. Be sure when sterilizing to fill the bubble trap completely, to sterilize the entire inner surface. Wear gloves at all times when handling the inlet and outlet tubes. Use sterile glassware for sample incubation, preparation, and collection.
18. Ethanol will produce bubbles. If the system gets blocked and overpressurizes, stop the centrifuge and shake tubing and chamber to dislodge those bubbles. Restart the centrifuge and continue ethanol recirculation. Two phases may form in the air trap; be sure to invert the air trap to ensure the phases mix.
19. Vegetative yeast are extremely sensitive to both ethanol and bleach; be sure to allow the entire liter of water to flush out the system.
20. If plating cells, use media containing ampicillin or carbenicilin to prevent bacteria from growing and out-competing the yeast.

***Treatment with Ionizing Radiation***

21. Samples may be exposed using a Faxitron X-ray cabinet or another ionizing radiation source. For exposure in a Faxitron cabinet, it is convenient to transfer 5-mL samples

to 35-mm Petri dishes and irradiate cells while suspended in liquid culture. Regardless of the source used, begin by exposing the sample to be irradiated to the greatest degree and then place samples to be exposed at lower doses at later time points during the irradiation. This strategy will stagger all of the irradiated samples to finish simultaneously. 0-, 50-, 100-, and 200-Gy doses are good starting points for healthy strains. As a control, cells may be exposed to 3 mU Bleomycin, which acts as an IR mimetic and will arrest checkpoint proficient strains.

***Treatment with Ultraviolet Radiation***

22. Filters should be prelabeled using pencil. Filter capacity is approximately 2 OD/cm$^2$. Thus, up to 10 and 40 OD yeast may be deposited on 25- and 47-mm filters, respectively.
23. A Stratalinker UV Crosslinker is a convenient way to deliver a defined UV dose. For repair proficient strains, doses up to 50 J/m$^2$ are useful.

## References


1. Sancar, A., Lindsey-Boltz, L. A., Unsal-Kacmaz, K., and Linn, S. (2004) Molecular mechanisms of mammalian DNA repair and the DNA damage checkpoints. *Annu Rev Biochem.* **73**, 39–85.
2. Stewart, E. and Enoch, T. (1996) S-phase and DNA-damage checkpoints: a tale of two yeasts. *Curr Opin Cell Biol.* **8**, 781–787.
3. Rhind, N. and Russell, P. (1998) Mitotic DNA damage and replication checkpoints in yeast. *Curr Opin Cell Biol.* **10**, 749–758.
4. Forsburg, S. L. and Rhind, N. (2006) Basic methods for fission yeast. *Yeast.* **23**, 173–183.
5. Edwards, R. J. and Carr, A. M. (1997) Analysis of radiation-sensitive mutants of fission yeast. *Methods Enzymol.* **283**, 471–494.
6. Sullivan, D. S. and Huffaker, T. C. (1992) Astral microtubules are not required for anaphase B in Saccharomyces cerevisiae. *J Cell Biol.* **119**, 379–388.
7. Walker, G. M. (1999) Synchronization of yeast cell populations. *Methods Cell Sci.* **21**, 87–93.

# Chapter 2

# Studying S-Phase DNA Damage Checkpoints Using the Fission Yeast *Schizosaccharomyces pombe*

Nicholas Willis and Nicholas Rhind

## Abstract

Slowing of replication in response to DNA damage is a universal response to DNA damage during S-phase. Originally discovered to be defective in checkpoint mutant cells in metazoans, this S-phase DNA damage checkpoint response has been extensively studied in yeast. Unlike other checkpoints that completely arrest cell cycle, the S-phase DNA damage checkpoint slows but does not completely halt replication in response to DNA damage. An analysis of mutants defective in the slowing response requires a sensitive assay to measure this quantitative effect. The use of centrifugal elutriation to synchronize cells and improved techniques in preparing cells for flow cytometry allow for more sensitive and accurate measurement of cells' ability to slow replication in the presence of DNA damage. This chapter describes the use of transient *cdc10-M17* temperature sensitive allele arrest and release combined with centrifugal elutriation to synchronize cells in G1. The S-phase progression of these cells is then assayed by flow cytometry of isolated nuclei, which allows sensitive determination of replication kinetics.

**Key words:** S-phase DNA damage checkpoint, DNA replication, fission yeast, *Schizosaccharomyces pombe*, flow cytometry, centrifugal elutriation, cell division cycle.

## 1. Introduction

Slowing of DNA replication is a hallmark of the S-phase DNA damage checkpoint response. Failure to reduce new synthesis in the presence of DNA damage is correlated with genomic instability and cancer development (1). In metazoans, new DNA synthesis may be measured in asynchronous cultures by the incorporation of radio-labeled nucleotides (2). However, in fission

W.X. Li (ed.), *Cell Cycle Checkpoints*, Methods in Molecular Biology 782,
DOI 10.1007/978-1-61779-273-1_2, © Springer Science+Business Media, LLC 2011

yeast only about 10% of cells in an asynchronous population are in S-phase, so kinetic measurement of replication in asynchronous cultures is not currently practical. Accurate measurement of replication progression requires the synchronization of cells, allowing one to follow the population through S-phase.

Many methods are available to synchronize yeast. Cells may be synchronized using cell division cycle (*cdc*) mutants, nutrient starvation, or drug-induced arrest. G1 synchronization may be achieved using temperature sensitive alleles of the transcriptional regulator *cdc10* or starvation in nitrogen-free media (3, 4). These techniques allow the manipulation of multiple cultures and the uses of large quantities of cells. However, these methods suffer from various disadvantages including checkpoint activation, inefficient release from starvation, and physiological disruption. In particular, we have observed that block-and-release protocols can prevent wild-type cells from properly responding to DNA damage (5).

To avoid synchronization artifacts, we combine *cdc10-ts* arrest with centrifugal elutriation (**Fig. 1a**). As mentioned above, prolonged arrest of *cdc10-ts* cells can compromise the S-phase DNA damage checkpoint. However, due to a peculiarity of the fission yeast cell cycle, G1 is cryptic, with cells beginning DNA replication before they finish cytokinesis. Therefore, in order to collect cells in G1, one must arrest cells in one way or another. Fortunately, one can briefly arrest *cdc10-ts* cells and then select the smallest, G1 arrested cells by centrifugal elutriation. This approach allows for the preparation of a synchronous G1 population that retains robust S-phase checkpoint response to DNA damage.

To accurately measure progression through S-phase and assess strains' ability to slow replication in response to DNA damage, we employ flow cytometry of isolated nuclei. Cytoplasmic background leads to an overestimation of nuclear DNA content due to the contribution of RNA and mitochondrial DNA to overall cellular nucleic acid content (6). Isolating nuclei reduces background due the contribution of mitochondrial DNA and cell size, greatly increasing sensitivity and simplifying data interpretation (7, 8).

Synchronization of cells in G1 is performed using the following scheme. Cells are arrested for 2 h at non-permissive temperature 35°C; meanwhile the elutriator is setup and temperature set to 35°C. The semi-synchronous culture is then loaded and the smallest and transiently arrested G1 cells are harvested. Harvested cells are released from the arrest to 25°C in the absence or presence of DNA damage. Samples are fixed at regular intervals for analysis by flow cytometry or frozen in liquid nitrogen for isolation of DNA, RNA, or protein.

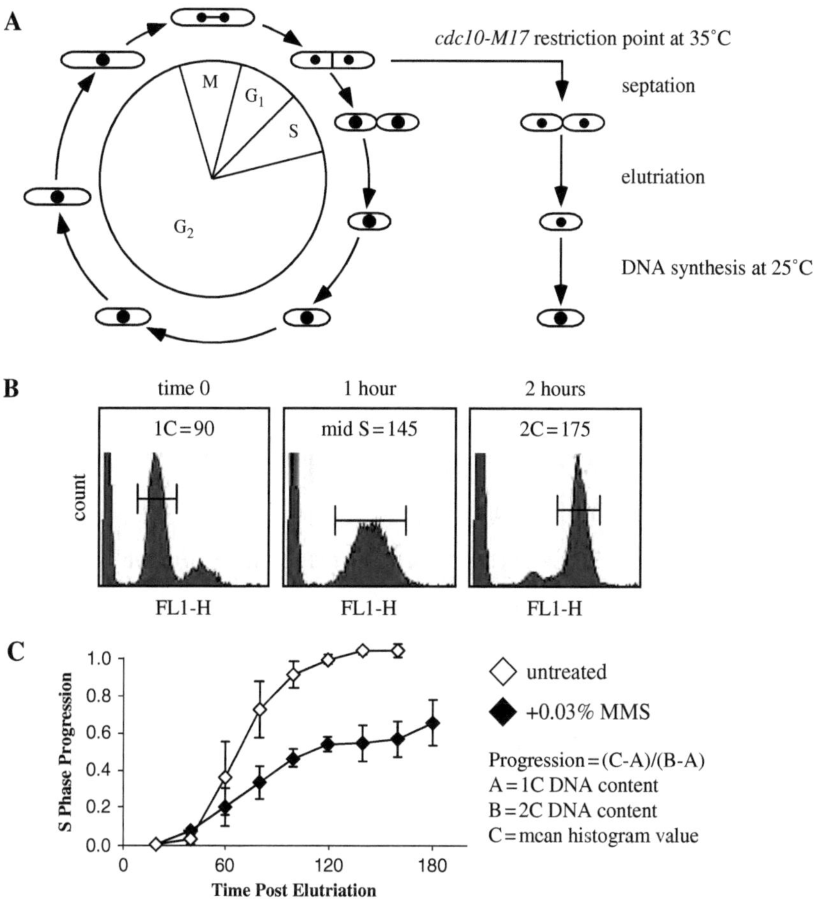


Fig. 1. S-phase DNA damage response. (**a**) Cytokinesis and replication normally occur almost simultaneously. However, use of the *cdc10-M17* temperature sensitive allele prevents replication while allowing cells to complete cytokinesis. Transient arrest of an asynchronous culture will produce small cells arrested in G1 containing an unreplicated genome. Centrifugal elutriation allows the harvesting of these cells. (**b**) Examples of flow cytometry profiles for 0, 1, and 2 h after release. Cells display unreplicated, mid-S-phase, and replicated nuclear DNA profiles, respectively. Brackets indicate sample portion measured to determine mean histogram peak value used for S-phase progression plots. (**c**) Example of S-phase progression plot in wild-type strain untreated or exposed to 0.03% MMS. The profile shown is an average of three experiments and error bar represent the standard error of the mean.

## 2. Materials

### 2.1. $G_1$ Synchronization

1. *cdc10-M17* fission yeast strains.
2. YES rich media (Yeast Extract + supplements): 5 g/L yeast extract, 30 g/L glucose, 75 mg/L leucine, 75 mg/L uracil, 75 mg/L adenine, 75 mg/L histidine, autoclaved.
3. Beckman J-20 with a JE-5 series rotor and 4-mL elutriation chamber, Beckman Instruments.

4. Two shaking water baths set to 25°C and 35°C, 200 rpm.
5. 70% ethanol.
6. Methyl methane sulfonate, Sigma, cat # M4016.

### 2.2. Nuclei Flow Cytometry

1. 0.6 M KCl.
2. 0.1 M KCl, 0.1% SDS.
3. 20 mM Tris, pH 8.0, 1 mM EDTA, pH 8.0.
4. 10 mg/ml RNAse A (Sigma, cat # R5503).
5. Branson 450 Analog Sonifier, #101-063-200 (VWR, cat # 33995-320, or equivalent).
6. Branson Sonifier 3 mm tapered tip (VWR cat # 33996-163).
7. BD FACScan flow cytometer (BD Sciences, or equivalent).
8. BD Sciences Cellquest software version 3.3.
9. Lysing enzymes from *Trichoderma harzianium* (Sigma, cat # L1412).
10. Zymolyase 20T (US Biological, cat# Z1000).
11. FACS sheath fluid, Diluent 2 (1× PBS) (VWR, cat # 45001-012).
12. Sytox Green nucleic acid stain diluted to 2 μm in FACS sheath fluid (Invitrogen, cat # S7020).
13. Falcon 100 × 17 mm 2054 tubes (Fischer, cat # 149595).

## 3. Methods

### 3.1. G1 Cell Elutriation

1. Grow cells to a final volume of 500 mL, mid-log phase ($OD_{600}$ ~1) (*see* **Note 1**).
2. Preheat shaking water baths to 25°C and 35°C, set to 200 rpm (*see* **Note 2**).
3. Pellet 0.5 OD cells in a 1.7-mL eppendorf tube, 16,000×*g*, for 2 min at room temperature and resuspend in 1 mL 70% ethanol to serve as the asynchronous control for flow cytometry (*see* **Note 3**).
4. Incubate 500 mL culture at 35°C in the shaking water bath for 2 h (*see* **Note 4**).
5. Setup the elutriator as described in Chapter 1, and set the chamber temperature to 35°C (*see* **Note 5**).

6. Load cells as described for G2 elutriation in Chapter 1.
7. Pellet and save 0.5 OD cells while loading to serve as the transiently arrested control (*see* **Note 6**).
8. Perform elutriation and harvest 150 mL cells. Visually check cells, homogenous small size with no divided pairs or septated cells is ideal (*see* **Notes 7** and **8**).
9. Sample cells as in #3 to serve as the time 0 samples.
10. Divide cells into three 50-mL cultures and treat as follows:
    a. Mock treatment.
    b. Add MMS to 0.03%.
    c. Add hydroxyurea to 10 mM (*see* **Note 9**).
11. Place cultures in 25°C shaking water bath (*see* **Note 10**).
12. Sample cells as in #3 at 20-min intervals for analysis by flow cytometry.
13. Store samples in 70% alcohol overnight at +4°C (*see* **Note 11**).

### 3.2. Preparation of Isolated Nuclei for Flow Cytometry

1. Pellet fixed cells at 5000×*g* for 3 min at room temperature, decant supernatant (*see* **Note 12**).
2. Wash once with 1 mL 0.6 M KCl.
3. Resuspend cells in 1 mL 0.6 M KCl, 1 mg/ml lysing enzyme, 0.5 mg/ml Zymolyase 20T (*see* **Note 13**).
4. Incubate cells for 30 min at 37°C.
5. Pellet cells and decant supernatant.
6. Resuspend cells in 1 mL 0.1 M KCl 0.1% Triton X-100, vortex well.
7. Pellet cells and decant supernatant.
8. Wash cells in 1 mL 20 mM Tris–HCl, 5 mM EDTA, pH 8.0.
9. Resuspend cells in 1 mL 20 mM Tris–HCl, 5 mM EDTA, pH 8.0 (*see* **Note 14**).
10. Add 25 μL 10 mg/ml RNase A to each sample (final conc. = 200 μg/ml), vortex well.
11. Incubate cells overnight at 37°C.
12. Chill cells at −20°C for 5–10 min (*see* **Notes 15–17**).
13. Sonicate samples using a Branson 450 Sonifier and a 3-mm tapered tip, maximum micro-tip power for 10 s (*see* **Note 18**).
14. Mix 300 μL of prepared nuclei with 300 μL of 2 μM Sytox in a Falcon 2054 tube. Briefly vortex samples.

**3.3. Analysis by Flow Cytometry**

1. Measure nuclear DNA content by FL1 signal for each sample according to flow cytometer instructions (*see* **Note 19**).
2. Calculate the position within S-phase for each sample taken using the following equation (**Fig. 1b** and *see* **Note 20**).

$$\% \text{ replicated} = (C - A)/(B - A)$$

$A = 1C$, $B = 2C$, and $C =$ mean histogram value.

3. Plot data as % replicated vs. time (*see* **Notes 21** and **22**).

## 4. Notes

**G1 Elutriation**

1. Grow 500 mL culture of *cdc10-M17* in YES at 25°C, 200 rpm to an optical density at 600 nm ($OD_{600}$) of about 1, starting from a mid-log 50-mL starter culture. $OD_{600}$ is used to follow yeast cell proliferation. Optical density is also used to approximate cell number; 1 OD unit (abbreviated OD) is the number of cells required to give 1 mL of culture an optical density of 1. Therefore, a 100-mL culture at $OD_{600}$ of 0.1, a 10-mL culture at $OD_{600}$ of 1, and a 1-mL culture at $OD_{600}$ of 10 all contain 10 ODs of cells. 1 OD is approximately $2 \times 10^7$ cells. The 4- and 40-mL chambers hold approximately 400 and 4000 OD yeast cells, respectively. For the large chamber, several liters of cells may be grown to an optical density of 2.0. Sick strains may require more cells to fill the chamber, in excess of 800 or 8000 ODs of cells for the small and large chambers, respectively. For cultures that are extremely sick or flocculate at low OD, sonicate the culture using 10–20 50 ms pulses at maximum power. Sonication also helps to break up divided pairs, thus increasing enrichment of small cells for any strain.
2. Water baths are required for rapid temperature shifts used for arrest and release. Air incubators are not suitable for quick temperature shifts.
3. The asynchronous sample serves as a control with mostly 2C nuclear DNA content. Keep the quantity of cells fixed for all time points approximately equal. Subsequent processing steps are sensitive to the total number of cells in each sample.
4. Culture density should increase by 50–100% depending on strain doubling time.
5. Allow about 40 min prior to cell loading to heat the chamber and recirculating media to 35°C.

6. Asynchronous cultures arrested for 2 h display two populations of cells, those containing 1C and 2C genomic DNA content. The 1C cells have encountered and arrested at the restriction point while the 2C cells have not.
7. At most, 10% of cells loaded may be synchronously harvested. Harvested culture density should be between 0.1 and 0.2 $OD_{600}$. Greater than 0.4 $OD_{600}$ results in far less synchronous population. For S-phase progression, 50 mL volume for each condition is sufficient.
8. The time zero control should contain mostly cells with 1C nuclear DNA content. Harvested cells should all appear small and homogenous in size. Sick, slow growing cultures may look far worse than wild-type control. Cultures may be sonicated to help separate cells. The larger 40-mL chamber may be employed to help select a healthy minority from sick cultures.
9. MMS induced replication slowing is dose dependent; 0.015% MMS causes very little slowing, 0.03% causes robust, checkpoint-dependent slowing, and 0.06% and greater concentrations induce checkpoint-independent slowing. Alternatives to 0.03% MMS include incubation with 0.5 μM 4NQO or exposure to 100–200 J/$m^2$ UV radiation. See Chapter 1 for details on UV exposure.
10. Multiple consecutive elutriations may be staggered approximately 30 min apart after sufficient experience operating the elutriator. Time courses may be synchronized by maintaining freshly harvested cells at 35°C for 5–15 min until release with the next time point of a previous and ongoing time-course.
11. Fixed cells may be stored at +4°C for days before processing. Overnight fixation in ethanol is ideal. Samples processed the same day as the time-course do not look as tight by flow cytometry as cells processed after an overnight fixation. Additionally, the longer cells are stored after 24 h, the worse the raw flow cytometry data will look.
12. G1 cells synchronized by centrifugal elutriation begin replication approximately 40 min after release and complete replication by 100 min. Most replication occurs about 80 min after release.

***Nuclei Preparation and Flow Cytometry***

13. Cell pellets are often soft and lost during supernatant aspiration. Simply decant supernatant by hand and do not attempt to remove more than 90% of the supernatant.
14. Properly digested cells appear dark by phase microscopy when mixed with 1% SDS. Lysing enzymes and zymolyase may be increased by 50% if needed. A titration of the

enzymes should be done if using an alternative vendor or the purified 100T fraction of zymolyase. Setup digests using 0.5×, 1×, 1.5×, and 2× concentration of enzymes used and digest as normal. Process all the samples in parallel for flow cytometry and compare. Select the enzyme concentration that displays sharp G1 and G2 peaks; a sub-G1 shoulder indicates over digestion and nuclear breakdown.

15. Be careful since the pellets at this step are extremely fragile.
16. Anecdotally, sonicating pelleted cells improves signal to noise of samples by flow cytometry. If necessary, pellet cells for 5 min at 5000×*g* prior to chilling.
17. Cooling of the samples prevents over heating during sonication. If the samples freeze, allow them to thaw on the benchtop before sonication.
18. Do not touch the side of the tube to the microtip, gently move the tip up and down in the tube while sonicating. Do not allow samples to froth up during sonication. This leads to a very poor sample flow profile. This nuclei preparation may be stored at 4°C for days.

***Calculate and Plot Population S-Phase Progression***

19. BD Sciences Cellquest software is used to measure the mean peak values of histograms displayed by populations exiting G1 and progressing through S-phase. For this analysis, all detectors are set to linear scale. Calibrate the FSC and SSC (forward and side-scatter) so that the samples analyzed are well within the cytometer detection range. Calibrate FL1 voltage and amperage gain to allow visualization of both the 1C and the 2C DNA populations prior to data collection. The FL1 signal representing the 1C population should be set to appear at around the 400th channel on a 1024-channel detection scale. Flow cytometry plots show a sub-1C peak, which is due to cell fragments lacking nuclei.
20. For calculation of the percent replicated, the 1C unreplicated value, A, is determined from the DNA content of the freshly elutriated, time 0 sample; the 2C fully replicated value B is determined from the DNA content of the later time points of the untreated sample and the 2C peaks of the asynchronous and transiently arrested samples (**Fig. 1**).
21. If sample signal drifts to the right for later time points, after cells have reached 2C DNA content, try resonicating the samples and collect data again and sonicate subsequent samples for longer than 10 s.
22. G1 cells synchronized by centrifugal elutriation begin replication approximately 40 min after release and complete replication by 100 min. Most replication occurs about 80 min after release.

## References

1. Kastan, M. B. and Bartek, J. (2004) Cell-cycle checkpoints and cancer. *Nature.* **432**, 316–323.
2. Painter, R. B. and Young, B. R. (1980) Radiosensitivity in ataxia-telangiectasia: a new explanation. *Proc Natl Acad Sci USA.* 77, 7315–7317.
3. Nurse, P., Thuriaux, P., and Nasmyth, K. (1976) Genetic control of the cell division cycle in the fission yeast *Schizosaccharomyces pombe. Mol Gen Genet.* **146**, 167–178.
4. Aves, S. J., Durkacz, B. W., Carr, A., and Nurse, P. (1985) Cloning, sequencing and transcriptional control of the *Schizosaccharomyces pombe* cdc10 'start' gene. *EMBO J.* **4**, 457–463.
5. Kommajosyula, N. and Rhind, N. (2006) Cdc2 Tyrosine Phosphorylation is not required for the S-Phase DNA damage checkpoint in fission yeast. *Cell Cycle* **5**, 2495–2500.
6. Sazer, S. and Sherwood, S. W. (1990) Mitochondrial growth and DNA synthesis occur in the absence of nuclear DNA replication in fission yeast. *J Cell Sci.* **97**, 509–516.
7. Carlson, C. R., Grallert, B., Bernander, R., Stokke, T., and Boye, E. (1997) Measurement of nuclear DNA content in fission yeast by flow cytometry. *Yeast.* **13**, 1329–1335.
8. Forsburg, S. L. and Rhind, N. (2006) Basic methods for fission yeast. *Yeast.* **23**, 173–183.

# Chapter 3

# Methods for Studying the G2 DNA Damage Checkpoint in Mammalian Cells


## Claudia Tapia-Alveal and Matthew J. O'Connell



## Abstract

In response to post-replicative DNA damage, cells activate the G2 DNA damage checkpoint to ensure mitosis is not attempted until the damage has been repaired. This is a common response to a variety of DNA damaging agents, including ionizing radiation and many chemotherapeutic agents used in the treatment of cancer. The G2 DNA damage checkpoint acts to inhibit the mitotic cyclin-dependent kinase, and thus cells are arrested in the G2 phase of the cell cycle. The kinetics of this checkpoint can be assayed by staining cells for markers of mitosis, which can then be quantified by flow cytometry or microscopy.



**Key words:** Checkpoint, mitosis, cell cycle, flow cytometry.


## 1. Introduction

Entry into mitosis is controlled by the mitotic cyclin-dependent kinase Cdc2 (1). During interphase, Cdc2 binds to the accumulating A- or B-type cyclins. However, these complexes are rapidly inactivated by phosphorylation on tyrosine-15 (Y15) by the Wee1 family of kinases. Once conditions appropriate for mitosis are achieved, Cdc2 is rapidly activated by Y15 dephosphorylation by the Cdc25 phosphatases (2). Downstream of Cdc2 during mitosis, other protein kinases are activated, including members of the Polo (3), Aurora (4), and NIMA-related (Nek) (5) kinase families, and there is a dramatic up-regulation of total protein phosphorylation (6). One abundant phospho-protein in mitotic cells is the core H3 histone, which is phosphorylated on Serine-10 (S10) by Aurora kinases (7, 8).


W.X. Li (ed.), *Cell Cycle Checkpoints*, Methods in Molecular Biology 782,
DOI 10.1007/978-1-61779-273-1_3, © Springer Science+Business Media, LLC 2011

Any form of DNA damage must be repaired prior to mitotic entry. To allow time for repair to occur, the G2 DNA damage checkpoint prevents Cdc2 activation, thus holding the cells in G2 phase (9). The effector of this checkpoint is the protein serine-threonine kinase Chk1. In response to DNA damage, complexes of checkpoint proteins assemble on single-stranded DNA at the sites of lesions. Chk1 is recruited into these complexes and is phosphorylated on C-terminal serines by the Ataxia Telangiectasia Mutated (ATM) and the ATM and Rad3-related (ATR) kinases. Activated Chk1 phosphorylates both Wee1 and Cdc25 proteins, and ensures that Cdc2 remains phosphorylated on Y15. The duration of the G2 DNA damage checkpoint-induced G2 cell cycle arrest is proportional to the amount of unrepaired DNA damage; that is, few (maybe as low as one) lesions will induce a potent checkpoint so long as they remain unrepaired. Once repair is complete, Chk1 is inactivated by dephosphorylation, and Cdc25 then activates Cdc2 and cells enter mitosis (10, 11).

Common activators of the G2 DNA damage checkpoint are factors used in cancer therapy – both ionizing radiation and DNA damaging drugs. The later includes inhibitors of type I and type II topoisomerases, DNA cross-linking agents, and anti-metabolites that interfere with dNTP metabolism. In the laboratory, each of these agents will activate Chk1 in cells grown in culture. There are several biochemical markers of checkpoint activation (e.g., activation of Chk1 – *see* Chapter 12). Physiologically, the checkpoint is assayed by accumulation of cells in G2 phase, and a concomitant loss of mitotic cells. The methods we describe can be applied to any mammalian cell type, and to any source of DNA damage.

## 2. Materials

### 2.1. Cell Culture and Checkpoint Activation

1. Media (Hepes-buffered Dulbecco's Modified Eagle's Medium, DMEM), phosphate-buffered saline (PBS), 1 × Trypsin/EDTA, fetal calf serum (FBS) (all from Invitrogen, Carlsband, CA), and tissue-culture grade 6-cm petri dishes (BD Falcon, Franklin Lanes, NJ) (*see* **Note 1**).
2. $^{137}Cs$ source irradiator (J.L. Sheppard and Associates, San Fernando, CA) (*see* **Note 2**).

### 2.2. Synchronization by Double Thymidine Block and Release

1. Thymidine (Sigma). 100 mM stock made in PBS, and sterilized by filtration through a 0.2-μM filter. Aliquot and store at –20°C.

### 2.3. Cell Harvesting and Fixation

1. Trypsin/EDTA and PBS (both from Invitrogen, Carlsband, CA).

2. Formaldehyde (Methanol free) 16% ultrapure EM grade (Polysciences, Inc., Warrington, PA) (*see* **Note 3**).
3. 90% (v/v) ice-cold methanol (HPLC grade, Fisher).
4. 15-mL Falcon tubes (BD Falcon, Franklin Lanes, NJ).
5. 1.5-mL microfuge tubes (Fisher).

#### 2.4. Nocodazole Trapping

1. Nocodazole (Sigma, St. Louis, MO). Stock solution is 5 mg/ml in dimethyl sulfoxide (DMSO). Aliquot and store at –20°C.

#### 2.5. Quantification of Mitotic Index

1. Blocking Solution: PBS plus 2% BSA (Fraction V, Fisher).
2. RNAase A (Sigma, St. Louis, MO). Stock solution is 10 mg/ml in water. Dissolve and heat to 95°C for 10 min, aliquot, and store at –20°C.
3. Primary antibody: Phospho-Histone H3 (Ser10) (6G3), (Cell Signaling Technology) and secondary antibody (FITC and Cy5-labelled anti-mouse IgG, Jackson ImmunoResearch Laboratories, Inc., West Grove, PA).
4. Propidium Iodide (Sigma, St. Louis, MO). Stock solution is 2.5 mg/ml in water. Store in the dark.
5. Flow Cytometry tubes (BD Falcon, Franklin Lanes, NJ).
6. DAPI/antifade: 1 × stock is 1 μ g/ml 4′,6′-diamidino-2-phenylindole (DAPI, Sigma, St. Louis, MO), plus 1 mg/ml ρ-phenylenediamine in 50% glycerol. Store at –20°C in the dark.
7. Microscope slides Superfrost/microscope cover glass (both Fisherbrand, Fisher).

## 3. Methods

Here we describe assays for checkpoint arrest, which in practice can be applied to any cell type – all that will vary is the culture conditions. As a general practice, it is important that the cells are no more than 70–80% confluent at the time of harvesting. The parameters described below are optimized for HeLa S3 and immortalized murine fibroblasts grown in 6-cm petri dishes, but can be scaled up if needed.

#### 3.1. Cell Culture and Checkpoint Activation

1. Grow cells to 70% density in DMEM + 10% FCS, at 37°C in the presence of 5% $CO_2$. Remove media, wash in PBS, and harvest cells with 0.5 mL trypsin/EDTA. Inactivate the trypsin with fresh media, and determine the cell density with

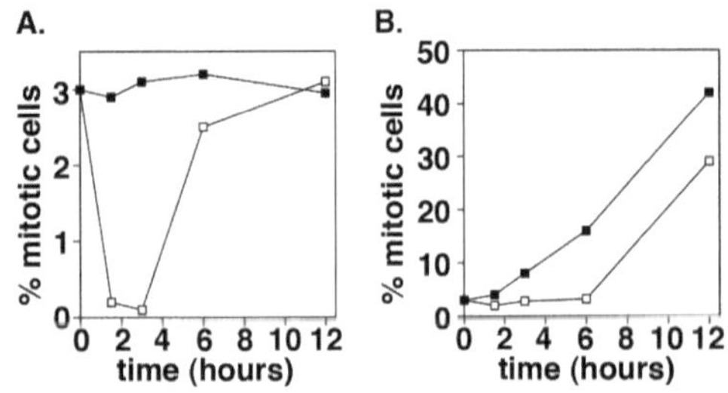


Fig. 1. Checkpoint activation assayed by mitotic index. (**a**) HeLa cells were grown and irradiated with 0 Gy (*filled squares*) or 2 Gy (*open squares*) as described in **Section 3.1**. Mitotic indices were determined as described in **Section 3.5**. Induction of the checkpoint is seen as a temporary loss of mitotic cells. (**b**) HeLa cells were grown and irradiated with 0 Gy (*filled squares*) or 2 Gy (*open squares*), and nocodazole was then added as described in **Section 3.3**. Checkpoint activation is seen as the delayed increase in mitotic cells in the irradiated population.

a hemocytometer or automated cell counter (e.g., Coulter Counter).

2. For each desired sample, plate $4.75 \times 10^5$ cells per 6 cm plate in 2.5-mL DMEM + 10% FCS and grow for 24 h at 37°C. A separate plate is required for each sample; samples can vary for dose of irradiation and time of recovery after irradiation – *see* **Fig. 1**. Two additional plates of cells are required for staining controls.
3. Irradiate cells with desired dose of ionizing radiation using the manufacturer's instructions. A robust checkpoint arrest is achieved by a dose of 2 Gy. (Alternatively, *see* **Note 2**.)

### 3.2. Synchronization by Double Thymidine Block and Release

In some cases, it is desired to synchronize the cells in G2 prior to irradiation. A simple and reliable method is a double thymidine block and release (*see* **Note 4**).

1. $2 \times 10^6$ cells are plated into a 10-cm petri dish and incubated at 37°C overnight.
2. Thymidine is then added to a final concentration of 2 mM for 17 h.
3. The media is removed by aspiration, the cells washed in PBS, and fresh media is added for 9 h.
4. Thymidine is then again added to 2 mM for 14 h.
5. The cells are again washed, and fresh media is then added. The cells progress out of an early S-phase arrest, and accumulate in G2 3–5 h later depending on the cell line used (*see* **Note 4**).

### 3.3. Nocodazole Trapping

As an alternative to synchronizing cells, the microtubule poison nocodazole can be added to cultures to trap the cells in mitosis; the nuclear envelope disappears, the chromosomes condense, but the spindle does not form and the cells are trapped in a pseudo-metaphase state. That is, there is a time-dependent accumulation of mitotic cells, but checkpoint activation delays the transition from G2 into mitosis – *see* **Fig. 1**.

1. Cells are grown and irradiated as described above in **Section 3.1**.
2. Following irradiation, nocodazole is added to a final concentration of 150 ng/ml.
3. Cells are then harvested at the desired time points (*see* **Note 5**).

### 3.4. Cell Harvesting and Fixation

1. At the desired time points, cells are harvested with trypsin as above, but importantly, the media must be retained to save cells that are not attached, including mitotic and apoptotic cells. Therefore, remove media to a test tube, wash cells in PBS, but retain that PBS and add to the saved media. Add 0.5 mL trypsin and incubate at 37°C until all cells are unattached. Add saved media/PBS to the plate and recover all cells to a single 15-mL falcon tube.
2. Spin tubes at 2000×*g* for 2 min. Wash cell pellet in 5 mL PBS and spin again at 2000×*g* for 2 min.
3. Resuspend cells in 1 mL of PBS containing 0.5% formaldehyde. Incubate for 10 min at 37°C. Spin at 2000×*g* for 2 min and remove fixative. Wash cells in 5 mL of PBS and spin again at 2000×*g* for 2 min.
4. Remove PBS and resuspend cells in 1 mL of ice-cold 90% methanol. Importantly, the methanol is added drop-wise as the cells are mixed gently with a vortex mixer. Cells need to be incubated in methanol for at least 30 min, but can be stored at −20°C in 1.5-mL microfuge tubes until desired (*see* **Note 6**).

### 3.5. Quantification of Mitotic Index

The most quantitative way to determine mitotic index is by flow cytometry. We describe methods using Becton Dickinson Flow cytometers using Cell Quest software, but the same principles can be applied to other machines. As an alternative, cells can be manually counted by microscopy.

1. Recover fixed cells from −20°C and spin in a microfuge at 1000×*g* for 2 min. Rotate the tubes 180°, and spin again at 1000×*g* for 2 min. Remove methanol by aspiration.
2. Resuspend cells in 200 μL of blocking solution and incubate for 1 h at room temperature with gentle agitation.
3. Add 1 μL of primary antibody (Phospho-Histone H3 (Ser10) (6G3), Cell Signaling Technology) to all tubes except one, which remains as a no primary antibody control. Add 2 μL of RNAaseA to each tube. Incubate

overnight at room temperature with gentle agitation (*see* **Note 7**).

4. Harvest the cells in a microfuge at 1000×*g* for 2 min. Rotate the tubes 180°, and spin again at 1000×*g* for 2 min. Remove blocking solution and primary antibody by aspiration. Wash cells three times in 200 μL of blocking solution, and resuspend in 200 μL of blocking solution.
5. Add 2 μL of secondary antibody and incubate for 2 h at room temperature with gentle agitation (*see* **Note 8**).
6. Repeat step 4 to wash the cells three times.
7. Resuspend the cells in 1 mL of PBS. Add 25 μL of Propidium Iodide (PI) to each tube except one, which remains as a no PI control – this tube must receive both primary and secondary antibody.
8. Transfer samples to flow cytometry tubes.
9. The data will be collected by flow cytometry, and will require training for both the hardware and software depending on the model and version used. As a general guide, under set-up mode, establish parameters using the two negative controls. PI will be collected on FL3 on a linear scale, and the secondary antibody signal will be collected on either FL1 (FITC) or FL4 (Cy5) using a log scale. Under the Detectors/Amps control panel, adjust the voltages applied to the detectors so that the no PI sample does not register on FL3, and the no primary antibody FL1 or FL4 signal is below 10. Set the 4 N DNA content to the approximately 400–500 on FL3, and the phospho-histone H3 positive mitotic cells (also 4 N) to above 100 on FL1 or FL4 (*see* **Fig. 2**).
10. The percent of total cells that are positive for histone H3 phosphorylation can then be calculated from the collected data. At least 10,000 cells should be recorded.

## 4. Notes

1. The culture conditions described are standard for many cell lines derived from human tumors and for cultures of fibroblasts of murine origin. The actual culture conditions are not critical for the subsequent assays, and other conditions optimized for a particular cell line can be substituted.
2. The source of DNA damage is not important. Other vendors for $^{137}Cs$ sources can be used, as can other isotopes. Ionizing

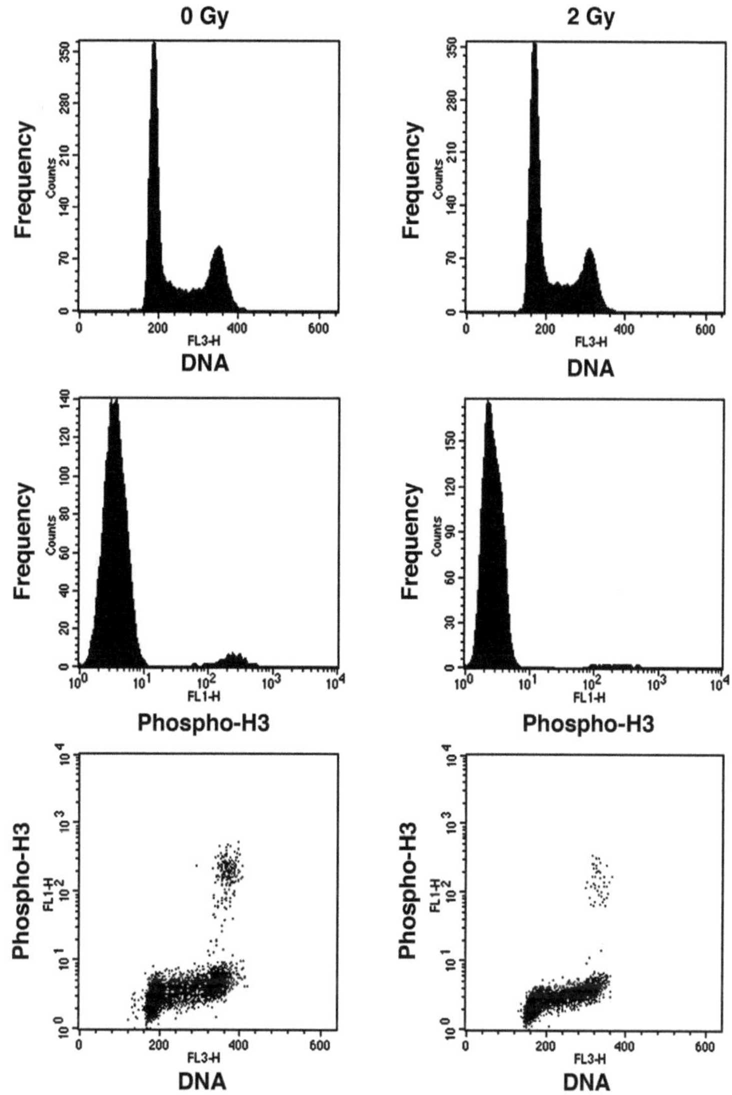


Fig. 2. FACS profiles used for mitotic indices. HeLa cells were grown and irradiated as described in **Section 3.1**. Mitotic indices were determined as described in **Section 3.5**. Shown are histograms for DNA (propidium iodide), phospho-histone H3 (phospho-H3, detected with an FITC labeled secondary antibody), and a 2-D plot of DNA and phospho-histone H3. Note the mitotic cells (between 100 and 1000 on the *y*-axis) also have a 4 N DNA content, and are lost in the irradiated cells.

radiation can be substituted for any radiomimetic drugs. We routinely use the Topoisomerase II inhibitor Etoposide (1–2 μM for 12–24 h) as an alternative. However, particularly for cells that express wild-type p53, it is important to harvest cells prior to widespread apoptosis.

3. Alternatively 16% formaldehyde can be prepared from paraformaldehyde powder. Dissolve 16 g of

paraformaldehyde in 80 mL of PBS, and heat (in a fume hood) to 65°C; this will make a white suspension. Gradually add drops of 5 N NaOH until the solution clears; be careful not to add too much NaOH as the pH will be too high. Filter solution through a 0.2-μm filter and adjust volume to 100 mL.

4. It should be noted that this protocol relies on two rounds of temporary checkpoint activation and will not work in cells that are defective in the checkpoint response. The time at which cells accumulate in G2 does vary between cell lines and will need to be verified by DNA content. In general, the G2 peak is 3–5 h after release from the second thymidine block, 5 h for HeLa cells and NIH3T3 fibroblasts, and 3 h for HCT116 cells.
5. Note that microtubule depolymerization caused by nocodazole will cause many cells to lose adherence to the plate, and so it is again critical to harvest both the cells floating in the media, and those that remain attached to the plate.
6. This method of fixation has been optimized for the antibodies described. If cells are expressing GFP, this fixation method preserves GFP fluorescence.
7. This amount of primary antibody saturates the signal in samples that are up to 50% mitotic cells. If mitotic indices are higher, add an additional 1 μL. Adding excessive primary antibody results in some non-specific staining. There are many commercial alternatives for antibodies that recognize phospho-Histone H3, though the suggested antibody is chosen because of high signal-to-noise ratios. The staining can also be carried out with the MPM-2 monoclonal antibody, which detects multiple phospho-epitopes in mitotic cells.
8. Using an FITC-labeled secondary antibody enables phospho-H3 and propidium iodide-stained DNA to be imaged with a single 488-nm excitation laser. If cells are GFP positive, a far red (Cy5)-labeled secondary antibody must be used, and the flow cytometry must be carried out using a machine equipped with a second 635-nm laser.

## Acknowledgments

This work was supported by NIH grant GM087326.

## References


1. Nurse, P. (1990) Universal control mechanism regulating onset of M-phase, *Nature 344*, 503–508.
2. Dunphy, W. G. (1994) The decision to enter mitosis, *Trends Cell Biol 4*, 202–207.
3. Barr, F. A., Sillje, H. H., and Nigg, E. A. (2004) Polo-like kinases and the orchestration of cell division, *Nat Rev Mol Cell Biol 5*, 429–440.
4. Carmena, M., Ruchaud, S., and Earnshaw, W. C. (2009) Making the Auroras glow: regulation of Aurora A and B kinase function by interacting proteins, *Curr Opin Cell Biol 21*, 796–805.
5. O'Connell, M. J., Krien, M. J., and Hunter, T. (2003) Never say never. The NIMA-related protein kinases in mitotic control, *Trends Cell Biol 13*, 221–228.
6. Davis, F. M., Tsao, T. Y., Fowler, S. K., and Rao, P. N. (1983) Monoclonal antibodies to mitotic cells, *Proc Natl Acad Sci USA 80*, 2926–2930.
7. Hsu, J. Y., Sun, Z. W., Li, X., Reuben, M., Tatchell, K., Bishop, D. K., Grushcow, J. M., Brame, C. J., Caldwell, J. A., Hunt, D. F., Lin, R., Smith, M. M., and Allis, C. D. (2000) Mitotic phosphorylation of histone H3 is governed by Ipl1/aurora kinase and Glc7/PP1 phosphatase in budding yeast and nematodes, *Cell 102*, 279–291.
8. Crosio, C., Fimia, G. M., Loury, R., Kimura, M., Okano, Y., Zhou, H., Sen, S., Allis, C. D., and Sassone-Corsi, P. (2002) Mitotic phosphorylation of histone H3: spatio-temporal regulation by mammalian Aurora kinases, *Mol Cell Biol 22*, 874–885.
9. O'Connell, M. J., Walworth, N. C., and Carr, A. M. (2000) The G2-phase DNA-damage checkpoint, *Trends Cell Biol 10*, 296–303.
10. Tapia-Alveal, C., Calonge, T. M., and O'Connell, M. J. (2009) Regulation of Chk1, *Cell Div 4*, 8.
11. Kuntz, K., and O'Connell, M. J. (2009) The G(2) DNA damage checkpoint: Could this ancient regulator be the achilles heel of cancer? *Cancer Biol Ther 8*, 1433–1439.

# Chapter 4

# Evaluating Spindle Assembly Checkpoint Competence in Mouse Oocytes Using Immunoblotting

Hayden Homer

## Abstract

The spindle assembly checkpoint (SAC) is a quality control mechanism for overseeing the fidelity of chromosome segregation. By modulating the activity of the anaphase-promoting complex or cyclosome (APC/C), the SAC sets the timing of anaphase-onset by co-ordinating the timely destruction of key proteins with the completion of chromosome alignment. How mammalian oocytes regulate chromosome segregation during the first meiotic division (meiosis I) is of immense importance as mis-segregation at this crucial stage in human oocytes underpins the majority of human aneuploidy and birth defects. In recent years, the SAC has been shown to be indispensable for the accuracy of meiosis I chromosome segregation. Here, I describe a technique based on immunoblotting for evaluating SAC competence during meiosis I in mouse oocytes.

**Key words:** Spindle assembly checkpoint (SAC), mouse oocyte, anaphase-promoting complex or cyclosome (APC/C), germinal vesicle breakdown (GVBD).

## 1. Introduction

Recent years have seen an explosion of interest into SAC function during meiosis I in mammalian oocytes, a stage recognised as the Achilles' heel of human reproduction and one that exhibits a peculiar vulnerability to aging (1). An extremely attractive hypothesis is that age-related decline in SAC integrity in oocytes could be a major contributor to the exponential increase in aneuploidy observed as women get older.

The focal point of the SAC is the APC/C acting in concert with its Cdc20 co-activator (APC/C$^{Cdc20}$) (2). Anaphase is brought about through APC/C$^{Cdc20}$-directed destruction of two pivotal proteins, securin and cyclin B (3, 4). The SAC monitors

W.X. Li (ed.), *Cell Cycle Checkpoints*, Methods in Molecular Biology 782,
DOI 10.1007/978-1-61779-273-1_4, © Springer Science+Business Media, LLC 2011

the completion of attachments between spindle microtubules and kinetochores, multiprotein complexes assembled on centromeric DNA (5). To prevent chromosome mis-segregation the SAC liberates APC/$C^{Cdc20}$ activity from under its inhibitory control only after kinetochores have become stably attached to microtubules in the correct configuration.

Therefore, in SAC-competent cells, experimentally perturbing kinetochore-microtubule attachments is expected to invoke SAC activation, a response which could be assayed as the persistent stabilisation of securin and cyclin B. One of the commonest means for disrupting kinetochore–microtubule interactions is by using drugs such as nocodazole to depolymerise the spindle. We have shown that under such conditions, mouse oocytes engage an SAC during meiosis I as measured by securin and cyclin B stabilisation (6). In contrast, failure to stabilise securin and cyclin B in the presence of nocodazole is indicative of SAC compromise. This sort of experimental approach has been used to demonstrate that proteins such as MAD2 and BUB1 are indispensable for SAC integrity during meiosis I in mouse oocytes (6, 7) and is invaluable for ascribing an SAC function to a given gene of interest. Here, I describe the use of immunoblotting for measuring the stability of securin and cyclin B in mouse oocytes treated with nocodazole.

A major drawback when working with mammalian oocytes is their limited availability. Unlike somatic cells, which can be propagated in vitro, oocytes must be derived from living hosts, each of which can only provide a finite number of oocytes. It is therefore imperative that techniques be optimised to yield maximal information from limited biological material. The technique described here enables two proteins of interest, securin and cyclin B, as well as actin as a loading control, to be assayed from a single blot by taking advantage of the differences in molecular weight among the three proteins.

## 2. Materials

### 2.1. Oocyte Harvesting, Culture and Lysis

1. M2 culture medium (with HEPES) (Sigma, Gillingham, Dorset UK).
2. 3-isobutyl-1-methyl-xanthine (IBMX) (Sigma) in dimethyl sulfoxide (DMSO) (*see* **Note 1**).
3. M16 culture medium (Sigma).
4. Nocodazole (Sigma) in DMSO (*see* **Note 2**).
5. 60 × 15 mm and 35 × 10 mm BD Falcon™ polystyrene cell culture dishes (Becton, Dickinson UK Ltd., Oxford, UK).

6. 1% Polyvinylpyrrolidone (PVP) in phosphate-buffered saline (PBS).
7. Sample lysis buffer (4×): 424 mM Tris–HCL, 564 mM Tris, 8% lithium dodecyl sulphate (LDS), 40% glycerol, 2.04 mM ethylenediamine tetraacetic acid (EDTA), 0.88 mM Coomassie® Blue (SERVA® Blue G250; SERVA, Heidelberg, Germany), 0.7 mM Phenol Red (*see* **Note 3**).

### *2.2. Sodium Dodecyl Sulphate (SDS)-Polyacrylamide Gel Electrophoresis (PAGE)*

1. Sample reducing agent: 500 mM dithiotreitol (DTT; 10×) (Invitrogen, Paisley, UK).
2. Prestained molecular weight markers: SeeBlue® Plus2 Prestained Standard (Invitrogen).
3. Precast 4–12% mini-gels (NuPAGE® Novex® 4–12% Bis-Tris Gel; Invitrogen).
4. Electrophoresis system compatible with precast Novex® mini-gels.
5. Running Buffer (20×): 1 M MOPS, 1 M Tris, 2% SDS, 20 mM EDTA (*see* **Note 3**).

### *2.3. Western Blotting*

1. Western blotting transfer buffer: 25 mM Tris, 0.192 M glycine and 20% methanol (*see* **Note 4**).
2. Tris-buffered saline (TBS; 10×): 0.25 M Tris, 1.5 M NaCl, pH 8.0.
3. TBS containing 0.05% Tween-20 (TBST).
4. Polyvinylidene fluoride (PVDF) membrane (Immobilon-P; Millipore, Bedford, MA).
5. Blotting paper (Whatman 3 MM Chr paper; Whatman Ltd., Kent, UK).
6. Electrophoretic Transfer Cell.
7. Blocking solution: 2% non-fat dried milk with 3% bovine serum albumin (BSA) in TBST.
8. Primary antibodies:
   a. Mouse anti-securin (Abcam, Cambridge, MA).
   b. Mouse anti-cyclin B (Abcam).
   c. Mouse anti-actin (Millipore).
9. Secondary antibody: Goat antimouse IgG conjugated to horse radish peroxidise (Bio-Rad).
10. Enhanced chemiluminescent (ECL) reagents (ECL Plus™ Western Blotting Detection System; GE Healthcare, Buckinghamshire, UK).
11. X-ray film (Hyperfilm ECL; GE Healthcare).

## 3. Methods

Morphologically, meiosis I incorporates the events occurring between germinal vesicle breakdown (GVBD) and first polar body extrusion (PBE, **Fig. 1a**). The duration of meiosis I ranges from 6 to 11 h depending upon the strain of mouse (8). In general, for all strains of mice, the levels of securin and cyclin B increase following GVBD and then decrease concomitant upon their destruction during the final ~2 h leading up to PBE (**Fig. 1b**). Our work has centred on the MF-1 strain in which meiosis I lasts about 9–10 h. In MF-1 oocytes, securin and cyclin B increase following GVBD to attain near-maximal levels by 6 h post-GVBD before beginning to decrease in the lead up to PBE (**Figs. 1** and **2a**) (6, 9, 10).

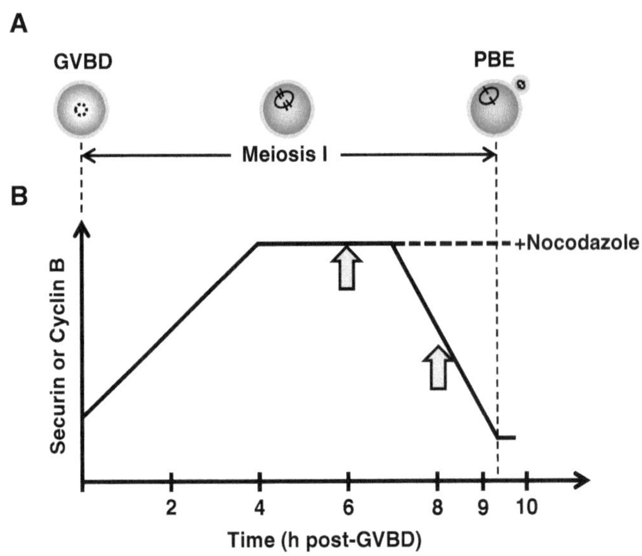


Fig. 1. Schematic of changes in securin and cyclin B levels during meiosis I in mouse oocytes. (**a**) Meiosis I begins with breakdown of the nuclear envelope, termed germinal vesicle breakdown (GVBD) in oocytes, and concludes with first polar body extrusion (PBE). (**b**) Securin and cyclin B undergo net synthesis following GVBD and attain maximal levels by mid-meiosis I. Thereafter, APC/C$^{Cdc20}$-mediated destruction leads to a decline in both proteins over a period of about 2 h, resulting in anaphase I and exit from meiosis I, the latter marked by PBE. To show that nocodazole-induced spindle depolymerisation activates the SAC and hence prevents APC/C$^{Cdc20}$-directed proteolysis, oocytes need to be sampled at time points that will encompass the interval in which a decline in securin and cyclin B would ordinarily occur. This figure is a schematic representation of the timing of events in the MF-1 strain of oocytes. The *vertical grey block arrows* illustrate time points at which oocytes might be sampled in order to detect protein destruction in this strain. The *dashed horizontal line* is a schematic illustration of protein stabilisation following nocodazole treatment, reflecting SAC activation. *See* also **Fig. 2**.

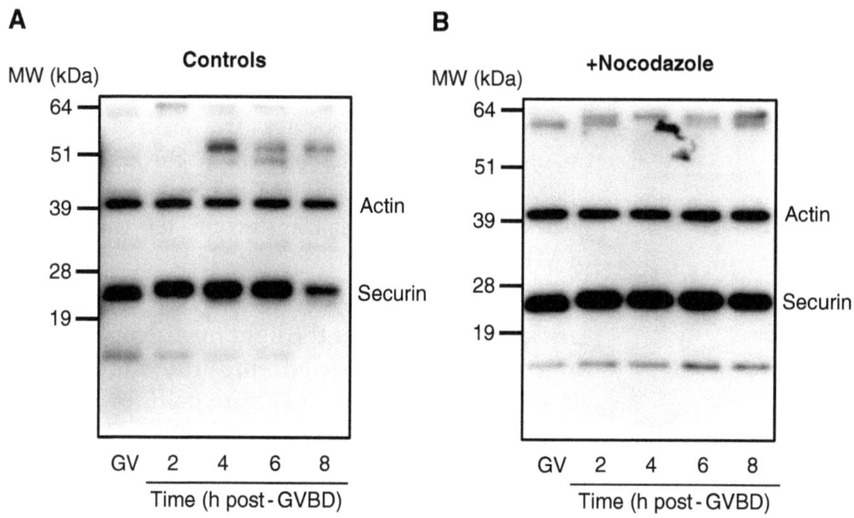


Fig. 2. Western blot of securin and actin in control and nocodazole-treated oocytes. Oocytes (50 oocytes per time point) of the MF-1 strain were immunoblotted at the GV-stage (GV) and at 2-h intervals for the first 8 h post-GVBD. Shown are full-length immunoblots. (**a**) Control oocytes. Note the decline in securin by 8 h post-GVBD. (**b**) Nocodazole-treated oocytes. Note that securin remains stable by 8 h post-GVBD. In this blot, after excising the segment of membrane above the 64-kDa marker, the membrane was first probed for securin and then directly reprobed for actin.

***3.1. Culture Media Preparation***

1. Filter 10 mL of M2 and M16 media (stored at 4°C) using a 22-μm syringe-driven filter unit and a 10-ml syringe.
2. Bring all media to a stable temperature of 37°C. For M2, this can be achieved with a pre-set bench-top hot-block. Apart from being at the correct temperature, M16 should also be equilibrated in an atmosphere of 5% $CO_2$ using an appropriate incubator, usually over a period of 2–3 h.
3. For IBMX-treated medium, add 2.5 μL of stock IBMX to 10 mL of M2 medium for a final concentration of 50 μM.
4. For nocodazole-treated medium, dissolve stock solution in 5 mL of M16 medium for a final concentration of 5 μM and equilibrate in a 5% $CO_2$ incubator (*see* **Note 5**).
5. Prepare 35 × 10 mm culture dishes (×5) with droplets (×4) under mineral oil, three of these dishes with M16 only and two with nocodazole-treated M16 (M16 + Noco).

***3.2. Oocyte Harvesting and Sample Collection***

1. Ovaries are dissected from hormonally primed female mice (*see* **Note 6**).
2. Oocytes are harvested into IBMX-treated M2 medium (*see* **Note 7**) in a 60 × 15 mm culture dish by puncturing ovarian follicles using a fine needle (e.g. 27G; BD Microlance$^{TM}$ 3) on the stage of a light microscope (*see* **Note 8**).

3. Fully grown cumulus-enclosed oocytes are identified and mechanically denuded using an appropriately sized glass pipette.
4. To induce GVBD, oocytes are washed free of IBMX by sequentially transferring them through four droplets of M16 in the first of the pre-equilibrated culture dishes before being transferred to the second dish of M16.
5. After 2 h of culture, oocytes that have undergone GVBD can be identified under the light microscope by the absence of a GV and transferred into the third dish of pre-equilibrated M16 for longer term culture.
6. The objective of the experiment is to obtain four groups of oocytes, two control untreated groups and two nocodazole-treated groups over the period when securin and cyclin B are being degraded in the lead up to PBE (**Fig. 1**; *see* **Note 9**).
7. For adequate signals using the antibodies listed above, we find that a minimum of 30 oocytes per group are required (*see* **Notes 10** and **11**).
8. After a further 4 h of culture (4 h post-GVBD), half of the oocytes are washed through droplets of nocodazole-treated medium in the first of the M16 + Noco dishes before being divided amongst the droplets in the second M16 + Noco dish (*see* **Note 12**). Thus, there will now be two groups of oocytes, one group in M16 medium (controls) and the other group in M16 + Noco medium (nocodazole-treated).
9. After a further 2 h of culture (6 h post-GVBD), two samples, one control and one nocodaole-treated, are obtained (*see* **Note 13**) (*see* **Fig. 1b**).
10. For sample collection, oocytes are briefly washed through pre-warmed 1% PVP and transferred in 10.4 μL volumes of 1% PVP into Eppendorfs containing 4 μL of sample lysis buffer (4×) and immediately snap-frozen in dry-ice. Samples can then be maintained at −80°C until ready for use.
11. After a further 2 h of culture (8 h post-GVBD), another two samples of oocytes are lysed and snap-frozen as above (*see* **Note 13**) (*see* **Fig. 1b**).

### 3.3. SDS-PAGE

1. This protocol assumes the use of an XCell *SureLock*™ Mini-Cell gel unit and Novex® pre-cast mini-gels.
2. Prepare 800 mL of running buffer by adding 40 mL of the 20× stock solution to 760 mL of ultra-pure water in a beaker and mixing gently using a magnetic stir-bar.

3. Prepare the mini-gel as per manufacturer's instructions by cutting open one of the individually sealed gel pouches, rinsing with deionised water and flushing the wells with running buffer (*see* **Note 14**).
4. Assemble the gel unit. Fill the upper chamber with 200 mL of running buffer and ensure that there is no leakage into the lower chamber before filling the latter with the remaining 600 mL of running buffer.
5. Thaw samples on ice and add 1.6 μL of reducing agent (10×) to each Eppendorf for a final sample volume of 16 μL.
6. Mix samples thoroughly and heat to 70°C for 10 min before loading wells.
7. Connect the gel unit to a power supply. Run at 200 V constant until the dye front reaches the edge of the gel (50–60 min is usually sufficient) (*see* **Note 15**).

### 3.4. Western Blotting for Securin, Cyclin B and Actin

1. Instructions for the electrophoretic transfer of proteins assume the use of a mini trans-blot cell.
2. Whilst the SDS-PAGE is underway, prepare the following:
   a. Blocking solution (*see* **Note 16**).
   b. Four sheets of blotting paper (9 × 7 cm).
   c. A segment of PVDF membrane (9 × 7 cm) (*see* **Note 14**). Pre-wet membrane in methanol for 30 s, rinse in water and then immerse in transfer buffer for 5–10 min.
   d. A rectangular tray containing transfer buffer into which fiber pads and sheets of blotting paper are submerged.
3. Immediately following SDS-PAGE, turn off the power supply and disconnect the power cables. Remove the gel cassette and prise apart the plastic plates encasing the gel using the gel knife. The gel will remain attached to one of the plastic plates. Rinse briefly with transfer buffer to remove traces of running buffer and trim the gel to remove unused segments using the gel knife.
4. Open the gel holder cassette and, with the black surface laid flat in the tray of transfer buffer, assemble the gel sandwich as follows. Place one fiber pad on the black surface of the open cassette followed by two sheets of blotting paper. Keeping the assembly together, gently apply the blotting paper to the gel on the plastic casing and invert so that the gel becomes adherent to the blotting paper. Cover the gel with the pre-wetted PVDF membrane and gently roll a glass rod over the membrane to extrude any trapped air bubbles.

5. Complete the assembly of the gel sandwich within the gel holder cassette using two further sheets of blotting paper over the membrane followed by the second filter pad. Ensure that the gel is towards the black surface of the gel holder cassette and that the membrane is towards the clear surface. Close and lock the cassette being careful to avoid movement of the gel and membrane.
6. Place a magnetic stir-bar into the transfer tank followed by the electrode module and cooling unit (*see* **Note 17**). Insert the gel holder cassette containing the gel sandwich into the module ensuring that the black surface is towards the black portion of the module (*see* **Note 18**). Fill the tank with pre-cooled transfer buffer and place the lid on the tank.
7. Place the tank on a magnetic stirrer and connect to the power supply. Circulate the transfer buffer by activating the stirrer to the fastest rate possible. Transfer at 400 mA constant for a total of 2 h (*see* **Note 19**).
8. Following transfer, switch off the power and remove the gel holder cassette. Carefully disassemble the gel sandwich by removing the fiber pad and blotting paper overlaying the membrane (*see* **Note 14**). With the membrane still overlaying the gel, the limits of the gel can be gently outlined on the membrane using a pencil. Trim off the excess membrane using a clean pair of scissors (*see* **Notes 20–22**).
9. To orientate the membrane, cut a corner out of the membrane above the lane having the prestained markers (*see* **Notes 14** and **23**).
10. Block the membrane in 10 mL of blocking solution for 1 h at room temperature (*see* **Note 24**), gently agitating on a rocking platform.
11. Following blocking, rinse the membrane in TBST.
12. Membranes can then either be sequentially incubated with antibodies (as done for **Fig. 2**) or the membrane can be divided transversely into two portions, each of which is then separately incubated with the relevant antibody (as done for **Fig. 3**) (*see* **Note 25**).
13. Whichever approach is taken, primary antibody dilutions in blocking solution (*see* **Note 25**) are as follows:
    a. Mouse anti-cyclin B (1:400).
    b. Mouse anti-securin (1:1000).
    c. Mouse anti-actin (1:3000).
14. Incubate in primary antibody overnight at 4°C with gentle agitation.

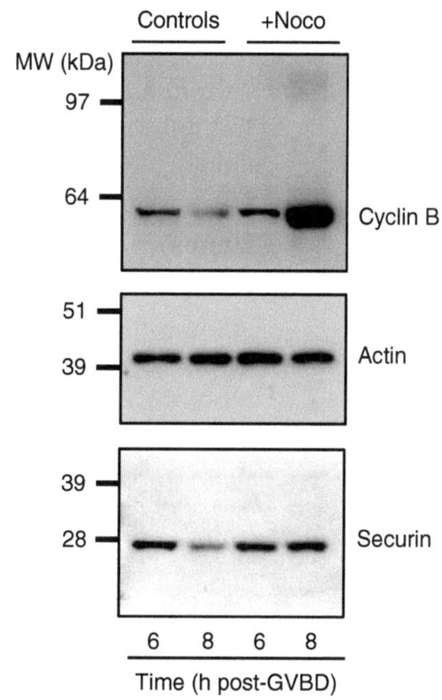


Fig. 3. Western blot of securin, cyclin B and actin in control and nocodazole-treated MF-1 oocytes. Control and nocodazole-treated (+Noco) oocytes (30 oocytes per time point) were immunoblotted for securin, cyclin B and actin at 6 and 8 h post-GVBD. Note the decline in signal intensity for both securin and cyclin B between 6 and 8 h post-GVBD in controls. In contrast, after nocodazole treatment both proteins are stabilised between 6 and 8 h post-GVBD. Unlike the membrane in **Fig. 2**, in the blot shown here, the membrane was divided after blocking around the 51-kDa band of the standard protein marker (*see* **Note 21**). The upper segment was then used to probe first for cyclin B (migrates at ~62 kDa) whilst the lower segment was simultaneously used to probe for securin (migrates at ~26 kDa). The lower segment was directly reprobed for actin (migrates at ~43 kDa) as a loading control at a later date.

15. Following overnight incubation, rinse in a large excess of TBST, then wash 3× with TBST, 5 min each time.
16. For detecting any of the three proteins, incubate in goat anti-mouse secondary at a dilution of 1:1000 in blocking solution for 1 h at room temperature on a rocking platform.
17. Wash as in Step 15.
18. Expose membrane to ECL Plus™ reagents for 5 min as per manufacturer's instructions. A total of 1 mL of ECL will be sufficient.
19. Drain off excess ECL reagents, and place the membrane between the leaves of a transparent acetate sheet protector that has been trimmed down to the appropriate size.

20. Detect bands by exposing X-ray film to the acetate-enclosed ECL-treated membrane using an X-ray film cassette in a dark room under safe light conditions. Usually, 1–2 min of exposure to film is sufficient to produce strong signals.
21. Following detection, wash the membrane in TBST for 10 min.
22. Directly reprobe the membrane (or segments of membrane) with primary and secondary antibodies (as per Steps 13–20 above) (11) until all proteins have been detected (*see* **Note 26**).

## 4. Notes

1. For 200 mM stock solution of IBMX, weigh 0.044 g of IBMX and dissolve in 1 mL of DMSO. This can then be divided into 15- to 20-μl aliquots and kept at −20°C. Stock solution should be renewed after 3–4 months as IBMX can lose potency beyond this time.
2. Nocodazole is solubilised in DMSO to make a stock solution of 5 mg/mL, divided into 10- to 20-μl aliquots and stored frozen at –20°C.
3. These reagents are commercially available from Invitrogen.
4. We find it is best if transfer buffer is freshly prepared on the day prior to transfer. For 2 L of transfer buffer, weigh 28.8 g of glycine and 6.06 g of Tris, dissolve in 1600 mL of water and make to 2 L with 400 mL of methanol. Divide into two 1 L glass bottles and store overnight at 4°C.
5. Micromolar concentrations of nocodazole have been shown to disrupt the spindle in mouse oocytes (6, 12–14). In contrast, nanomolar concentrations of nocodazole do not depolymerise the spindle although it does activate an SAC-dependent meiosis I delay (14–17). We suggest using nocodazole at a concentration of 5 μM, which we have previously shown to completely depolymerise the spindle and to induce a robust meiosis I arrest in MF-1 oocytes (6).
6. To stimulate ovarian follicle recruitment and so increase the yield of oocytes per animal, mice are hormonally primed by intra-peritoneally injecting 7.5–10 I.U. of pregnant mares' serum gonadotrophin (PMSG) about 44–46 h prior to sacrificing the animal to obtain ovaries.
7. To synchronise GVBD and hence collect samples of oocytes at comparable stages of meiosis I, it is important to use drugs such as IBMX, which inhibit GVBD.

8. Before commencing follicle puncture for releasing oocytes, it is important to dissect off any remnants of oviduct and para-ovarian tissue from around the ovaries.
9. In evaluating SAC competence, the objective is to determine whether securin and cyclin B are stabilised when the spindle is depolymerised. By means of immunoblotting, one would anticipate stable signals for securin and cyclin B in the nocodazole-treated group during the interval when signals for both proteins show a decline in controls. For MF-1 oocytes, both proteins undergo destruction between 6 h post-GVBD and 9 h post-GVBD (*see* **Figs. 1b** and **2a**).
10. For 4 samples of oocytes, and given that 30 oocytes per sample are needed for an adequate signal (*see* **Note 11**), a minimum of 120 GVBD oocytes will be required. Given an ~80% GVBD rate by 2 h after release from IBMX-treated medium, about 150 GV-stage oocytes should be present after ovarian puncture. In our hands, at least 40 fully grown meiotically competent oocytes can be obtained from one 4- to 6-week-old PMSG-primed mouse. Overall therefore, four mice should suffice for this experiment. Note that oocyte yield may differ with other strains of mice and under different animal housing conditions.
11. Other labs may find that their threshold number of oocytes for signal detection may differ from that presented here.
12. In mouse oocytes, bipolar spindle assembly is in its advanced stages by mid-meiosis I (8, 12, 18). For most mouse strains, 4 h post-GVBD is an appropriate stage for oocytes to be transferred into nocodazole as this is late enough that the bipolar spindle is mostly formed but early enough that securin and cyclin B proteolysis is not yet underway.
13. For the protocol detailed here, the time points set for sample collection (6 h post-GVBD and 8 h post-GVBD) pertain to MF-1 oocytes as this is when protein degradation becomes evident in this strain of mouse (6, 9, 10) (*see* **Figs. 1b**, **2a** and **3**). For other mouse strains, the dynamics of securin and cyclin B should be characterised in order to ascertain the most appropriate times for sample collection.
14. Always wear gloves when handling the gel cassette and membranes to avoid contamination.
15. It is important to prevent the dye front from completely running off the gel as the dye forms a useful guide for subsequently trimming off the unused gel portions.
16. Blocking solution is prepared fresh. Weigh 0.6 g of non-fat dried milk and 0.9 g of BSA and dissolve in 30 mL of TBST.

17. After each use, ensure that the ice-tray cooling unit of the blotting module is replaced in the −20°C freezer so that an ice-filled cooling unit is available for subsequent experiments.
18. The correct orientation is vitally important to prevent protein being transferred into the buffer rather than on to the PVDF membrane.
19. The buffer will gradually warm up during transfer. Consequently, we divide the transfer into two equal portions of ~1 h each thereby allowing warm transfer buffer to be exchanged for cooled buffer, hence the reason for preparing two separate bottles of transfer buffer (*see* **Note 4**). Also, after 1 h, the ice in the cooling unit would have melted and can be replaced with ice blocks or preferably, a spare cooling unit with ice can be used.
20. The segment of membrane above the 64-kDa marker can be excised as the heaviest protein, cyclin B, migrates at 62 kDa (*see* **Notes 21** and **22**).
21. These markers pertain to the use of SeeBlue® Plus2 with MOPS running buffer.
22. Relatively high concentrations of antibodies are often required for detecting cell cycle proteins in mouse oocytes and for this reason, trimmed down membrane portions that allow comparatively small volumes of antibody solution are advantageous.
23. Occasionally, after transfer, portions of the membrane may become dried out. The membrane can be re-wet by immersing in TBST for about 1 h.
24. Blocking can be undertaken in a 90-mm Petri dish (Scientific Laboratory Supplies, Yorkshire, UK), which will accommodate the membrane after it is trimmed down.
25. Dividing the membrane into two portions enables both securin and cyclin B to be simultaneously probed thereby producing faster results (*see* **Fig. 3**). Dividing the membrane also allows smaller volumes of antibody solution to be used (*see* **Note 22**). For four test samples with an adequately trimmed down membrane, 3 mL of antibody solution in a 60 × 15 mm culture dish will suffice for each membrane segment (as was the case for **Fig. 3**). Otherwise, double this volume in a 90-mm Petri dish will be required when the membrane is left undivided (as for **Fig. 2**).
26. The three proteins of interest migrate at ~62 kDa (cyclin B), ~43 kDa (actin) and ~26 kDa (securin) and are therefore sufficiently separated from one another to avoid the need for membrane stripping.

## Acknowledgements

This work was funded by a Wellcome Trust Clinical Fellowship (082587/Z/07/Z).

### References


1. Homer, H. (2007) Ageing, aneuploidy and meiosis: eggs in a race against time. *Yearbook of Obstetrics and Gynaecology – Volume 12*, ed. Hillard, T. (RCOG Press, London), pp. 139–158.
2. Musacchio, A. & Salmon, E. D. (2007) The spindle-assembly checkpoint in space and time. *Nat Rev Mol Cell Biol* **8**, 379–393.
3. Peters, J. (2006) The anaphase promoting complex/cyclosome: a machine designed to destroy. *Nat Rev Mol Cell Biol* 7, 644–656.
4. Reis, A., Madgwick, S., Chang, H. Y., Nabti, I., Levasseur, M., & Jones, K. T. (2007) Prometaphase APCcdh1 activity prevents non-disjunction in mammalian oocytes. *Nat Cell Biol* **9**, 1192–1198.
5. Cheeseman, I. M. & Desai, A. (2008) Molecular architecture of the kinetochore-microtubule interface. *Nat Rev Mol Cell Biol* **9**, 33–46.
6. Homer, H., McDougall, A., Levasseur, M., Murdoch, A., & Herbert, M. (2005) Mad2 is required for inhibiting securin and cyclin B degradation following spindle depolymerisation in meiosis I mouse oocytes. *Reproduction* **130**, 829–843.
7. McGuinness, B. E., Anger, M., Kouznetsova, A., Gil-Bernabq, A. M., Helmhart, W., Kudo, N. R., Wuensche, A., Taylor, S., Hoog, C., Novak, B., et al. (2009) Regulation of APC/C activity in oocytes by a Bub1-dependent spindle assembly checkpoint. *Curr Biol* **19**, 369–380.
8. Polanski, Z., Ledan, E., Brunet, S., Louvet, S., Verlhac, M.-H., Kubiak, J. Z., & Maro, B. (1998) Cyclin synthesis controls the progression of meiotic maturation in mouse oocytes. *Development* **125**, 4989–4997.
9. Homer, H., Gui, L., & Carroll, J. (2009) A spindle assembly checkpoint protein functions in prophase I arrest and prometaphase progression. *Science* **326**, 991–994.
10. Homer, H., McDougall, A., Levasseur, M., Yallop, K., Murdoch, A., & Herbert, M. (2005) Mad2 prevents aneuploidy and premature proteolysis of cyclin B and securin during meiosis I in mouse oocytes. *Genes Dev* **19**, 202–207.
11. Liao, J., Xu, X., & Wargovich, M. J. (2000) Direct reprobing with anti-b-actin antibody as an internal control for western blot analysis. *BioTechniques* **28**, 216–218.
12. Brunet, S., Santa Maria, A., Guillaud, P., Dujardin, D., Kubiak, J. Z., & Maro, B. (1999) Kinetochore fibers are not involved in the formation of the first meiotic spindle of mouse oocytes, but control the exit from the first meiotic M phase. *J Cell Biol* **146**, 1–12.
13. Brunet, S., Pahlavan, G., Taylor, S., & Maro, B. (2003) Functionality of the spindle checkpoint during the first meiotic division of mammalian oocytes. *Reproduction* **126**, 443–450.
14. Homer, H. (2006) Mad2 and spindle assembly checkpoint function during meiosis I in mammalian oocytes. *Histol Histopathol* **21**, 873–886.
15. Li, M., Li, S., Yuan, J., Wang, Z. B., Sun, S. C., Schatten, H., & Sun, Q. Y. (2009) Bub3 is a spindle assembly checkpoint protein regulating chromosome segregation during mouse oocyte meiosis. *PLoS One* **4**, e7701.
16. Wassmann, K., Niault, T., & Maro, B. (2003) Metaphase I arrest upon activation of the MAD2-dependent spindle checkpoint in mouse oocytes. *Curr Biol* **13**, 1596–1608.
17. Yin, S., Wang, Q., Liu, J. H., Ai, J. S., Liang, C. G., Hou, Y., Chen, D. Y., Schatten, H., & Sun, Q. Y. (2006) Bub1 prevents chromosome misalignment and precocious anaphase during mouse oocyte meiosis. *Cell Cycle* **5**, 2130–2137.
18. Schuh, M. & Ellenberg, J. (2007) Self-organization of MTOCs replaces centrosome function during acentrosomal spindle assembly in live mouse oocytes. *Cell* **130**, 484–498.

# Chapter 5

# Analysis of Changes in Protein Level and Subcellular Localization During Cell Cycle Progression Using the Budding Yeast *Saccharomyces cerevisiae*

**Xiaorong Wu, Lili Liu, and Mingxia Huang**

## Abstract

Methods are described here to monitor changes in protein level and subcellular localization during the cell cycle progression in the budding yeast *Saccharomyces cerevisiae*. Cell synchronization is achieved by an α-factor-mediated block-and-release protocol. Cells are collected at different time points for the first two cell cycles upon release. Cellular DNA contents are analyzed by flow cytometry. Trichloroacetic acid protein precipitates are prepared for monitoring levels of cell cycle regulated proteins by Western blotting. The dynamic changes in protein subcellular localization patterns are examined by indirect immunofluorescence microscopy.

**Key words:** Cell cycle, checkpoint, cell synchronization, flow cytometry, immunofluorescence microscopy, protein stability, subcellular localization, *Saccharomyces cerevisiae*.

## 1. Introduction

Cell cycle checkpoints are regulatory pathways that ensure the proper order and timing of phase-specific events in order to maintain genomic stability (1–4). Extensive studies from numerous laboratories have established that the basic mechanisms of cell cycle checkpoints are evolutionarily conserved in all eukaryotes (2, 5–7). The budding yeast *Saccharomyces cerevisiae* is a popular model eukaryotic organism for the study of various cellular

W.X. Li (ed.), *Cell Cycle Checkpoints*, Methods in Molecular Biology 782,
DOI 10.1007/978-1-61779-273-1_5, 

processes including cell cycle regulation (8). In fact, the concept of checkpoint was initially developed in *S. cerevisiae* over two decades ago (9). The ease of propagation and maintenance, combined with its powerful genetic, genomic, molecular, and biochemical tools, has made the budding yeast an organism of choice for studying the mechanisms underlying cell cycle checkpoints.

A critical technique in cell cycle experiments is cell synchronization. Two methods are routinely used to obtain synchronized yeast populations: centrifugal elutriation and block-and-release protocols. While elutriation can collect homogenously dividing cells without much perturbation (10), the block-and-release methods have the advantage of easily generating much larger populations of synchronized cells for biochemical characterization of proteins of interest at different time points of the cell cycle (11, 12). Among various block-and-release protocols the yeast pheromone α-factor is frequently used to arrest *MATa* cells at the G1/S-phase boundary (START) by inhibiting Cdc28-Cln activity (13). The arrested cells have an 1 N DNA content and a pear-shaped ("schmoo") morphology (14). Upon washing off α-factor in the medium, yeast cells recover quickly and progress synchronously through the first couple of cell cycles. The progression of cells through cell cycle can be monitored by examining changes in cellular morphology (budding index analysis) and by measuring cellular DNA contents using fluorescence-activated cell sorting (FACS). In this chapter, we compiled protocols of a series of time-course experiments of cell cycle analysis. Cells are synchronized in G1/S-phase boundary by α-factor-mediated arrest and released back into the cell cycle, whereupon samples are collected at 10-min intervals during the first two cell cycles and processed for FACS and protein analyses.

The ribonucleotide reductase (RNR) catalyzes conversion of ribonucleoside diphosphates to deoxyribonucleoside diphoasphates, a rate-limiting step in deoxyribonucleotide (dNTP) biosynthesis (15). RNR is regulated by both the cell cycle and DNA damage checkpoints. Two regulatory processes are responsible for the increase in RNR activity and dNTP production during S phase. One event is the nucleus-to-cytoplasm redistribution of the RNR small subunit, which allows co-localization of the large and small subunits of RNR in the cytoplasm and increases holoenzyme formation (16). Another event is phosphorylation and proteolysis of the Rnr1 inhibitor Sml1, which further enhances RNR catalytic activity (17). In the following, we use the RNR small subunit protein Rnr2 as an example for analysis of protein subcellular localization pattern changes by using indirect immunofluorescence (IMF), and Sml1 for protein level fluctuation by using Western blotting.

## 2. Materials

The water for making media and solutions should be double distilled ($ddH_2O$) or from a Milli-Q still (Millipore). Yeast cultures are grown at 30°C.

### 2.1. Culture Growth and Synchronization

1. Yeast strains
   *MATa ade2 his3 leu2 trp1 ura3 can1,*
   *MATa ade2 his3 leu2 trp1 ura3 can1 Myc-RNR2.*
2. YEPD medium: 1% yeast extract, 2% peptone, and 2% dextrose.
3. α-factor is from GenScript and Sigma-Aldrich. Make 5 mg/mL stock solution in 0.1 M HCl and keep aliquots at −20°C.

### 2.2. Flow Cytometry

1. 70% Ethanol.
2. 1× PBS buffer, pH 7.4: 40 mM $K_2HPO_4$, 10 mM $KH_2PO_4$, 0.15 M NaCl, pH 7.4.
3. FACS buffer: 0.2 M Tris–HCl, pH 8.0, 20 mM EDTA, pH 8.0, with 0.1% RNAse A to be added immediately before use.
4. 500 μg/mL × propidium iodide (10× PI) stock solution: dissolve 10 mg of PI (Sigma-Aldrich) in 20 mL of 1× PBS, pH 7.4 buffer in a 50-mL conical tube, wrap the tube in aluminum foil, and store at 4°C. The 1× PI working solution is made freshly before use by dilution of one part 10× PI stock solution with nine parts 1× PBS, pH 7.4.
5. Sonicator (Branson Sonifier 250).

### 2.3. Protein Extraction and Western Blotting

1. 20% trichloroacetic acid (TCA).
2. Glass beads, 0.5 mm size (Research Products International Co).
3. Bullet Blender™ (Next Advance Inc.).
4. 18-gauge syringe needle.
5. Sonicator (Branson Sonifier 250).
6. 1 M Tris-base (pH not adjusted).
7. Sodium dodecyl sulfate polyacrylamide gel electrophoresis (SDS-PAGE):
   4× resolving gel buffer : 1.5 M Tris–HCl, pH 8.8, 0.4% SDS;
   4× stacking gel buffer 0.5 M Tris–HCl, pH 6.8, 0.4% SDS;

2× SDS protein loading buffer: 125 mM Tris–HCl, pH 6.8, 20% glycerol, 4% SDS, 2% β-mecaptomethanol, 0.01% bromophenol blue;

1× TGS (Tris–Glycine–SDS) running buffer: 25 mM Tris-base, 200 mM glycine, 0.1% SDS;

10% ammonium persulfate;

TEMED (*N,N,N,N′*-tetramethyl-ethylenediamine).

30%/0.8% acrylamide/bis-acrylamide solution.

8. 1× Transfer buffer: 25 mM Tris-base, 195 mM glycine, and 15% methanol.
9. 10× Tris-buffered saline (10× TBS) buffer: 200 mM Tris–HCl, pH 7.6, 1.37 M NaCl.

   For 1× TBST, mix 100 mL 10× stock with 900 mL $ddH_2O$ and add 0.5 mL Tween-20 to 0.05%.
10. Blocking buffer: 5% nonfat dry milk in 1× TBST.
11. Antibody dilution buffer: 1× TBST supplemented with 1% nonfat dry milk.
12. Antibodies: the polyclonal anti-Sml1 is a gift from JoAnne Stubbe lab (Massachusetts Institute of Technology, Department of Chemistry); antibodies against glucose-6-phosphate dehydrogenase (anti-G6DPH) is from Sigma-Aldrich. Horse Radish Peroxidase (HRP) conjugated goat-anti-rabbit IgG is from Jackson ImmunoResearch.
13. Enhanced chemiluminescence reagents (Western Lightening Plus-ECL from PerkinElmer).

### 2.4. Indirect Immunofluorescence (IMF)

1. Formaldehyde 37% (w/v).
2. K buffer: 0.1 M $KPO_4$ pH 6.5, 5 mM $MgCl_2$.
3. P buffer: 0.1 M $KPO_4$ pH 6.5, 5 mM $MgCl_2$, 1.2 M sorbitol.
4. β-mecaptoethanol.
5. Zymolyase 100T (Fisher Scientific): make 10 mg/mL stock solution in 0.1 M sorbitol, aliquot and store at –20°C.
6. Poly-L-lysine (Sigma-Aldrich): make 1% stock solution with $ddH_2O$ and store aliquots at –20°C.
7. Methanol: prechilled at –20°C in a Coplin jar.
8. Acetone: prechilled at -20°C in a Coplin jar.
9. PBS/BSA buffer: 10 mg/mL BSA in 1 × PBS, pH 7.4, with 0.01% $NaN_3$.
10. Antibodies: the monoclonal anti-Myc antibody 9E10 is from Covance, FITC goat-anti-mouse IgG conjugate is from Jackson ImmunoResearch .

11. 2 μg/mL DAPI (4,6-diamidino-2-phenylindole, Sigma-Aldrich): make 1000× stock (2 mg/mL) solution in $ddH_2O$, store aliquots in the dark at –20°C.
12. VECTASHIELD mounting medium (Vector Laboratories).
13. 25 × 75 × 1 mm microscope slides with hydrophobic coating, 14-well, 5 mm diameter (Thermo Scientific Erie) and 22 × 60 × 0.15 mm coverslips.
14. Rubber cement.

## 3. Method

### 3.1. Synchronization of Cells by α-Factor

1. Pick a single colony into 5-mL YEPD liquid to grow overnight at 30°C.
2. Dilute the overnight culture with YEPD to a cell density of 2–4 × $10^6$ cells/mL and grow at 30°C to early-to-mid log phase (0.6–1.0 × $10^7$ cells/mL, *see* **Note 1**).
3. Remove a sample of the asynchronous culture before adding α-factor. Take 0.5–1 × $10^7$ cells for FACS, 2–4 × $10^7$ cells for IMF and 0.6~1.0 × $10^8$ cells for protein extraction and Western blotting).
4. Add α-factor to 5–10 μg/mL and keep incubating at 30°C. It takes 1.5–3 h to arrest cells at G1/S-phase boundary at 30°C (*see* **Note 2**).
5. Begin assessing cell cycle arrest efficiency at 1.5 h after α-factor addition. Take 5 μL cell culture and examine under a microscope. Cells that were prior to START at the time of α factor addition will arrest at the G1/S-phase boundary with a "schmoo" morphology and continue to grow larger. Cells past START at α-factor addition will divide (i.e. exhibiting a budded morphology) and arrest at the G1/S-phase boundary of the subsequent cell cycle. An additional dose of 5 μg/mL α-factor may be added to the culture after the initial 1.5 h if necessary. Examine cells every half hour until they are > 95% "schmoo" (i.e. <5% budded).
6. Remove an α-factor-arrested faction for assessing synchrony (*see* **Notes 3–5**). Release cells from α-factor arrest by centrifugation (4000×$g$ for 3–5 min), washing twice with prewarmed (30°C) YEPD, and resuspend in prewarmed (30°C) YEPD to a final concentration of 0.5–1 × $10^7$ cells/mL.
7. Remove a time-zero fraction. Return the synchronized population to 30°C and begin taking time points to collect fractions of culture every 10 min for 120–180 min.

### 3.2. Processing Cells for Flow Cytometry Analysis

1. Fixation. Collect 0.5–1.0 × $10^7$ cells (*see* **Note 5**) by centrifugation and resuspend in 0.5–1.0 mL of 70% ethanol, incubate from 30 min to overnight on bench top. The cells are stable at this point for up to 2 weeks at 4°C.
2. Rehydration. Collect ethanol-fixed cells by centrifugation and resuspend in 1 mL of 1× PBS, pH 7.4. Rehydration can take approximately 1 h if the cells have been in 70% ethanol for extended time. The cells are stable at this point for up to 2 weeks at 4°C. Add $NaN_3$ to a final concentration of 0.1% if the cells are to be kept for a while.
3. RNAse A treatment. Collect rehydrated cells by centrifugation and resuspend into 100 μL of FACS buffer supplemented with 0.1% RNase A. Incubate at 37°C from 4 h to overnight.
4. PI staining. Collect cells by centrifugation and resuspend in 100 μL of 1× PBS supplemented with 50 μg/mL of PI. Staining occurs rapidly at room temperature (within 1 h), although this is often a convenient point to leave the cells overnight at 4°C in the dark. Cells can kept for 1–2 weeks before FACS analysis (**Fig. 1A**).
5. Sonication. Add 900 μL 1 × PBS to each sample for a dilution of 1:10. Sonicate briefly (10–12 s at 20% output) to break up cell aggregates before proceeding to flow cytometry.

### 3.3. Monitoring Protein Level Changes by Western Blotting

#### 3.3.1. Preparing Total Protein Extracts by TCA Precipitation

1. Collect 0.6–1 × $10^8$ cells by centrifugation and wash once with 1 mL of 20% TCA.
2. Resuspend cells in 500 μL of 20% TCA in a 1.5-mL microtube, add equal volume of 0.5 mm diameter glass beads, and disrupt cells by vortexing in the Bullet Blender at speed level 9 for 2.5 min at 4°C.
3. Puncture a hole at the bottom of the microtube with an 18-gauge syringe needle, insert it into another 1.5-mL tube and collect the liquid through the hole into the new tube by centrifugation at 2000×*g* for 30 s.
4. Centrifugate the collected liquid at 10,000×*g* for 10 min at 4°C.
5. Discard the supernatant, add 100 μL of 2× SDS loading buffer to the pellet, and then add 50 μL of 1 M Tris-base to adjust the pH, the dye will change from yellow to blue as the pH is being adjusted.
6. Sonicate each sample on the Branson Sonifier 250 for 5–10 s at 20% output to solubilize the pellet.
7. Heat the samples at 95–100°C for 5 min and chill on ice. Centrifugate at 2000×*g* for 10 min before loading.

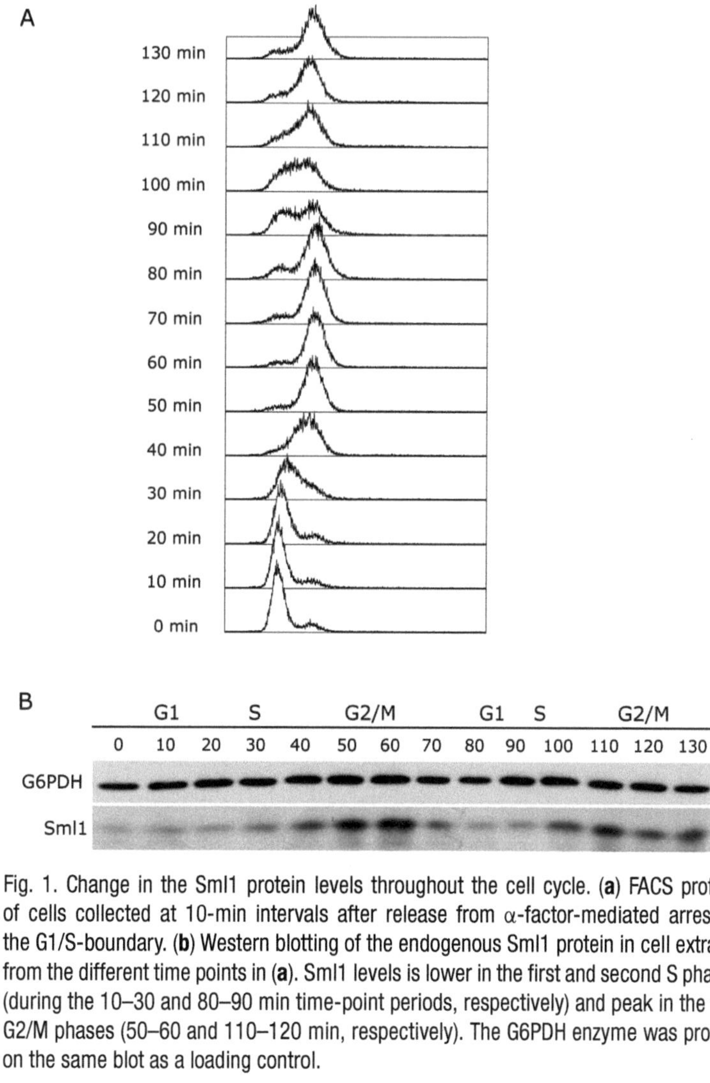


Fig. 1. Change in the Sml1 protein levels throughout the cell cycle. (**a**) FACS profiles of cells collected at 10-min intervals after release from α-factor-mediated arrest at the G1/S-boundary. (**b**) Western blotting of the endogenous Sml1 protein in cell extracts from the different time points in (**a**). Sml1 levels is lower in the first and second S phases (during the 10–30 and 80–90 min time-point periods, respectively) and peak in the two G2/M phases (50–60 and 110–120 min, respectively). The G6PDH enzyme was probed on the same blot as a loading control.

*3.3.2. SDS-PAGE and Western Blotting*

We use the Bio-Rad Mini-Protean and Mini Trans-blot systems for SDS-PAGE and protein transfer to membrane for immunoblotting. For SDS-PAGE of Sml1, 10–12% gel is recommended because of the small size of the protein. Prepare 5 mL separating gel mixture (1.2 mL of 4× separating gel buffer, 2 mL of 30%/0.8% acrylamide/bis-acrylamide , 1.8 mL of $ddH_2O$, 17 μL of 10% ammonium persulfate and 3.5 μL of TEMED), and 1 mL of stacking gel mixture (0.25 mL of 4× stacking gel buffer, 0.17 mL of 30%/0.8% acrylamide/bis-acrylamide, 0.6 mL of $ddH_2O$, 8 μL of 10% ammonium persulfate, and 1.5 μL of TEMED) for a 0.75-mm mini-gel. TCA protein suspension equivalent to ~2 × $10^6$ cells is loaded per lane. Electrophoresis conditions are 80 V for ~2.5 h or 20 V for 10 h, till the dye front reaches the bottom of the gel. Slower voltage and longer electrophoresis time can help better separation of different mobility species of the same protein due to phosphorylation.

Transfer of proteins to either PVDF or nitrocellulose membrane is done in 1× transfer buffer in the Mini Trans-blot tank at 300 mA for 1 h at 4°C. After transfer, the membrane is incubated first in blocking buffer for at least 1 h, then in antibody dilution buffer containing anti-Sml1 at 1:10,000 dilution for 2 h, followed by three washes with 1× TBST. The membrane is then incubated with goat anti-rabbit HRP conjugate at 1:10,000 dilution for 1 h, washed in TBST, and processed for chemiluminescence (**Fig. 1B**).

### 3.4. Immunostaining of Yeast Cells on a Microscope Slide

1. Fixing cells. Add 110 μL of 37% (w/v) formaldehyde directly into 1 mL of cell culture (2–4 × $10^7$ cells) in a 1.5-mL microtube so the final concentration of formaldehyde is 3.7%. Incubate the cells at 30°C for 15 min on a rolling drum.
2. Collect cells by centrifugation at 2000×*g* for 2 min. Wash cells first with 1 mL of K buffer and then with 1 mL of P buffer.
3. Spheroplasting cells. Resuspend cells in 1 mL of P buffer in a 1.5-mL microtube. Add 0.5 μL of β-mecaptoethanol and 5 μL of 10 mg/mL zymolyase 100T stock. Incubate in a 37°C water bath for 5–10 min (*see* **Note 6**). Start monitoring cell wall removal after 5 min of zymolyase treatment: place 5 μL of cells on a microscope slide and mount a coverslip over the cells, add 10 μL of water with a pipette from one corner of the coverslip while examining the cells under a microscope. Spheroplasts will explode in water due to osmotic shock because they lack cell wall; the burst spheroplasts also appear dimmer under the microscope comparing to intact cells.
4. Collect the spheroplasts by centrifugation at 800×*g* for 2 min (*see* **Note** 7) and resuspend the cells in 1 mL of P buffer.
5. Preparing slides. Use one 14-well microscope slide for simultaneous processing of cell samples collected from all time points of a cell cycle time course experiment. The small well sizes (5 mm) also cut down the amount of antibodies used in IMF. Add 15 μL of 0.1% poly-L-lysine to each well and incubate at room temperature for 10 min. Aspirate off the polylysine by using a Pasteur pipette linked to a vacuum. Wash each well three times by adding a drop of $ddH_2O$ and then aspirating it off. Let the slide air-dry after the final wash.

6. Add 15 μL of the spheroplast suspension from each time point sample to each well and wait 5–10 min for cells to settle. Aspirate the excess liquid.
7. Permeabilizing cells. Immerse the slide in prechilled methanol (–20°C) for 6 min and then in prechilled acetone (–20°C) for 20 s. Let the slide air-dry completely. Check cell density in each well under a microscope.
8. Blocking cells. Wash each well three times with PBS/BSA buffer.
9. Primary antibody staining. For detection of Myc-Rnr2, dilute the 9E10 monoclonal antibody 1:200 in PBS/BSA. Add 15 μL of the diluted antibody to each well. Incubate the slide in a humid chamber (e.g. a covered petri dish with a piece of wet paper towel) at room temperature for 2 h or at 4°C overnight.
10. Wash each well ten times with PBS/BSA buffer.
11. Secondary antibody staining. Add 15 μL secondary antibody diluted in PBS/BSA to each well. For goat-anti-mouse FITC conjugate, use 1:1000 dilution.
12. Incubate the slide in a dark and humid chamber at room temperature for 2 h.

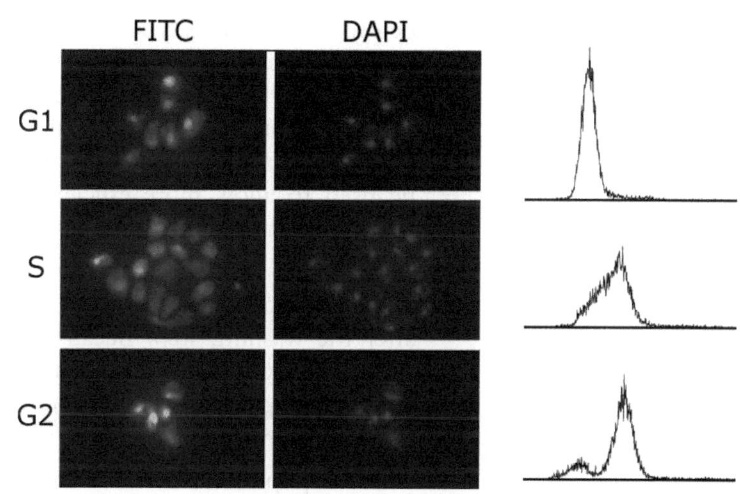


Fig. 2. Change of Rnr2 subcellular localization patterns at different phases of the cell cycle. Cells harboring N-terminally Myc-tagged *RNR2* that was integrated in its own chromosomal locus were synchronized by α-factor-mediated arrest-and-release. Samples were taken from the G1/S-boundry (0 min), the first S phase (30 min) and the first G2 phase (60 min) for IMF analysis. FITC images show signals from Myc-Rnr2; DAPI images show signals from nuclear and mitochondrial DNA. FACS profiles of cells are shown on the *right*. Rnr2 is predominantly localized in the nucleus in cells from the G1 and G2 phases, but become ubiquitously localized between the nucleus and the cytoplasm in cells from the S phase.

13. DNA staining. Add 15 μL of 2 μg/mL DAPI solution to each well and leave 3 min.
14. Wash each well ten times with PBS/BSA; aspirate the excess liquid after each wash.
15. Add 3 μL of VECTASHIELD mounting medium to each well, and mount a coverslip onto the slide. Seal the coverslip with rubber cement.
16. Examine the cells immediately with a fluorescence microscope (**Fig. 2**). The cells can be kept at –20°C in the dark for up to 1 week.

## 4. Notes

1. Cells of an overnight culture usually have reached stationary phase and need 3–4 h to resume log-phase growth. Cell density can be measured in a UV spectrophotometer at $OD_{600}$ as well as counted with a hemocytometer. An $OD_{600}$ reading of 1.0 is usually equivalent to ~$2 \times 10^7$ cells/mL but the correlation may vary depending on cell size.
2. Cell density is critical to the success of α-factor arrest of *BAR1* strains. It is recommended that α-factor is added before the cell density reaches $1 \times 10^7$ cells/mL. In contrast, efficient α-factor arrest can be achieved at higher cell densities with *bar1* mutant strains that lack the secreted aspartyl protease and are much more sensitive to the pheromone relative to *BAR1* strains. YEPD of pH 3.9 can be used to allow more efficient arrest by α-factor, as the acidic condition inhibits the protease activity that degrades α-factor.
3. For *BAR1* cells, release from arrest is quite homogenous after the washing step. Because the *bar1* mutant cells are very sensitive to residual α-factor, 0.1 mg/mL of Pronase E (Sigma-Aldrich) is used in the washing media to facilitate α-factor degradation and quick recover from the arrest.
4. The cell population released from α-factor-mediated arrest usually proceeds through the first two cell cycles synchronously; cells will lose synchrony after the first two cell cycles.
5. The $0.5 \sim 1.0 \times 10^7$ cells collected at each time point provide a good number of cells to track through the staining procedure and gives reasonable flow rates on the flow cytometer. For comparison of FACS profiles between samples, it is recommended that the cell density (cell

number/mL) be kept fairly constant for all samples; the positions of fluorescence peaks may drift from one sample to another if there is a high variability in cell density between samples.

6. Spheroplasting. Under the conditions described above, 5–10 min of zymolyase digestion is usually sufficient to achieve complete spheroplasting. Zymolyase concentration and incubation time can be adjusted to achieve optimal result.

7. Spheroplasts are more fragile than intact yeast cells and need to be handled with more care from this step and forward. Lower centrifugation speed for spheroplast collection protects them from disintegration but may also result in some cell loss because they are lighter than intact cells. It is recommended that cell densities be checked in Step 7 before proceeding to antibody staining to make sure there is a sufficient number of cells in each well.

## References


1. Murray, A. (1994) *Cell cycle checkpoints.* Curr Opin Cell Biol. **6**(6): pp. 872–6.
2. Elledge, S.J. (1996) *Cell cycle checkpoints: preventing an identity crisis.* Science **274**(5293): pp. 1664–72.
3. Hartwell, L., et al. (1994) *Cell cycle checkpoints, genomic integrity, and cancer.* Cold Spring Harb Symp Quant Biol. **59**: pp. 259–63.
4. Kastan, M.B. and J. Bartek (2004) *Cell-cycle checkpoints and cancer.* Nature. **432**(7015): pp. 316–23.
5. Sclafani, R.A. and T.M. Holzen (2007) *Cell cycle regulation of DNA replication.* Annu Rev Genet. **41**: pp. 237–80.
6. O'Connell, M.J., N.C. Walworth, and A.M. Carr (2000) *The G2-phase DNA-damage checkpoint.* Trends Cell Biol. **10**(7): pp. 296–303.
7. Gardner, R.D. and D.J. Burke (2000) *The spindle checkpoint: two transitions, two pathways.* Trends Cell Biol. **10**(4): pp. 154–8.
8. Forsburg, S.L. (2001) *The art and design of genetic screens: yeast.* Nat Rev Genet. **2**(9): pp. 659–68.
9. Hartwell, L.H. and T.A. Weinert (1989) *Checkpoints: controls that ensure the order of cell cycle events.* Science **246**(4930): pp. 629–34.
10. Johnston, L.H. and A.L. Johnson (1997) *Elutriation of budding yeast.* Methods Enzymol. **283**: pp. 342–50.
11. Walker, G.M. (1999) *Synchronization of yeast cell populations.* Methods Cell Sci. **21**(2–3): pp. 87–93.
12. Breeden, L.L. (1997) *Alpha-factor synchronization of budding yeast.* Methods Enzymol. **283**: pp. 332–41.
13. Dirick, L., T. Bohm, and K. Nasmyth (1995) *Roles and regulation of Cln-Cdc28 kinases at the start of the cell cycle of Saccharomyces cerevisiae.* EMBO J. **14**(19): pp. 4803–13.
14. Perlman, R., et al. (1989) *Rapid intracellular alkalinization of Saccharomyces cerevisiae MATa cells in response to alpha-factor requires the CDC25 gene product.* Cell Signal. **1**(6): pp. 577–86.
15. Jordan, A. and P. Reichard (1988) *Ribonucleotide reductases.* Annu Rev Biochem. **67**: pp. 71–98.
16. Yao, R., et al. (2003) *Subcellular localization of yeast ribonucleotide reductase regulated by the DNA replication and damage checkpoint pathways.* Proc Natl Acad Sci USA. **100**(11): pp. 6628–33.
17. Zhao, X., et al. (2001) *The ribonucleotide reductase inhibitor Sml1 is a new target of the Mec1/Rad53 kinase cascade during growth and in response to DNA damage.* EMBO J. **20**(13): pp. 3544–53.

# Chapter 6

# Studying Cell Cycle Checkpoints Using *Drosophila* Cultured Cells

## Katarzyna Siudeja, Jannie de Jong, and Ody C.M. Sibon

## Abstract

*Drosophila* cell lines are valuable tools to study a number of cellular processes, including DNA damage responses and cell cycle checkpoint control. Using an in vitro system instead of a whole organism has two main advantages: it saves time and simple and effective molecular techniques are available. It has been shown that *Drosophila* cells, similarly to mammalian cells, display cell cycle checkpoint pathways required to survive DNA damaging events (de Vries et al. 2005, *Journal of Cell Science 118*, 1833–1842; Bae et al. 1995, *Experimental Cell Research 217*, 541–545). Moreover, a number of proteins involved in checkpoint and cell cycle control in mammals are highly conserved among different species, including *Drosophila* (de Vries et al. 2005, *Journal of Cell Science 118*, 1833–1842; Bae et al. 1995, *Experimental Cell Research 217*, 541–545; LaRocque et al. 2007, *Genetics 175*, 1023–1033; Sibon et al. 1999, *Current Biology 9*, 302–312; Purdy et al. 2005, *Journal of Cell Science 118*, 3305–3315). Because of straightforward and highly efficient methods to downregulate specific transcripts in *Drosophila* cells, these cells are an excellent system for genome-wide RNA interference (RNAi) screens. Thus, the following methods, assays and techniques: Drosophila cell culture, RNAi, introducing DNA damaging events, determination of cell cycle arrest, and determination of cell cycle distributions described here may well be applied to identifying new players in checkpoint mechanisms and will be helpful to investigate the function of these new players in detail. Results obtained with studies using in vitro systems can subsequently be extended to studies in the complete organism as described in the chapters provided by the Su laboratory and the Takada laboratory.

**Key words:** *Drosophila* cells, flow cytometry, RNAi, mitotic index, DNA content, G2/M checkpoint.

## 1. Introduction

Similar to other organisms, *Drosophila* cells respond to DNA insults via organized and tightly regulated pathways. Checkpoint activation, cell death, and compensatory proliferation were all

W.X. Li (ed.), *Cell Cycle Checkpoints*, Methods in Molecular Biology 782,
DOI 10.1007/978-1-61779-273-1_6, © Springer Science+Business Media, LLC 2011

shown to significantly contribute to the survival of *Drosophila* whole organisms after the exposure to DNA damage inducing agents (1, 7). Cultured *Drosophila* cells can be applied to study some, although not all, of these responses. *Drosophila* Schneider's cells (commonly known as S2 cells) (8) are often used for these type of studies. Although cell death and apoptosis are difficult to detect in *Drosophila* cells following exposure to DNA damaging agents, it has been proven that S2 cells delay entry into mitosis and display a G2/M checkpoint arrest after exposure to DNA damaging events and compounds such as ionizing radiation or hydroxyurea (1).

The advantages of using an in vitro system are listed below:

1. Cell lines are homogenous, easy to handle, and show stable behavior over time when good culturing conditions are applied.
2. The treatment of cell cultures with DNA damage inducing agents (chemicals, ionizing radiation) is easier compared to the intact organism. This ensures an equal exposure and response of the whole cell population, which can additionally be assayed with the use of relatively straightforward tests and reliable quantification methods.
3. It is straightforward and relatively cheap to perform RNA interference (RNAi) experiments in *Drosophila* cells. RNAi is highly effective in these cells and can be applied to analyze consequences of downregulation of specific checkpoint proteins or can be useful to identify new players involved in checkpoint function.

In this chapter we first introduce the basic techniques in culturing *Drosophila* cells. Handling insect cells follows, to a large extend, the standard protocols developed for mammalian cell lines. Up to date a number of *Drosophila* cell lines have been established. Initially, *Drosophila* cell lines were derived from spontaneously immortalized cells obtained from mechanically dissociated embryos. Among these lines, S2 and Kc cells (8, 9) are currently the best characterized and the most commonly used cell lines by *Drosophila* researchers.

Second, we describe the methods to induce a checkpoint response in *Drosophila* cell cultures. A number of DNA damaging insults can be used in *Drosophila* cells. We and others have been successfully using hydroxyurea (HU) or ionizing radiation (IR) to investigate checkpoint function and cell survival after the induction of DNA damage (1, 2, 6). Both HU and IR introduce a cell cycle arrest in wild-type cells. HU is a DNA replication inhibitor that decreases the pool of dNTPs, resulting in stalled replication forks (10). Ionizing radiation, on the other hand, damages DNA by direct deposition of energy, as well as producing hydroxyl radicals that attack the DNA. IR induces multiple forms of DNA

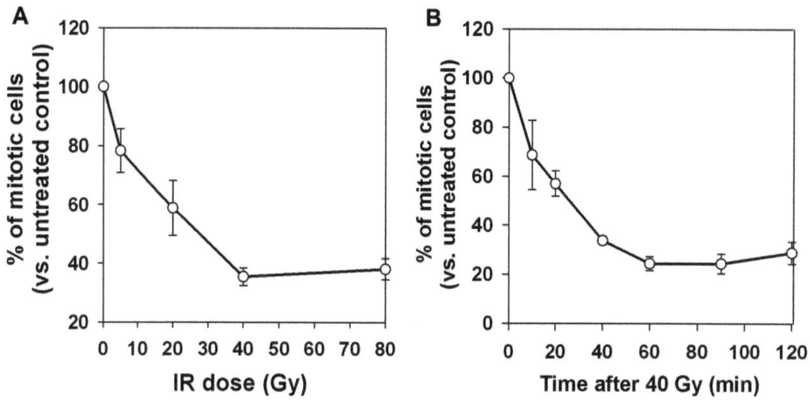


Fig. 1. G2/M checkpoint response in wild-type *Drosophila* S2 cells. *Drosophila* S2 cells exposed to ionizing radiation (IR) were stained for mitosis specific phosphorylated histone 3 and the percentage of mitotic cells was determined by flow cytometry. (**a**) Dose–response curve. Cells were exposed to increasing doses of IR, allowed to recover for 45 min, and the mitotic index was determined. (**b**) Time kinetics of the mitotic arrest. S2 cells were exposed to 40 Gy of IR, allowed to recover for the indicated times, and the mitotic index was determined.

damage including double-strand breaks, which are considered to be the most lethal form of DNA damage (11, 12). After exposure to DNA damaging agents wild-type cells activate a cell cycle checkpoint pathway and a decreased fraction of cells which enter mitosis is observed (**Fig. 1**). Mitotic cells can be stained using an antibody against mitosis specific phosphorylated histone 3 (13). The amount of cells in mitosis can then be determined using immunofluorescence techniques: fluorescence microscopy or flow cytometry. Additionally, cell cycle distribution can also be studied in *Drosophila* cells with the use of a well known technique of DNA content measurements using propidium iodide staining.

Finally, given that *Drosophila* cell lines are especially useful for RNA interference (RNAi) mediated downregulation of specific genes, we will also include the protocols for designing and performing an RNAi knockdown experiment. These will be useful for all the researchers interested in the effect of the downregulation of a specific protein on cell cycle checkpoint functioning. RNAi has already been widely used in *Drosophila* cells to dissect signaling pathways and cellular processes (14–18). In *Drosophila* long double-stranded RNA molecules (dsRNA) can be introduced into the cells without triggering an interferon response and without causing other side effects that are commonly observed in mammalian systems (19–21). dsRNA molecules taken up by the cells are cleaved by the cellular RNAi machinery and a pool of gene specific short silencing RNAs is generated, resulting in a highly efficient down-regulation of the target mRNA. The RNAi technique applied to cultured *Drosophila* cells has already been proven to be useful in studying, among others, the involvement of *Drosophila* checkpoint kinases, Chk1 homologue *grapes* (grp/Dchk1) and Chk2 homologue DmChk2, in cell cycle progression after exposure to DNA damaging insults (1).

Downregulation of a specific gene product involved in cell cycle checkpoint function will result in an increased fraction of mitotic cells after introducing DNA damage to the cells. The above techniques were used to show that in *Drosophila* cells grp/Dchk1 is required for G2/M checkpoint activation, whereas Dmnk/DChk2 is dispensable (1). Thus, RNAi knockdown of grp/Dchk1 can serve as a control for checkpoint studies in studies to identify new potential genes involved in DNA damage response.

For a fast and reliable data analysis the protocols provided here were in part optimized for flow cytometry analysis. Cells fixed and stained according to the protocols provided in this chapter can be analyzed with the use of standard protocols available for a range of cytometers. Many research centers are equipped with a dedicated cytometry facility providing technical support and knowledge. Detailed protocols for the acquisition and analysis of the flow cytometry data exceed the limits of this chapter. A comprehensive methodology of the flow cytometry of *Drosophila* cells has been described elsewhere (22).

In summary, we provide an easy to follow strategy to downregulate a specific gene of interest in *Drosophila* cultured cells and to investigate whether this downregulation affects cell cycle distribution and/or checkpoint functioning.

## 2. Materials

### 2.1. Cell Culture

1. Schneider's *Drosophila* Medium with L-glutamine (Invitrogen) supplemented with 10% heat inactivated fetal bovine serum (FBS; Gibco/Invitrogen or other suppliers, *see* **Note 1**), 100 U/mL penicillin, and 100 U/mL streptomycin (various suppliers).
2. Phosphate buffer saline (PBS), pH 7.4 (various suppliers).
3. 0.25% Trypsin with 1 mM EDTA (Invitrogen).
4. 0.4% Trypan Blue solution (Sigma) and hemocytometer for counting the cells.
5. Tissue culture flask, multi-well plates, disposable pipettes.

### 2.2. RNAi

1. PCR reaction: 5 mM oligonucleotides (*see* **Section 3** for guidelines to design primers), Platinum Blue PCR SuperMix (Invitrogen), PCR template (cDNA).
2. PCR Purification Kit (Qiagen).

3. In vitro transcription: MEGAscript RNAi Kit (Ambion, Catalog #1626).
4. Agarose gel electrophoresis: 0.8–1% agarose in TAE buffer (TAE : 40 mM Tris–acetate, 1 mM EDTA), 0.5 μg/mL ethidium bromide (from 10 mg/mL stock, Invitrogen).
5. Schneider's *Drosophila* Medium with L-glutamine (Invitrogen) without FBS (*see* **Note 2**).

#### 2.3. DNA Damaging Agents

1. Hydroxyurea (HU, Sigma): HU is toxic and should be handled with extra care. Working solution of HU should be prepared fresh for every experiment.
2. Cesium-137 source IBL 637 irradiator (CIS Bio-International) or other source of ionizing radiation (X-rays, gamma-rays).

#### 2.4. Immunolabeling of Cells for Flow Cytometry

1. PBS for washing steps.
2. 3.7% Formaldehyde in PBS. Dilute always freshly before use from commercially available 37% stock.
3. Methanol (absolute).
4. Incubation buffer: 1% bovine serum albumin (BSA) in PBS (solution should be stored at 4°C).
5. Primary antibody: polyclonal rabbit anti-phospho-histone H3 (Cell Signaling, *see* **Note 3**).
6. Fluorophore conjugated secondary antibody: Alexa Fluor 488 goat anti-rabbit IgG (Molecular Probes/Invitrogen, *see* **Note 3**).
7. Round-bottom 6-mL conical tubes suitable for a cytometer.

#### 2.5. Immunolabeling of Cells for Fluorescent Microscopy

1. Poly-L-lysine-coated cover slips. Cover the slips with sterile 0.001% poly-L-lysine (Sigma) solution and incubate for 10 min. Wash the cover slips three times with water and allow to dry.
2. PBS for washing steps.
3. 0.1% Tween in PBS (PBST).
4. 3.7% Formaldehyde in PBS. Dilute always freshly before use from commercially available 37% stock.
5. 0.2% Triton X-100 in PBS.
6. 3% BSA in PBS (Solution should be stored at 4°C).
7. Primary antibody: polyclonal rabbit anti-phospho-histone H3 (Cell Signaling, *see* **Note 3**).
8. Fluorophore conjugated secondary antibody: Alexa Fluor 488 goat anti-rabbit IgG (Molecular Probes/Invitrogen, *see* **Note 3**).

9. 0.2 μg/mL DAPI in PBS (Invitrogen).
10. Citifluor mounting medium (Agar Scientific or other suppliers).

#### 2.6. DNA Staining

1. PBS for washing steps.
2. 80% ethanol:acetone (1:1).
3. DNase-free RNaseA (make 10 mg/mL RNaseA stock solution in water and boil for 10 min to destroy DNase activity; store at –20°C in aliquots).
4. Propidium iodide: 25 μg/mL solution in PBS (solution should be kept in the dark, at 4°C).
5. Round-bottom 6-mL conical tubes suitable for cytometer.

## 3. Methods

### 3.1. Cell Culture

#### 3.1.1. Choosing a Cell Line to Study

All *Drosophila* cell lines, together with their descriptions, are available through the Drosophila Genomics Resource Center (https://dgrc.cgb.indiana.edu/cells/). The collection currently includes 108 lines. In general, cells of an embryonic origin are easy to handle, fast-growing, and very convenient for RNAi experiments. The collection includes also a range of cell lines derived from the central nervous system (CNS) or imaginal discs. These lines are characterized by the presence of tissue specific markers and often posses a distinct morphology but, at the same time, are usually more demanding in culture. The protocols described below were developed mainly for Schneider's cells (S2 cells), nevertheless they may as well be applicable to other cell lines with similar characteristics (like Kc cells).

#### 3.1.2. Guidelines for Maintaining Drosophila Cells in Culture

1. Commonly used fast-growing cell lines, like S2 or Kc, should be maintained by splitting the cells twice a week in a dilution 1/10 to 1/20. Other *Drosophila* cell lines can be sensitive to low cell density and should only be diluted four- to sixfold at each passage.
2. Cell viability should be assayed by a Trypan Blue exclusion test (23). Mix the cell suspension with Trypan Blue solution in 1:1 ratio. Incubate the cells with the Trypan Blue for 5 min and transfer the cells in solution to the chambers of a hemocytometer. Cells excluding Trypan Blue are viable; cells containing the Trypan Blue dye are considered to be not viable.
3. Most *Drosophila* cells adhere loosely to the culturing surfaces and therefore can be easily detached by pipetting the

medium over the cell layer. For more strongly adhered cells, trypsin–EDTA solution should be used in order to detach the cells.

4. *Drosophila* cells are grown in a cell culture incubator without $CO_2$ at temperatures ranging from 22 to 25°C. Culturing at 25°C may be beneficial in reducing the probability of yeast contaminations, since yeast cells prefer to grow at lower temperatures.

### 3.2. RNAi in Drosophila Cells

dsRNA can easily be obtained in large amounts in a two-step procedure. First a gene-specific DNA template is created by a PCR reaction. Subsequently, the PCR product will be used for an in vitro transcription reaction to generate dsRNA (**Fig. 2**). Using a simple transfection protocol, *Drosophila* cells take up dsRNA and no expensive transfection agents are required.

#### 3.2.1. Generating a DNA Template for In Vitro Transcription

1. Designing primers. For in vitro transcription a gene-specific transcription template with T7 RNA polymerase promoter overhangs is needed. Templates of 300–800 bp, with the least possible homology to other targets, should be used (*see* **Note 4**). Design the forward and reverse primers such that the 18–24 gene specific sequence is proceeded by the T7 promoter sequence: 5′-TAATACGACTCACTATAGGGA-gene specific-3′ (the underlined G will be the transcription start site). As a positive control for cell cycle checkpoint studies, a gene-specific transcription template can be generated to downregulate grp/Dchk1. For this the following T7-primers should be used: TAATACGACTCACTATAGGGATGTGCGTGTGTGTGCCG and

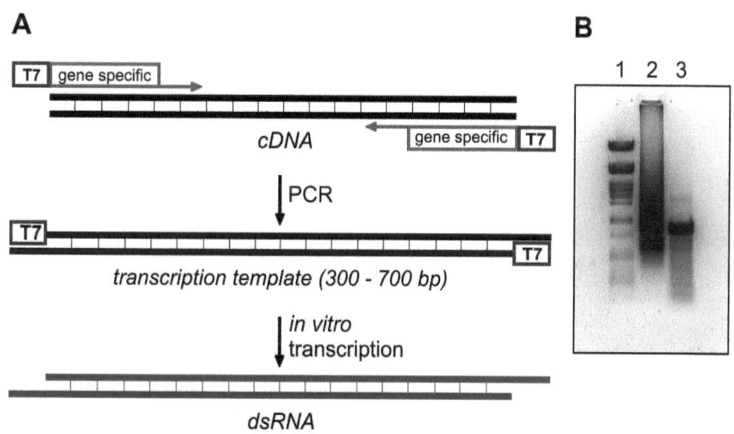


Fig. 2. Generation of long dsRNA for RNAi in *Drosophila* cells. (**a**) Double-stranded DNA template is obtained by standard PCR reaction with gene specific primers flanked with T7 promoter sites. The template is used for an in vitro transcription to produce dsRNA. (**b**) An agarose gel illustrating the steps of generating dsRNA: 1 – DNA Ladder; 2 – 1 μL of an in vitro transcription reaction; 3 – purified dsRNA of approximately 700 bp.

TAATACGACTCACTATAGGGAGGATATGCTTATC-CTG (corresponding to the 493–1200 bp fragment of the GeneBank sequence with the accession number: AF057041).

2. Amplify the template with a PCR reaction. In a PCR tube mix 2 μL of 5 mM solution of each primer, 50 ng of cDNA and Platinum Blue PCR SuperMix to a final volume of 100 μL. For cycles 1–5 use an annealing temperature by calculating the $T_m$ of the sequence specific parts of the primers ($T_a1$). For cycles 6–30, the annealing temperature should be calculated based on the $T_m$ of the entire PCR primers ($T_a2$). Set the thermocycler and run the amplification reaction: 94°C for 30 s, $T_a1$ for 30 s (cycles 1–5), $T_a2$ for 30 s (cycles 6–30), 72°C for 1 min.
3. Analyze 5 μL of the reaction by agarose gel electrophoresis to verify the product yield. The product should appear as a sharp single band of an expected size.
4. Purify the PCR product using a PCR purification kit (Qiagen) and elute the DNA in RNase-free water.
5. Determine the concentration of the PCR product by measuring the $A_{260}$ absorbance. The following formula can be used for calculating the DNA concentration: DNA [μg/mL] = $A_{260} \times 50 \times$ dilution factor.
6. A clean PCR product of a concentration of at least 0.1 μg/μL should be obtained at this step (*see* **Note 5**).

*3.2.2. Generating dsRNA*

1. Perform an in vitro transcription reaction with the use of the MEGAScript RNAi Kit (Ambion) according to manufacturer's instructions. Use 1 μg of DNA template to ensure a high transcription yield. Incubate the transcription reaction at 37°C overnight.
2. Check 0.5 μL of the transcription reaction on an agarose gel. At this step the product is usually visible as a smear and not as a distinct band (lane 2, **Fig. 2b**).
3. Proceed with nuclease digestion and purification of dsRNA according to the protocol of the kit.
4. Elute purified dsRNA from the column with pre-warmed elution buffer (two times 50 μL).
5. Determine the concentration of dsRNA by measuring the $A_{260}$ absorbance. Dilute dsRNA 20–100 times in nuclease-free water or elution buffer. Use the following formula to calculate dsRNA concentration: dsRNA [μg/mL] = $A_{260} \times 40 \times$ dilution factor.
6. Check 0.5 μL of eluted dsRNA on an agarose gel. At this step the product should be visible as a distinct band (lane 3, **Fig. 2b**).

#### 3.2.3. RNAi "Bathing" Technique in Drosophila Cells

1. Use only healthy, exponentially growing cultures, with a viability of more than 97%.
2. Suspend the cells, transfer them to a conical tube, count, and spin the cells down for 5 min at 1000 rpm. Resuspend the cells in serum-free medium with glutamine to a density of $1 \times 10^6$ cells/mL.
3. Transfer 1 mL of cell suspension to each 35-mm dish (or to each well of a 6-well plate).
4. Immediately add 5–15 μg of dsRNA (*see* **Note 6** and **Table 1**) and mix by gently rocking the dish (plate) back and forth.
5. Incubate the cells with dsRNA for 1 h.
6. Add 2 mL of complete Schneider's S2 medium to the cells.
7. Test the efficiency of RNAi on the protein or RNA level after 48–72 h.
8. If examination of the cells at later time points is required, split the cells into the density of $0.5 \times 10^6$/mL and continue the culture until days 4–9 (*see* **Note 7**).

### 3.3. Exposing Cells to DNA Damaging Insults

In general *Drosophila* cells tend to be more resistant to IR than most commonly used mammalian cells. We routinely use doses of 10–50 mM HU in cell culture medium for 6–12 h and an exposing dose of 20–160 Gy of radiation.

1. Resuspend the cells in culture medium, count the cells, and determine the cell viability as described under **Section 3.1.2**. Only exponentially growing cultures with a viability of more than 97% should be used for an experiment.
2. Seed the cells to a density of $1 \times 10^6$/mL in tissue culture dishes or 15-mL conical tubes.

**Table 1**
**The amount of dsRNA (in μg) to be added in 1 mL of serum-free medium in an RNAi experiment**

| | Final concentration | | |
|---|---|---|---|
| **dsRNA length (bp)** | **10 nM** | **20 nM** | **40 nM** |
| 300 | 2.0 | 4.0 | 8.0 |
| 400 | 2.7 | 5.3 | 10.7 |
| 500 | 3.3 | 6.7 | 13.3 |
| 600 | 4.0 | 8.0 | 16 |
| 700 | 4.7 | 9.3 | 19 |

3. Allow the cells to settle down for 30 min under standard growth conditions.
4. Add HU to the medium or expose the cells to ionizing radiation. Incubate the cells with HU containing medium for 6–12 h and replace the medium with a fresh standard medium.
5. Return the cells to the incubator for a recovery period of a desired time.

### 3.4. Immunolabeling of Mitotic Cells – Flow Cytometry

For determining the mitotic index of cells after exposure to checkpoint inducing agents we recommend staining of mitotic chromosomes in formaldehyde fixed cells. The antibody against mitosis-specific phosphorylated form of histone 3 can be used in combination with a fluorescently labeled secondary antibody. The percentage of mitotic cells is then identified by flow cytometry.

1. Transfer 3–5 million cells in a 15-mL conical tube (*see* **Note 8**).
2. Spin the cells for 5 min at 1000 rpm at 4°C. Remove the medium.
3. To fix the cells resuspend the cell pellet in 1 mL of 3.7% formaldehyde in PBS and incubate at room temperature for 15 min.
4. Spin the cells for 5 min at 1000 rpm at 4°C.
5. Remove the fixative and resuspend the cells in 250 μL of PBS.
6. Pre-chill tubes on ice for approximately 2 min.
7. Permeabilize the cells by slowly adding 2.25 mL of ice-cold methanol and vortexing at the same time.
8. Incubate on ice for 30 min (*see* **Note 9**).
9. Spin the cells for 5 min at 1000 rpm at 4°C.
10. Resuspend in 1 mL of incubation buffer and allow the cells to rehydrate for 5 min at room temperature.
11. Spin the cells for 5 min at 1000 rpm at 4°C.
12. Repeat Step 10, but incubate the cells in the incubation buffer for 20 min.
13. Spin the cells for 5 min at 1000 rpm at 4°C.
14. Incubate the cells with primary antibody against phosphorylated histone H3 overnight at 4°C (*see* **Note 3**). Dilute the antibody (1:250) in the incubation buffer and resuspend the cell pellet in 50–60 μL of antibody solution.
15. Wash the cells twice with 1 mL of incubation buffer and spin the cells down after each wash (5 min, 1000 rpm, 4°C).

16. Incubate the cells in the dark in 50 μL of fluorophore conjugated secondary antibody diluted in the incubation buffer.
17. Wash the cells twice with 1 mL of incubation buffer and spin the cells down after each wash (5 min, 1000 rpm, 4°C).
18. Resuspend the cell pellet in 500 μL of PBS and store the cell suspension a 4°C in the dark until flow cytometry analysis.
19. Analyze the samples on flow cytometer. Using Cell Quest or Flow Jo software determine the percentage of cells within the analyzed population which stained positively for mitotic marker (green channel) (**Fig. 3**).

### *3.5. Immunolabeling of Mitotic Cells – Fluorescent Microscopy*

The mitotic cells after exposure to DNA damaging insults can also be observed and counted manually under a fluorescent microscope. For this purpose we include the protocol for immunolabeling of cells fixed on microscope slides.

1. Insert poly L-lysine coated microscope cover slips into the wells of a 6-well plate or 35-mm dishes.
2. Seed the cells on coated cover slips and expose to DNA damaging agents as described in **Section 3.3**.
3. Wash the wells twice with PBS.
4. Add 1 mL of formaldehyde solution to each well. Incubate for 15 min at room temperature to fix the cells.
5. Remove the fixative and wash the wells three times for 5 min with PBS.
6. Add 1 mL of 0.2% Triton X-100 to each well. Incubate for 15 min at room temperature to permeabilize the cells.
7. Incubate the coverslips with 3% BSA solution for 1 h.
8. Dilute the primary antibody against phosphorylated histone H3 1:250 in 3% BSA. Apply a droplet (25–40 μL) of antibody solution on each coverslip and cover with a piece of parafilm.
9. Incubate overnight at 4°C in a humid chamber.
10. Remove parafilm pieces and wash the wells three times 5 min with PBST.
11. Dilute the Alexa Fluor 488 conjugated secondary antibody (1:1000) in 3% BSA. Apply a droplet (25–40 μL) of antibody solution on each coverslip and cover with a piece of parafilm.
12. Incubate 1–2 h at room temperature.
13. Remove parafilm pieces and wash the wells three times 5 min with PBST.

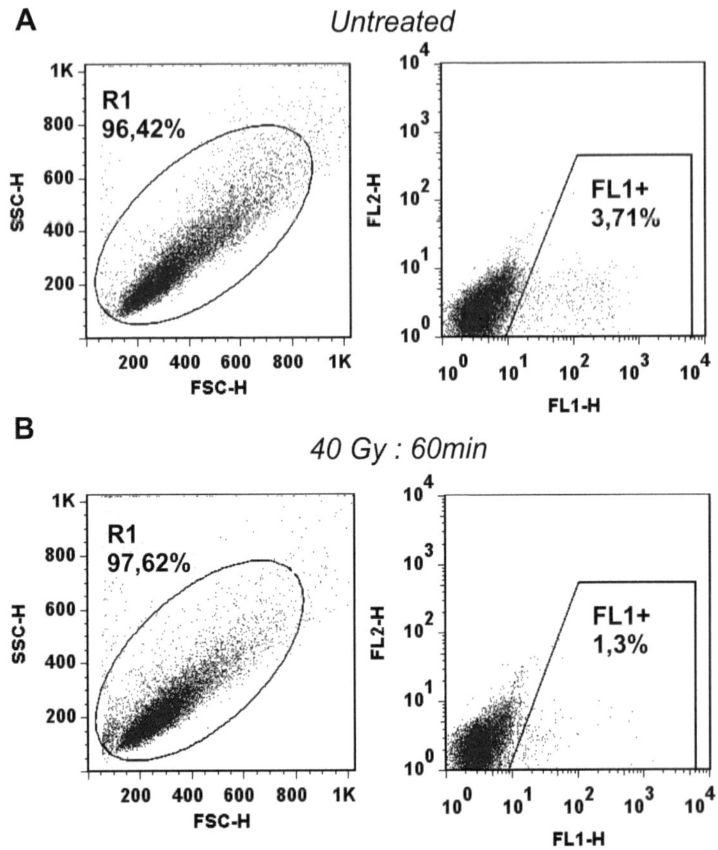


Fig. 3. Flow cytometry analysis of *Drosophila* S2 cells stained for mitosis specific phosphorylated histone H3. S2 cells were fixed and stained with anti phosphorylated histone H3 antibody in combination with Alexa 488 conjugated secondary antibody. (**a, b**) *Left images*: Forward Scatter/Side Scatter plot shows homogeneous distribution of cells gated within one population (marked by R1; 98.51% of all cells). *Right images*: FL1/FL2 plot shows the fluorescence signal of all cells selected in R1. The *boxed area* indicates a distinct population of mitotic cells, which stained positively and give a signal in FL1 (FITC signal) channel (marked FL1+). **a** shows the amount of mitotic cells under control culturing conditions (3.71% of all cells show a FITC signal above threshold and therefore are considered as cells in mitosis). **b** shows the amount of mitotic cells after ionizing radiation (40 Gy, and 60 min recovery), (1.3% of all cells show a FITC signal above threshold and therefore are considered as cells in mitosis).

14. Apply a droplet (25–40 μL) of DAPI solution on each coverslip and cover with a piece of parafilm. Incubate for 15 min to satin DNA in the cells.
15. Remove parafilm pieces and wash the wells three times with PBS.
16. Gently lift the cover slips from the wells and mount on microscope slides with a droplet of Citifluor mounting solution. Apply slight pressure on the cover slips to remove

the excess of mounting medium. Cover the edges of each cover slip with a layer of nail polish.

17. Analyze the slides using a fluorescent microscope. To calculate the percentage of cells in mitosis, at least 1000 cells should be counted for each condition.

#### 3.6. DNA Staining – Determination of the Cell Cycle Distribution

1. Transfer 1–2 million cells in a conical tube.
2. Wash cell pellet twice with 1 mL of PBS by spinning down the cells after each wash (5 min, 1000 rpm, 4°C).
3. Fix the cells in 2 mL of 80% ethanol:acetone (1:1). Add the fixative drop-wise vortexing the tube at the same time.
4. Incubate overnight at 4°C.
5. Spin the cells for 5 min at 1000 rpm at 4°C.
6. Wash the cell pellet once with 5 mL PBS.
7. Resuspend the cell pellet in 20 μL of DNase-free RNase A and incubate for 30 min in 37°C.
8. Add 400 μL of propidium iodide solution.
9. Incubate at room temperature for 1 h or overnight at 4°C.
10. Analyze samples by flow cytometry in the presence of the dye.
11. Using FlowJo or ModFit software, determine the distribution of cells in G1, S, and G2/M phase.

## 4. Notes

1. Serum purchased from different suppliers may differ greatly. It is recommended to test each new batch of serum for possible cell toxicity. Serum should be heat inactivated in 65°C for 30 min and stored in aliquots at –20°C. Repeated freezing and thawing should be avoided.
2. We have tried a number of serum-free media designed for serum-free insect cell cultures for our RNAi experiments. Optionally to classical Schneider's *Drosophila* Medium with L-glutamine, Express Five SFM Medium (Invitrogen) can also be used with a similar transfection efficiency. Add 9 mL of 200 mM L-glutamine to each 100 mL of Express Five SFM Medium prior to use.
3. A wide selection of primary antibodies against phosphorylated histone H3 as well as fluorophore conjugated secondary antibodies is commercially available. Alexa 488

(green) may as well be replaced with other fluorophore if a change of the color is required.

4. Choosing an appropriate gene fragment for a design of dsRNA is crucial for a good RNAi experiment. Online software tools, such as E-RNAi (http://e-rnai.dkfz.de/) can be very helpful by designing such a template (24). Special attention should be paid to avoiding off-target effects. If possible a dsRNA should not contain any 19-mer homology to a gene other than the gene of interest. Two non-overlapping sequences acting on distant places of the target mRNA can also be used.
5. In case of a low yield multiple PCR reactions should be combined, purified, and eluted with one volume.
6. The optimal dsRNA concentration should always be determined experimentally; however, the given ranges should, in most cases, result in an efficient knockdown. We recommend to use 10–40 nM dsRNA in 1 mL of serum-free medium (*see* **Table 1**). The lowest effective concentration should always be used to avoid nonspecific gene targeting.
7. A successful knockdown is usually observed after 3 days of RNAi. However, this time may differ greatly depending on the stability and half-life of the target protein. We routinely perform very efficient RNAi experiments in which more than 90% knockdown can be observed during a period of 3–9 days of RNAi treatment.
8. This number of cells is more than enough for the flow cytometry analysis. We do not recommend decreasing this number since after centrifugation less cells will produce only a small pellet which can be easily lost (especially by inexperienced hands) during the 2-day staining protocol.
9. Cells in 90% methanol can be stored for at least a couple of weeks in –20°C.

## References

1. de Vries, H. I., Uyetake, L., Lemstra, W., Brunsting, J. F., Su, T. T., Kampinga, H. H., and Sibon, O. C. M. (2005) Grp/DChk1 is required for G(2)-M checkpoint activation in Drosophila S2 cells, whereas Dmnk/DChk2 is dispensable. *Journal of Cell Science 118*, 1833–1842.
2. Bae, I., Smith, M. L., and Fornace, A. J. (1995) Induction of P53-Like, Mdm2-Like, and Waf1/Cip1-like molecules in insect cells by DNA-damaging agents. *Experimental Cell Research 217*, 541–545.
3. LaRocque, J. R., Jaklevic, B., Su, T. T., and Sekelsky, J. (2007) Drosophila ATR in double-strand break repair. *Genetics 175*, 1023–1033.
4. Sibon, O. C. M., Laurencon, A., Hawley, R. S., and Theurkauf, W. E. (1999) The Drosophila ATM homologue Mei-41 has an essential checkpoint function at the midblastula transition. *Current Biology 9*, 302–312.
5. Purdy, A., Uyetake, L., Cordeiro, M. G., and Su, T. T. (2005) Regulation of mitosis in response to damaged or incompletely replicated DNA require different levels of Grapes (Drosophila Chk1). *Journal of Cell Science 118*, 3305–3315.

6. Jaklevic, B., Uyetake, L., Lemstra, W., Chang, J., Leary, W., Edwards, A., Vidwans, S., Sibon, O., and Su, T. T. (2006) Contribution of growth and cell cycle checkpoints to radiation survival in Drosophila. *Genetics 174*, 1963–1972.
7. Song, Y. H. (2005) Drosophila melanogaster: a model for the study of DNA damage checkpoint response. *Molecules and Cells 19*, 167–179.
8. Schneider, I. (1972) Cell lines derived from late embryonic stages of drosophila-melanogaster. *Journal of Embryology and Experimental Morphology 27*, 353–365.
9. Echalier, G., and Ohanessi, A. (1969) Isolation in cultivation in vitro of diploid cell strains of drosophila melanogaster. *Comptes Rendus Hebdomadaires Des Seances De L Academie Des Sciences Serie D 268*, 1771–1773.
10. Koc, A., Wheeler, L. J., Mathews, C. K., and Merrill, G. F. (2004) Hydroxyurea arrests DNA replication by a mechanism that preserves basal dNTP pools. *Journal of Biological Chemistry 279*, 223–230.
11. Dianov, G. L., O'Neill, P., and Goodhead, D. T. (2001) Securing genome stability by orchestrating DNA repair: removal of radiation-induced clustered lesions in DNA. *Bioessays 23*, 745–749.
12. Povirk, L. R. (2006) Biochemical mechanisms of chromosomal translocations result ing from DNA double-strand breaks. *DNA Repair 5*, 1199–1212.
13. Hendzel, M. J., Wei, Y., Mancini, M. A., VanHooser, A., Ranalli, T., Brinkley, B. R., BazettJones, D. P., and Allis, C. D. (1997) Mitosis-specific phosphorylation of histone H3 initiates primarily within pericentromeric heterochromatin during G2 and spreads in an ordered fashion coincident with mitotic chromosome condensation. *Chromosoma 106*, 348–360.
14. Clemens, J. C., Worby, C. A., Simonson-Leff, N., Muda, M., Maehama, T., Hemmings, B. A., and Dixon, J. E. (2000) Use of double-stranded RNA interference in Drosophila cell lines to dissect signal transduction pathways. *Proceedings of the National Academy of Sciences of the United States of America 97*, 6499–6503.
15. Muller, P., Boutros, M., and Zeidler, M. P. (2008) Identification of JAK/STAT pathway regulators-Insights from RNAi screens. *Seminars in Cell & Developmental Biology 19*, 360–369.
16. Bjorklund, M., Taipale, M., Varjosalo, M., Saharinen, J., Lahdenpera, J., and Taipale, J. (2006) Identification of pathways regulating cell size and cell-cycle progression by RNAi. *Nature 439*, 1009–1013.
17. Boutros, M., Kiger, A. A., Armknecht, S., Kerr, K., Hild, M., Koch, B., Haas, S. A., Paro, R., Perrimon, N., and Heidelberg Fly Array, C. (2004) Genome-wide RNAi analysis of growth and viability in Drosophila cells. *Science 303*, 832–835.
18. Kiger, A. A., Baum, B., Jones, S., Jones, M. R., Coulson, A., Echeverri, C., and Perrimon, N. (2003) A functional genomic analysis of cell morphology using RNA interference. *Journal of Biology 2*, 27.
19. Hammond, S. M., Bernstein, E., Beach, D., and Hannon, G. J. (2000) An RNA-directed nuclease mediates post-transcriptional gene silencing in Drosophila cells. *Nature 404*, 293–296.
20. Caplen, N. J., Fleenor, J., Fire, A., and Morgan, R. A. (2000) dsRNA-mediated gene silencing in cultured Drosophila cells: a tissue culture model for the analysis of RNA interference. *Gene 252*, 95–105.
21. Kennerdell, J. R., and Carthew, R. W. (1998) Use of dsRNA-mediated genetic interference to demonstrate that frizzled and frizzled 2 act in the wingless pathway. *Cell 95*, 1017–1026.
22. de la Cruz, A. F. A., and Edgar, B. A. (2008) Flow cytometric analysis of Drosophila cells. *Methods in Molecular Biology 420*, 373–389.
23. Phillips, H. J., and Terryberry, J. E. (1957) Counting actively metabolizing tissue cultured cells. *Experimental Cell Research 13*, 341–347.
24. Arziman, Z., Horn, T., and Boutros, M. (2005) E-RNAi: a web application to design optimized RNAi constructs. *Nucleic Acids Research 33*, W582–W588.

# Chapter 7

# In Vivo Live-Analysis of Cell Cycle Checkpoints in *Drosophila* Early Embryos

**Saeko Takada and Byeong J. Cha**

## Abstract

Live-imaging of cells has been an excellent technique to provide us with highly accurate and valuable information about cell cycle checkpoint regulation and DNA damage responses. Early stage *Drosophila* embryos have several advantages to be studied by live-imaging. Fly embryos are much tougher than cultured cells and stand up to relatively rough manipulation, such as protein/chemical microinjection followed by time-lapse imaging. Cell cycles in the embryonic cleavage stage progress rapidly (9–20 min/cycle) and nuclear divisions are synchronous, allowing observation of multiple nuclei/cell cycles in a short period of time. Somatic precursor nuclei form a monolayer at the cortex of the embryo during the syncytial blastoderm stage (cell cycles 10–13). Thus the nuclei in this stage are particularly accessible by various microscopic techniques (Sullivan and Theurkauf, 1995, *Curr. Opin. Cell Biol.* 7, 18–22). Live-imaging of embryos complements the versatility of the *Drosophila* embryonic system, in which we can utilize various approaches, including genetics and biochemistry, to obtain comprehensive understanding of biological processes. In this chapter, we will describe basic methods of microinjection and live-imaging during early embryogenesis by differential interference contrast (DIC) or confocal microscopy, and the use of such methods to study cell cycle checkpoints.

**Key words:** Embryogenesis, cell cycle, checkpoint, DNA replication, DNA damage, spindle, syncytial blastoderm.

## 1. Introduction

*Drosophila melanogaster* embryos start their development with the cleavage stage that contains 13 rapid nuclear divisions occurring synchronously in a common cytoplasm. Each cell cycle consists of only S and M phases and initially lasts about 9 min. Dividing nuclei migrate to the cortex and form a monolayer during the cycle 10 interphase (S phase) and undergo additional four

W.X. Li (ed.), *Cell Cycle Checkpoints*, Methods in Molecular Biology 782,
DOI 10.1007/978-1-61779-273-1_7, © Springer Science+Business Media, LLC 2011

divisions. Cell cycles/nuclear divisions in the syncytial embryo are driven by materials (mRNA and protein) maternally provided during oogenesis. In interphase 14, a dynamic developmental transition called the midblastula transition (MBT) happens, where maternal control of the cell cycle is terminated and zygotic gene transcription is fully activated (1, 2) (**Fig. 1**).

Despite non canonical S M cell cycles and unusually fast syncytial divisions, the *Drosophila* early embryo shares some important mechanisms of cell cycle checkpoints with somatic cells. Among them, the DNA replication checkpoint ensures that cell cycle progression is delayed or arrested in intra S phase or G2/M transition when replication blocks/defects exist in somatic cells. Key players of this checkpoint mechanism include ATR (*A*taxia *T*elangiectasia Mutated and *R*ad3-related) and Chk1 (*Ch*eckpoint *K*inase *1*) (3, 4). The same ATR and Chk1-dependent replication checkpoint appears to coordinate S phase completion and mitotic entry during early embryogenesis and disruption of ATR or Chk1 causes early embryonic lethality (5, 6). In *Drosophila* embryos during the final four cortical cleavage divisions (cycles 10–13), interphase (S phase) length progressively increases because of spontaneous activation of the DNA replication checkpoint that

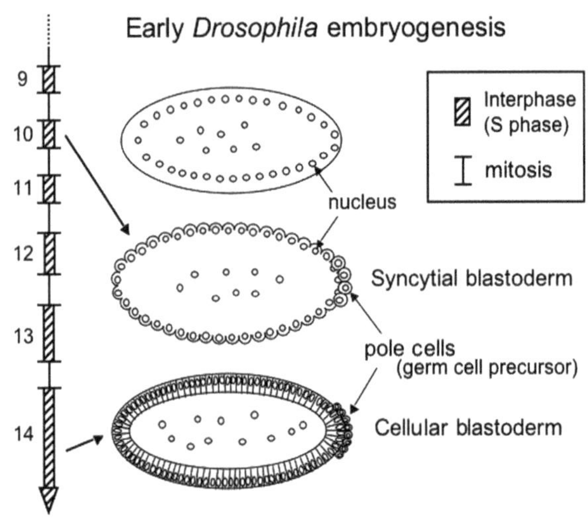


Fig. 1. A schematic view of *Drosophila melanogaster* early embryogenesis. Thirteen cleavage divisions in *Drosophila* embryo are syncytial, rapid (about 9 min initially), and consist of only S and M phases. Pole cell nuclei migrate to the posterior pole during cycle 9 interphase. Somatic cell precursor nuclei migrate to the periphery during interphase 10 and form the blastoderm monolayer. During the final four cleavage divisions (the syncytial blastoderm stage cycle 10–13), interphase length gradually increases from 5 to 15 min. The midblastula transition (MBT) occurs after cycle 13, which features termination of cleavage stage, cell cycle remodeling, zygotic transcription activation, and cellularization. An embryo goes through the 10th division approximately 90 min after egg deposition. Interphase 14 starts about 120 min after egg deposition and lasts about 60–70 min (2).

delays mitotic entry. Mutant embryos, such as *mei-41* (*atr*) and *grp* (*chk1*) that lack this replication checkpoint function, fail to increase the interphase length. As a result, these mutant embryos accumulate DNA lesions, do not go through the MBT, and die after undergoing extra cleavage divisions (7–9). Thus, the DNA replication checkpoint that slows down the late cleavage divisions is essential to *Drosophila* embryogenesis. As in somatic cells, the replication checkpoint can be activated in syncytial embryos by exposure to DNA replication inhibitors such as aphidicolin or hydroxyurea (7, 8).

In the canonical cell cycle, DNA damage including DNA double-strand break triggers the G1, S, or G2/M checkpoints to inhibit progression of the cell cycle (10). However, in *Drosophila* early embryos, DNA damage, in contrast to DNA replication defects, does not appear to delay mitotic entry. Compensating for the lack of the DNA damage cell cycle checkpoint, fly embryos have effective ways to remove damaged genomes from the system. These mechanisms include mitotic centrosome inactivation leading to chromosome segregation failures, and dropping of damaged-nuclei from the embryo cortex (11, 12). Intriguingly, the fly homologue of Chk2 (*Ch*eckpoint *K*inase *2*) is essential for both centrosome inactivation and nuclear dropping (12). Chk2 is one of the major components of the DNA damage signaling pathway in somatic cells and plays a role in cell cycle checkpoint activation and apoptosis (13). Despite the different outcomes of DNA damage response, fly embryos likely maintain genome integrity by utilizing the same DNA-damage signaling pathways used in somatic cells.

Another checkpoint mechanism that operates during the *Drosophila* early embryogenesis is the spindle checkpoint during mitosis. The spindle checkpoint ensures that the microtubule-chromosome attachments (kinetochores) properly form and align at the metaphase plate before committing to anaphase (14). Several proteins are involved in the spindle checkpoint mechanism and localize to the kinetochore (14). Disrupting microtubules by treating cells with nocodazole or colchicine triggers this checkpoint. The spindle checkpoint operates in syncytial embryos and is required for normal synchrony of nuclear divisions and cell cycle arrest when perturbation of spindle assembly occurs (15).

Approximately 90 min after egg deposition, embryogenesis progresses to the syncytial blastoderm stage (cycles 10–13) (2). During this stage, nuclei make a monolayer at the embryonic cortex. Differential interference contrast (DIC) microscopy allows us to distinguish between interphase and mitotic phase based on the presence or absence of the nuclear envelope in live embryos (**Fig. 2**). In addition, expanding and contracting movements are visible on the cortex at the end of mitosis due to anaphase spindle elongation and sister nuclei formation. Therefore, it is possible to

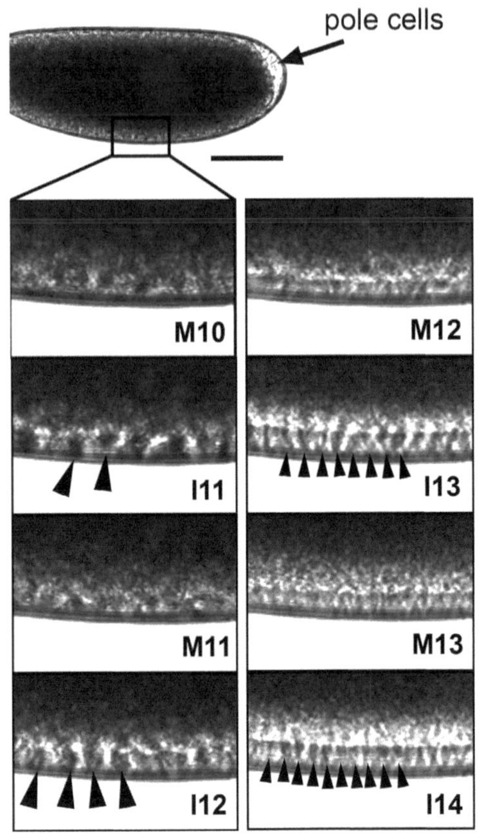


Fig. 2. DIC live-imaging of syncytial blastoderm divisions in *mnk* (*chk2*) mutant embryo. *mnk* mutant embryo shows wild-type cell cycle timing (**9**). Still images from a time-lapse recording are shown. A part of the embryo is magnified and mitosis 10 to interphase 14 are shown. *Arrowheads* indicate positions of nuclei. Note that the nuclei (nuclear envelope) are visible in interphase, but disappear in mitosis. Bar: 20 μm.

determine duration of interphase (S phase) and mitosis (M phase) in living embryos by DIC time-lapse recordings (9). An advantage of this method is that only short exposures to regular bright-field light are necessary for imaging; thus, it is not as invasive as confocal microscopy that uses laser illumination. Time-lapse DIC recording is also useful for capturing developmental progression in later embryogenesis (9, 16).

If microtubules and/or chromosomes are marked with fluorescently labeled Tubulin and/or fluorescent DNA dye by microinjection, cell cycle progression can be examined in detail by confocal laser scanning microscopy. This may be the fastest and most informative way of investigating cell cycle progression/nuclear divisions during early stages in fly embryos. Transgenic flies that express GFP-tagged proteins are also highly useful for investigating cell cycle progression by live-imaging (e.g., His2AvD-GFP or GFP-Tubulin, (17, 18)).

## 2. Materials

### 2.1. Drosophila Culture and Embryo Collection

1. *Drosophila* culture medium. Regular laboratory fly food in vials and bottles.
2. Bakers yeast (e.g., RED STAR Active Dry Yeast).
3. Grape juice agar plates. To prepare 1 L of grape juice agar solution, put 0.75 L of $dH_2O$, 20.3 g of granulated agar (e.g., Difco), and a magnetic stir bar into a 2-L Erlenmeyer glass flask and place the flask on a heated-magnetic stir plate. Heat and stir until the agar dissolves and the solution just starts to boil (when small bubbles form and rise from the bottom). Turn off the heat and continue stirring. Mix 25 g of glucose (Sigma) and 250 mL of 100% grape juice (e.g., Welch Food, Inc.) (*see* **Note 1**) in a beaker and stir until glucose completely dissolves. Add the grape juice/glucose mixture to the agar solution. Prepare Tegosept (anti-fungal reagent: methylparaben) solution by dissolving 1.5 g of Tegosept (e.g., Genesee Scientific) in 6 mL of ethanol. When grape juice/agar/glucose solution cools down to 60°C, add Tegosept solution and stir. Pour the grape juice/agar/glucose/Tegocept solution into 60 × 15-mm polystyrene petri dishes (e.g., BD Bioscience). When the plates completely solidify, wrap them securely with the original dish sleeves or plastic wrap to prevent drying and store at 4°C. This recipe makes 80–100 grape juice plates.
4. 25°C Incubator with 12/12 h light/dark cycle.
5. Collection cages. To make a cage, remove the bottom of a 100-mL Tri-cornered polypropylene beaker using heated-blade (*see* **Note 2**). Cut a disc with the size of the beaker bottom from stainless steel mesh (Genesee Scientific) and attach the disc over the bottom of the beaker to cover the opening. Permanently fix with Epoxy-bond.
6. *Drosophila* Stocks. Most of the fly stocks described here are available from the Bloomington *Drosophila* Stock Center at Indiana University (http://flystocks.bio.indiana.edu/).

### 2.2. Mounting Embryos

1. Cover glasses 40 × 22 mm #1.5 (e.g., Fisher Scientific).
2. Microscope slide glasses 25 × 75 mm (e.g., Fisher Scientific).
3. Scotch permanent double-sided tape 12.7 mm width (3M).
4. Sticky octane solution. Place about 50 cm of Scotch double-sided tape and 50–75 mL of octane together in a 100-mL glass media bottle with cap. Keep the bottle at room tem-

perature for 2 days and let the adhesive on the tape dissolve into octane. The solution will become slightly cloudy. It is not necessary to remove the remaining tape from the bottle. Store the sticky octane solution at room temperature. Supplement more tape and octane as needed.

5. Pasteur pipettes and small rubber bulbs.
6. Art paintbrush (small).
7. Dumont forceps #5 (Fine Science Tools).
8. Halocarbon oil 700 (Sigma) stored in a 30- or 60-mL dispensing polyethylene squeeze bottle.
9. A stereo microscope with lighting source, appropriate for fly pushing or dissection.

### 2.3. DIC Imaging

1. Inverted DIC microscope (e.g., Zeiss Axiovert 100 from Carl Zeiss) equipped with a computer-operated shutter (e.g., Uniblitz) and a CCD camera (e.g., Coolsnap HQ from Photometrics).
2. A computer and software for image acquisition and analysis (e.g., MetaMorph).

### 2.4. Embryo Injection

1. Drierite (WA Hammond).
2. 100 mm petri dish (either plastic or glass).
3. Mechanical micromanipulator with an injection pipette holder (e.g., Narishige Cat. # M-152 or MN-153, WPI Cat.# M3301).
4. Needle puller (e.g., Sutter Instruments or Narishige).
5. Injection needles. Pull needles from glass capillary tubing (e.g., Borosil 1.2 mm OD × 0.9 mm ID, 100 mm, FHC) using a needle puller (*see* **Note 3**).
6. Injection syringe. Cut the sharp end of an 18-G disposable needle (BD Medical) to make a blunt end. Tightly attach the needle to a 10-mL disposable syringe (BD Medical) using a lure lock. Connect about 50 cm of thin Teflon tubing to the needle. At the other end of the Teflon tubing, attach a 1- to 2-cm piece of soft silicon tubing that can tightly fits onto an injection glass needle making it easy to attach the glass needle.
7. Rhodamine–Tubulin (Cytoskeleton Inc., Cat. # TL590M) dissolved in BRB80 (80 mM Na-PIPES, pH 6.9, 1 mM $MgCl_2$, 1 mM EGTA, and 1 mM GTP) at 10 mg/mL and stored as aliquots of 0.65 μL at –80°C (*see* **Note 4**).

8. OliGreen ssDNA reagent in dimethylsulfoxide (DMSO) (Invitrogen Cat. # O7582) stored as small volume aliquots at –20°C.

### 2.5. Reagents to Activate Cell Cycle Checkpoint or to Induce DNA Damage

These reagents may be toxic. Handle them with caution, following manufacture's instructions. If a reagent would be co-injected with Rhodamine–Tubulin, make injection mixture with BRB80 buffer (*see* **Section 2.4**); otherwise, use PBS (137 mM NaCl, 2.7 mM KCl, 10 mM $Na_2HPO_4$, 1.76 mM $KH_2PO_4$, pH 7.4). Use the final concentration for each drug indicated below as your reference; however, the least and effective concentration should be determined individually.

1. Hydroxyurea (Sigma) dissolved in $dH_2O$ at 2 M and stored as small aliquots at –80°C. Dilute to 0.1–0.2 M with BRB80 or PBS for injection.
2. Aphidicolin (Sigma) dissolved in $dH_2O$ at 200 mM and stored as small aliquots at –80°C. Dilute to 100–200 μM with BRB80 or PBS for injection.
3. Nocodazole (Sigma) dissolved in $dH_2O$ at 20 mM and stored as small aliquots at –80°C. Dilute to 12.5 μM with BRB80 or PBS for injection.
4. Colchicine (Sigma) dissolved in $dH_2O$ at 100 mg/mL and store at –80°C. Dilute to 100 μg/mL with BRB80 or PBS for injection.
5. Camptothecin (Sigma) dissolved in DMSO at 20 mM and stored as small aliquots at –80°C. Dilute to 20 μM with BRB80 or PBS for injection.
6. Bleomycin (Sigma) dissolved in $dH_2O$ at 50 mg/mL and stored as small aliquots at –80°C. Dilute to 50 μg/mL with BRB80 or PBS for injection.

### 2.6. Confocal Imaging

1. Confocal laser scanning microscopy system with an inverted microscope (e.g., Leica TCS SP2 DM IRE2 inverted scanning confocal microscope).
2. Software for image acquisition (e.g., Leica Confocal Software) and a computer.

### 2.7. Software for Image Data Processing and Analysis

ImageJ (http://rsb.info.nih.gov/ij/)

## 3. Methods

Before performing in vivo live-imaging of *Drosophila* early embryos, young wild-type or mutant flies are placed in small collection cages (*see* **Note 5**). The cages are kept in a 25°C incubator equipped with a programmed light/dark cycle to synchronize fly circadian rhythms and control the speed of embryogenesis. On the day of live-imaging, 0- to 2-h-old embryos are collected and the chorion membrane is manually removed. The dechorionated embryos are mounted on a coverslip and covered with Halocarbon oil so that they can be kept alive during microinjection and live-recordings. Time-lapse live-imaging can be performed by DIC microscopy without any microinjection.

By microinjecting Rhodamine–Tubulin and/or OliGreen into embryos, cell cycle progression can be followed by laser-scanning confocal microscopy. Rhodamine-labeled Tubulin visualizes microtubules throughout the cell cycle. Because of the dynamic nature of microtubules, some of injected Rhodamine–Tubulin molecules are quickly incorporated into microtubules, whereas the remaining molecules freely diffuse into the cytoplasm. At interphase, Rhodamine–Tubulin stains microtubule asters nucleated from two centrosomes adjacent to the nucleus. The nuclear envelope totally excludes Rhodamine–Tubulin molecules from the nucleus; therefore, the shape of the nucleus is shown as a dark circle (**Fig. 4b**, **g–j**). Breakdown of the nuclear envelope at the beginning of mitosis is readily detected because the Rhodamine–Tubulin molecules in the cytoplasm diffuse into the nuclear region at the moment of breakdown (**Fig. 4c**). In mitosis, Rhodamine–Tubulin stains spindle microtubules and asters nucleated from the spindle poles (centrosomes). Mitotic progression and formation of sister nuclei at the end of mitosis can be examined in detail (**Fig. 4d–e**). OliGreen is a DNA-binding fluorescent dye that stains chromatin throughout the cell cycle. The cell cycle progression, which is marked by chromatin condensation in prophase, chromosome segregation in mitosis, and chromatin de-condensation at the end of mitosis, can be followed in detail (11, 12).

### 3.1. Collecting Embryos

1. To set up a collection cage, 50–150 young adult females and 1/3 to 1/2 as many young adult males should be placed together in a cage, preferably 2 days before planned live-imaging experiment (*see* **Note 6**).
2. Smear a small amount of yeast paste (*see* **Note 7**) on a grape juice plate and put the plate upside down on top of the cage that contains flies. Place the cage in a 25°C incubator with 12/12 h light/dark cycle.

3. Replace the grape juice plate once or twice a day even if you are not collecting embryos on that day. Flies should be fed well so that they develop ovaries to produce many eggs.
4. In the morning of your experiment, place a new plate on the cage and wait at least a half hour to let females lay any leftover eggs from the previous night.
5. Collect 0- to 2-h-old embryos for live-imaging.

### 3.2. Preparing Embryos for Cell Cycle Measurement

#### 3.2.1. Preparing Coverslips to Mount Embryos

For inverted-microscope observations, embryos are mounted on a #1.5 coverslip. It is necessary to make good contact between the embryo and the surface of the coverslip so that the embryos do not move or float. In order to make an approximately 5 × 25 mm sticky surface on a cover glass, spread some sticky octane solution (*see* **Section 2**) with a Pasteur pipette and let the octane completely evaporate. This may take 5–10 min. After the octane evaporates, the adhesive substance that was dissolved in octane should be left and dried on the glass and makes the glass surface sticky.

#### 3.2.2. Hand-Dechorionating and Mounting Embryos

1. After 2 h of embryo collection, remove the grape juice plate. Put a new plate on the cage to start another collection. Put a 5-cm piece of double-sided tape on a glass slide and gently transfer 30–40 embryos from the grape juice plate with a paintbrush.
2. Place the slide glass with the embryos under a stereo microscope and illuminate with light. Looking through the microscope eyepiece, scratch adhesive substance on an empty region of double-sided tape with fine Dumont forceps (#5) and make a small sticky ball. Hold the sticky ball with the forceps.
3. To remove the chorion membrane, gently touch an embryo and push and roll with the sticky ball. The chorion membrane may be easily torn open. Gently touch the dechorionated embryo (It is covered with the vitelline membrane.) with the sticky ball to lift and transfer onto a prepared sticky coverslip as described in **Section 3.2.1**.
4. Embryos should be aligned longitudinally in the middle of the sticky area on the coverslip. The whole process to align 10–20 embryos should not take more than 10 min.
5. Cover the embryos with a small amount of Halocarbon oil 700 to prevent drying.

### 3.3. Cell Cycle Measurement by DIC Microscopy

1. Place the coverslip with the embryos on the stage of an inverted DIC microscope. Make sure that all the necessary optical elements for DIC, such as prisms, polarizer, and

analyzer, are in the light path. Tweak the objective prism to achieve the highest contrast at the periphery of an embryo.

2. Focus on the embryos with 20× objective lens and look for an embryo that is just entering the syncytial blastoderm stage. It is best to start recording during cell cycle 9 interphase when pole bud formation occurs at the posterior pole of the embryo to capture the entire four divisions from cycles 10–13 (*see* **Note 8**).
3. Start time-lapse recording every 10 s.
4. Calculate cell cycle timing by counting the number of frames between nuclear-envelope formation and breakdown. This number gives the duration of interphase (S phase) and mitotic phase (M phase) (**Fig. 2**) (8).
5. Cell cycle timing can be analyzed either using an image-data acquisition software (e.g., MetaMorph) or ImageJ that is publicly available (*see* **Section 2**).

### 3.4. Microinjection and Live-Recording to Study Cell Cycle Progression by Confocal Microscopy

#### 3.4.1. Preparing Injection Mixture

1. Thaw a stock of Rhodamine–Tubulin, OliGreen, or any other drugs to be injected into embryos and keep them on ice. Make injection mixture that contains the appropriate final concentration of each component (*see* **Note 9**).
2. Centrifuge the injection mixture at maximum speed in a microcentrifuge for 15 min at 4°C to precipitate any aggregate. Keep the mixture on ice and use the supernatant.

#### 3.4.2. Microinjection

1. Collect, manually dechorionate, and align embryos on a coverslip as described in **Sections 3.1** and **3.2**.
2. Before covering the embryos with Halocarbon oil, place the embryos/coverslip on top of a bed of Drierite in a 100-mm petri dish and replace the lid. Wait for 3–4 min and remove the embryos/coverslip from the dish. Cover the embryos with Halocarbon oil. The purpose of this process is to partially dehydrate the embryos, so that the embryo can take up a small volume of injection solution without bursting from internal pressure (*see* **Note 10**).
3. Set a glass injection needle connected to the injection syringe (*see* **Section 2**) on the micromanipulator on the confocal microscope with a shallow angle (20–30°). Turn the bright-field light on. Bring the tip of the needle to the center of the light path. Break open the needle tip, if the needle puller produces a closed end (**Fig. 3a**, also *see* **Note 3**).
4. Place the embryo/coverslip on the stage of the confocal microscope. The embryos should be placed longitudinally and vertically on the stage so that the embryos can be injected from their side (**Fig. 3e**).

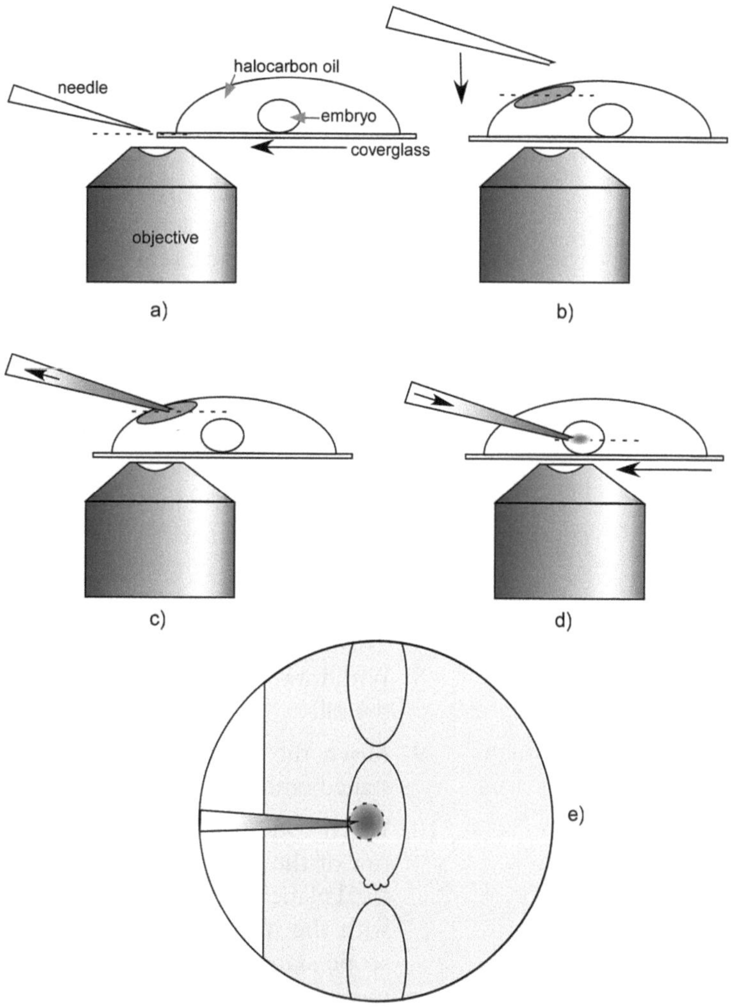


Fig. 3. A schematic drawing of microinjection. Embryos are aligned longitudinally and microinjection is performed from the side to the middle part of the embryo in order to obtain efficient diffusion of injected materials as quickly as possible into the large part of the embryo, unlike microinjection for germline transformation, which targets the posterior pole of the embryo. (**a**) Place the coverslip with the embryos on the microscope stage. If the needle puller produces closed end needles, break and open the needle tip. Move the stage to bring the coverslip to the view center and focus onto the edge of coverslip (*dashed lines* indicate the focal plane). Bring the needle tip into the same focal plane near the glass edge. By moving the coverslip, gently bump the needle tip to the glass edge. Move the needle up. (**b**) Place a droplet of injection solution onto the Halocarbon oil and focus on the middle of the droplet. Lower the needle tip to insert to the droplet. (**c**) Draw the injection solution by slowly pulling the syringe plunger. Remove the tip from the solution, but keep it in the halocarbon oil. (**d**) Focus on the periphery of embryos and find an embryo at the right stage. Bring the needle tip to the same focal plane as the embryo's equator and insert the needle by moving the stage. Inject solution gently and withdraw by moving the stage away from the needle tip. (**e**) Aligned embryos that are looked through the eyepiece of microscope. The middle embryo is being injected with injection mixture from the left side and the mixture quickly diffuses in the syncytial cytoplasm.

5. Using a P2 pipette, dispense 0.5 μL of injection mixture onto the Halocarbon oil on the left side of the row of embryos and away from any embryo. The injection mixture containing Rhodamine–Tubulin should make a pink droplet on the oil (**Fig. 3b**).
6. Using the micromanipulator, bring the needle tip very close to the droplet, first, looking with your eyes. Then, through the microscope eyepiece and 20× objective, focus on the droplet and bring the needle tip further down to the focal plane of the droplet.
7. Insert the needle tip into the droplet using the micromanipulator. Pull the syringe plunger slowly and draw the injection mixture into the needle (**Fig. 3c**). When enough amount of injection mixture is drawn, but before half the droplet is drawn into the needle, unscrew the lure lock of the syringe needle to disconnect the syringe from the Teflon tubing connecting to the injection needle. This process quickly releases the sucking pressure and prevents drawing the Halocarbon oil into the injection needle.
8. Withdraw the needle tip from the droplet, but keep it in the oil.
9. Move the microscope stage and search for a correctly staged embryo to inject (*see* **Note 8**).
10. When you find an embryo in the right stage, focus on the rim of the embryo and bring the tip of the needle to the focal plane and to the side of the embryo as close as possible with the micromanipulator. Then, by moving the microscope stage (*x* axis direction), not the micromanipulator, insert the tip of the needle into the embryo. Inject the solution by applying gentle pressure to the injection syringe (**Fig. 3d**). Then, withdraw the needle tip by moving the microscope stage in the opposite direction. Lift the needle tip slightly up to avoid any accidental poking of an embryo during subsequent imaging.
11. Turn objective lens to higher magnification (40×, 63×, or 100×). If the lens is water or oil immersion, put appropriate media between the lens and the coverslip. With fluorescent viewing, you should see Rhodamine–Tubulin (with red filter) or OliGreen (with green filter) quickly diffuses into the embryo. The volume of injected solution can be roughly estimated by the size of the circular area formed by the injected solution (**Fig. 3e**).

*3.4.3. Live-Imaging*

1. Before injection, open the image acquisition software and activate appropriate laser lines (i.e., 488-nm laser for OliGreen and 561-nm laser for Rhodamine–Tubulin). Select

scanning mode as xyt and set up a time-lapse image acquisition condition. Select appropriate dichroic beam splitter and emission detectors (or optical filters) to collect fluorescent emissions of OliGreen at 495–540 nm range (FITC/EGFP detection setting) and of Rhodamine at 575–630 nm range (Cy3/TRITC detection setting).

2. Immediately after injection, change the mode of microscope to confocal scanning. If you used DIC for bright-field viewing for injection, the DIC analyzer should be out of the light path in order to permit the laser (polarized) into the light path. Test scan the embryo to determine focal plane to see cortical nuclei. Examine whether the embryo was injected at the right timing (i.e., cycles 9–12).
3. If everything worked out well (the microinjection was done at the correct timing, the embryo was not damaged by injection, etc.), start scanning every 10 s to make time-lapse recording (*see* **Notes 11** and **12**).
4. Analyze cell cycle timing as described in **Section 3.3**.

*3.4.4. Cell Cycle Checkpoint Activation*

To examine whether a cell cycle checkpoint operates in a specific mutant embryo, a drug that should activate the checkpoint is injected together with Rhodamine–Tubulin. To activate DNA replication checkpoint function, aphidicolin or hydroxyurea can be used. Aphidicolin is an inhibitor of DNA polymerase α and stalls replication forks. Hydroxyurea is an inhibitor of ribonucleotide reductase that is required for dNTP production. These reagents should be injected during mitosis or at the very beginning of S phase to have sufficient time to effectively cause a delay in next mitotic entry (**Fig. 4f–j**). To induce DNA damage, Camptothecin or Bleomycin can be injected. Neither drug causes any delay of mitotic entry but instead causes Chk2-dependent centrosome inactivation and dropping of damaged-nuclei in fly embryo (12). Camptothecin is a Topoisomerase I inhibitor and Bleomycin is a radiomimetic drug. Both reagents cause DNA double-strand breaks. To activate the spindle checkpoint, microtubule-disrupting reagents, nocodazole or colchicin are injected before metaphase. The cell cycle should completely arrest or partially delay at metaphase in wild-type embryos.

1. Make injection mixture containing appropriate concentration of a checkpoint-activating drug (*see* **Section 2.5**) and 5 mg/mL of Rhodamine–Tubulin. Inject the mixture at the correct timing indicated above (also *see* **Note 13**) and start time-lapse recording every 10 s.
2. After finishing live-recordings, analyze cell cycle timing as described in **Section 3.3**.

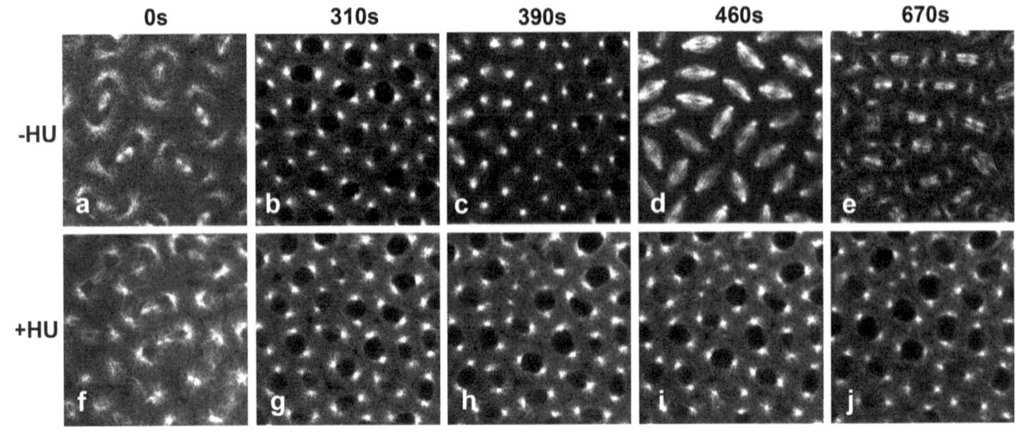


Fig. 4. Hydroxyurea injection causes a delay of mitotic entry in wild-type embryo. 5 mg/mL Rhodamine–Tubulin was injected without (**a–e**) or with (**f–j**) 100 mM hydroxyurea in wild-type embryos during mitosis 11. Numbers at the *top of the panels* indicate time after the beginning of interphase 12 in seconds. **a** and **f** show the beginning of interphase12 (0 s), where nuclear envelope forms but newly divided two sister nuclei are still connected by microtubules. *Dark circles* excluding Rhodamine–Tubulin molecules show interphase nuclei. Two centrosomes adjacent to each nucleus nucleate microtubule asters in interphase (**b, g–j**). At the beginning of mitosis, the nuclear envelope breaks down (**c**). Without hydroxyurea injection, nuclei entered cycle 12 mitosis at 390 s (6.5 min) (**c**), and exit mitosis12 at 670 s. (11.2 min) (**e**). The hydroxyurea injection significantly delayed mitotic entry (**h–j**). This embryo finally entered mitosis12 at 1200 s (20 min) (data not shown). Bar: 20 μm.

## 4. Notes

1. Purple color of grape juice makes it easy to see embryos on agar plates. Quality of grape juice plates (moisture and aroma) affects the yield of embryos.
2. Any plastic beaker that has a ridge around the opening would work. The 60-mm petri dish should tightly sit upside down over the ridge.
3. The injection mixture containing Rhodamine–Tubulin tends to be viscous and easily clogs injection needles. Therefore, slightly wider needle diameter and tip opening than those for germline transformation is preferred. Some trial-and-error is necessary to obtain reasonably good needles. If the needle puller produces closed tip, break open the tip by gently bumping against an edge of a cover glass under an inverted microscope equipped with a manipulator (**Fig. 3a**). The opening of the tip should be about 3–5 μm in diameter.
4. Rhodamine–Tubulin stock solution at 10 mg/mL is viscous and it is a challenge to make aliquots of less than 1 μL each. However, once the Rhodamine–Tubulin solution is thawed, it is useful only a couple of hours on ice. Thus, it

is recommended to make the smallest volume aliquots as possible to avoid wasting the expensive material.

5. The syncytial divisions/cell cycles in *Droosphila* early embryos are driven by maternal products. Full activation of zygotic transcription does not occur until cycle 14 (*see* **Section 1**). To obtain homozygous maternal-effect mutant embryos, embryos must be collected from adult females that are homozygous for the mutation (19). Separately collect homozygous mutant adult females and wild-type males (or *w* males depending on the mutant background and what should be the control) and place the females and the males together in a cage. Since paternal contribution is not relevant during early embryogenesis, females do not need to be virgins and crossing with any adult males should not affect early embryonic phenotypes. Some maternal-effect mutations also affect male fertility; therefore, we recommend using wild-type males. If the mutation causes zygotic lethality and homozygous females are simply not available, an alternative approach, for example, making germline clones, or inducing RNAi in oogenesis or in early embryogenesis to knockdown gene expression, would be necessary. The germline clone technique allows you to obtain homozygous mutant eggs from heterozygous adult females by inducing FLP/FRT-driven recombination in the female germline (20).
6. Flies usually take at least 2 days to adjust to a new environment (cage) in order to produce a good number of embryos.
7. Grape juice plate should be supplied with fresh yeast paste (mix water and dry active yeast to make peanut butter consistency) and warmed-up to room temperature before usage.
8. Look for pole bud formation (cycle 9) or clearing of the rim, which indicates that nuclei have migrated to the cortex and are going through cortical divisions.
9. Rhodamine–Tubulin should be 5 mg/mL in injection mixture. OliGreen stock is dissolved in 100% DMSO and should be diluted 10× with water to bring the DMSO concentration down to 10%. Fly embryos can tolerate injection of up to 10% DMSO. Given that the injected solution would be diluted approximately 100× in the embryo, this corresponds to approximately 0.1% final DMSO concentration in the embryo.
10. Time to dehydrate the embryos on Drierite should be determined empirically. Do not over-dry the embryos. If you see wrinkles on the embryo surface, it is over-dried.

11. Illuminating embryos with laser light could potentially damage cellular components or structures. This could easily affect cell cycle timing. It is thus a good idea to find the optimal imaging condition by compromising the laser power level, scan speed, and amplification of signal intensity. Recent new confocal microscopy systems, both point scanning and spinning disk from major brands are well equipped with sensitive photomultiplier detectors or a CCD camera that allow high-quality live-imaging without damaging objects. Nevertheless, when measuring cell cycle timing using confocal microscopy, it is important to have appropriate controls for each experiment. Inject least effective amount of fluorescent markers and keep laser energy output as low as possible. Always keep in mind that some mutants or transgenic embryos may be more sensitive to laser illumination than wild-type embryo.
12. When Rhodamine–Tubulin is injected, try to maintain focusing on centrosomes (microtubule asters) in interphase and spindles in mitosis by manually adjusting the focal plane (z position) throughout the recording (*see* **Fig. 4** for an example). When OliGreen is injected or a Histone-GFP line is used, keep focusing on chromosomes. If both structures are monitored simultaneously, try focusing on both as much as possible.
13. It is especially important to perform checkpoint-activating drug injection at the right time in the cell cycle to achieve sufficient effects. With some practice, you may be able to distinguish between mitosis and interphase with bright-field viewing using the 20× objective. However, at the beginning, it is better to inject randomly during the syncytial blastoderm stage and immediately scan with confocal to find out in which phase the injection was just done. If the timing was not right, then try injecting another embryo. One way to perform injection at the right timing is to use GFP-tagged protein expressing transgenics like Histone-GFP (16) or GFP-Tubulin (17). However, to analyze mutant embryos, the transgene should be genetically brought into the mutant background. Another method is performing "double-injection." This may be feasible when you become an expert at microinjection. First, a fluorescent marker like Rhodamine–Tubulin is injected and the embryo is scanned with confocal to check the timing. Second, a checkpoint-activating drug is injected at the correct timing.

## Acknowledgments

The authors would like to thank Professor William Theurkauf for providing us with good opportunities and environment to learn live-imaging techniques in flies, and MaryJane O'Connor, Aidan Peterson, and Hideki Aihara for critical reading of the manuscript.

### References


1. Sullivan, W., and Theurkauf, W.E. (1995). The cytoskeleton and morphogenesis of the early *Drosophila* embryo. *Curr. Opin. Cell Biol.* 7, 18–22.
2. Foe, V., Odell, G., and Edgar, B. (1993). Mitosis and Morphogenesis in the Drosophila Embryo: Point and Counterpoint. In The Development of *Drosophila melanogaster*, M. Bate and A. Martinez Arias, eds. (New York, NY: Cold Spring Harbor Laboratory Press), pp. 149–300.
3. Liu, Q., Guntuku, S., Cui, X.S., Matsuoka, S., Cortez, D., Tamai, K., Luo, G., Carattini-Rivera, S., DeMayo, F., Bradley, A., Donehower, L.A., and Elledge, S.J. (2000). Chkl is an essential kinase that is regulated by Atr and required for the G2/M DNA damage checkpoint. *Genes Dev.* **14**, 1448–1459.
4. Feijoo, C., Hall-Jackson, C., Wu, R., Jenkins, D., Leitch, J., Gilbert, D.M., and Smythe, C. (2001). Activation of mammalian Chkl during DNA replication arrest: a role for Chkl in the intra-S phase checkpoint monitoring replication origin firing. *J.Cell Biol.* **154**, 913–923.
5. Brown, E.J., and Baltimore, D. (2000). ATR disruption leads to chromosomal fragmentation and early embryonic lethality. *Genes Dev.* **14**, 397–402.
6. Takai, H., Tominaga, K., Motoyama, N., Minamishima, Y.A., Nagahama, H., Tsukiyama, T., Ikeda, K., Nakayama, K., and Nakanishi, M. (2000). Aberrant cell cycle checkpoint function and early embryonic death in Chkl(-/-) mice. *Genes Dev.* **14**, 1439–1447.
7. Sibon, O.C., Stevenson, V.A., and Theurkauf, W.E. (1997). DNA replication checkpoint control at the *Drosophila* midblastula transition. *Nature* **388**, 93–97.
8. Sibon, O.C., Laurencon, A., Hawley, R., and Theurkauf, W.E. (1999). The *Drosophila* ATM homologue Mei-41 has an essential checkpoint function at the midblastula transition. *Curr. Biol.* **9**, 302–312.
9. Takada, S., Kwak, S., Koppetsch, B., and Theurkauf, W.E. (2007). *grp(chk1)* replication-checkpoint mutations and DNA damage trigger a Chk2-dependent block at the *Drosophila* midblastula transition. *Development* **134**, 1737–1744.
10. Abraham, R.T. (2001). Cell cycle checkpoint signaling through the *ATM* and *ATR* kinases. *Genes Dev.* **15**, 2177–2196.
11. Sibon, O.C., Kelkar, A., Lemstra, W., and Theurkauf, W.E. (2000). DNA-replication/DNA-damage-dependent centrosome inactivation in *Drosophila* embryos. *Nat. Cell Biol.* **2**, 90–95.
12. Takada, S., Kelkar, A., and Thuerkauf W.E. (2003). *Drosophila* checkpoint kinase 2 couples centrosome function and spindle assembly to genomic integrity. *Cell* **113**, 87–99.
13. Bartek, J., Falck, J., and Lukas, J. (2001). CHK2 kinase–a busy messenger. *Nat. Rev. Mol. Cell Biol.* **2**, 877–886.
14. Musacchio, A., and Hardwick, K. G. (2002). The spindle checkpoint: structural insights into dynamic signalling. *Nat. Rev. Mol. Cell. Biol.* **3**, 731–741.
15. Pérez-Mongiovil, D., Malmanchel, N., Bousbaa, H., and Sunkell, C. (2005). Maternal expression of the checkpoint protein BubRl is required for synchrony of syncytial nuclear divisions and polar body arrest in *Drosophila melanogaster*. *Development* **132**, 4509–4520.
16. Weigmann, K., Klapper, R., Strasser, T., Rickert, C., Technau, G., Jäckle, H., Janning, J., and Klämbt, C. (2003). FlyMove – a new way to look at development of *Drosophila*. *Trends Genet.* **19**, 310. http://flymove.uni-muenster.de
17. Clarkson, M., and Saint, R. (1999). A His2AvDGFP fusion gene complements

a lethal His2AvD mutant allele and provides an in vivo marker for Drosophila chromosome behavior. *DNA Cell Biol.* **18**, 457–462.

18. Rogers, S.L., Rogers, G.C., Sharp, D.J., and Vale, R.D. (2002). *Drosophila* EB1 is important for proper assembly, dynamics, and positioning of the mitotic spindle. *J. Cell Biol.* **158**, 873–884.
19. Sullivan, W., Fogarty, P., and Theurkauf, W. (1993). Mutations affecting the cytoskeletal organization of syncytial *Drosophila* embryos. *Development* **118**, 1245–1254.
20. Chou, T.B., and Perrimon, N. (1996). The autosomal FLP-DFS technique for generating germline mosaics in *Drosophila* melanogaster. *Genetics* **144**, 1673–1679.

# Chapter 8

# Using *Drosophila* Larval Imaginal Discs to Study Low-Dose Radiation-Induced Cell Cycle Arrest

**Shian-Jang Yan and Willis X. Li**

## Abstract

Under genotoxic stress, activation of cell cycle checkpoint responses leads to cell cycle arrest, which allows cells to repair DNA damage before continuing to cycle. *Drosophila* larval epithelial sacs, called imaginal discs, are an excellent in vivo model system for studying radiation-induced cell cycle arrest. Larval imaginal discs go into cell cycle arrest after being subjected to low-dose irradiation, are subject to easy genetic manipulation, are not crucial for survival of the organism, and can be dissected easily for further molecular or cellular analysis. In this chapter, we describe methods for assessing low-dose irradiation-induced cell cycle arrest. Mitotic cells are identified by immunofluorescence staining for the mitotic marker phosphorylated histone H3 (phospho-histone H3 or pH3). When wandering third-instar control larvae, without transgene expression, are exposed to 500 rads of X-ray or γ-ray irradiation, the number of pH3-positive cells in wing imaginal discs is reduced from hundreds before irradiation to approximately 30 after irradiation, with an equal distribution between the anterior and posterior compartments (Yan et al., 2011, FASEB J). Using the GAL4/UAS system, RNAi, cDNA, or microRNA sponge transgenes can be expressed in the posterior compartment of the wing disc using drivers such as *engrailed (en)-Gal4*, while the anterior compartment serves as an internal control. This approach makes it possible to do genome-wide genetic screening for molecules involved in radiation-induced cell cycle arrest.

**Key words:** Drosophila, larvae, imaginal discs, radiation, checkpoint, cell cycle arrest, mitosis.

## 1. Introduction

The fruitfly *Drosophila melanogaster* was first introduced into the laboratory one hundred years ago by Thomas H. Morgan (2, 3) and has been used since then as a model organism for studies in genetics, animal development, and many other biological processes. The use of simple, yet elegant and powerful genetics tools

W.X. Li (ed.), *Cell Cycle Checkpoints*, Methods in Molecular Biology 782,
DOI 10.1007/978-1-61779-273-1_8, © Springer Science+Business Media, LLC 2011

in *Drosophila* has twice resulted in Nobel awards (4, 5). Today, we enjoy the insights that *Drosophila* has brought us regarding fundamental biological principles, and molecular mechanisms of human diseases. Each day we learn about new applications in biomedical research that are possible with this model organism.

*Drosophila* has been a great asset in studying cell cycle checkpoints (6). Many gene products involved in these checkpoints were first identified in *Drosophila* and were subsequently shown to have evolutionarily conserved functions in DNA damage responses and in cell cycle arrest in many other organisms including humans (7). It is likely that more genes which play important roles in cell cycle arrest will be discovered, since the *Drosophila* genome encodes many proteins and non-coding RNAs whose functions are still unknown.

Development of *Drosophila* larval wing epithelial sacs, called imaginal wing discs, has been studied intensively and comprises an excellent in vivo system for understanding organogenesis (8) and for investigating how different conserved signal transduction pathways function and interact during organ growth (9). During larval growth, a small population of wing progenitor cells proliferates rapidly, resulting in up to 50,000 cells at the late third-instar stage (10). Their rapid cellular division rate and their cell cycle arrest in response to low doses of radiation are among the advantages of using larval imaginal wing discs to study radiation-induced cellular responses (1). Furthermore, wing discs are dispensable for *Drosophila* survival, so they can be manipulated genetically with ease. For example, the Gal4/UAS system can be used to drive transgene expression in particular regions of discs (**Figs. 1** and **2**) (1, 11).

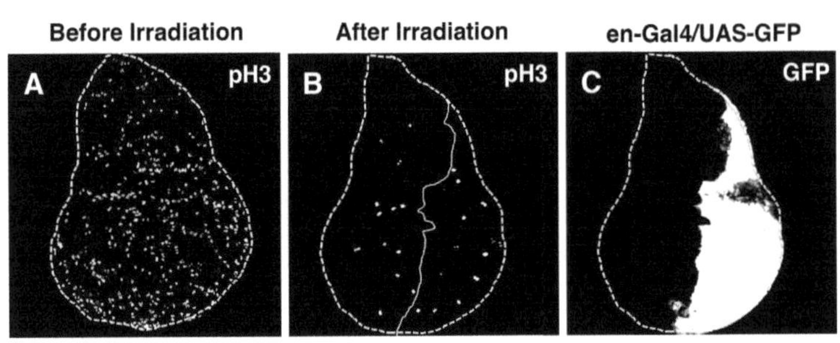


Fig. 1. Low-dose radiation-induced cell cycle arrest. (**a**) A representative confocal image of a wild-type control, unirradiated imaginal wing disc immuno-stained with anti-pH3 and Alexa 546 (as the secondary antibody). Note >200 pH3-positive cells (*white dots*) in the disc. Imaginal wing discs are shown with anterior to the left and dorsal up. *Dashed lines* outline the wing discs. (**b**) Larvae were fixed 1 h post-irradiation (500 rads of X-rays or $\gamma$ rays) and were immuno-stained with anti-pH3. Note that the number of pH3-positive cells is reduced to approximately 30, with a nearly equal distribution between the anterior and posterior compartments (about 15 in each). *Solid lines* demarcate the anterior–posterior boundaries. (**c**) en-Gal4 driven UAS-GFP expression in the posterior compartment. Note that the images in (**b**) and (**c**) come from the same disc.

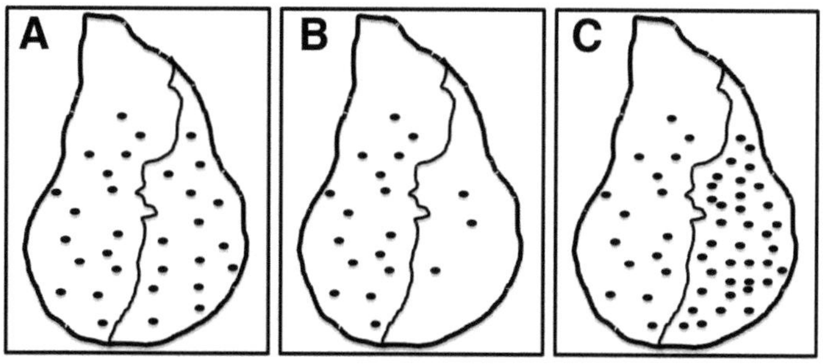


Fig. 2. A schematic of different responses to radiation-induced cell cycle arrest. Following irradiation (500 rads of γ rays), the total number of pH3-positive cells is similar in the anterior and posterior compartments when only GFP was expressed in the posterior compartment (**a**), whereas fewer (**b**) and more (**c**) pH3-positive cells are found in the posterior compartments expressing particular UAS-cDNAs or UAS-RNAi transgenes, respectively, indicating that such genes may control cell cycle arrest. Note that the anterior compartment of the discs serves as an internal control. It is possible to scale up such an assay for genome-wide genetic screening of molecules involved in the radiation-induced cell cycle arrest response.

Recently, UAS-cDNA transgenes from many fly laboratories and stock centers have become available at little or no cost, UAS-RNAi transgenes are available from the Vienna *Drosophila* RNAi Center (12), and an emerging *Drosophila* collection of UAS-microRNA sponge transgenes (13) is becoming available from Perrimon's lab. The vast collection of UAS-RNAi trangenes has made it possible to do genome-wide genetic screening in order to study certain biological processes (14, 15). Such powerful tools also provide an opportunity to those who have not previously worked with *Drosophila* to begin using it in their research.

Here, we describe methods for studying low-dose radiation-induced cell cycle arrest. Our goal is to introduce these methods to researchers outside of as well as within the *Drosophila* community.

## 2. Materials

### 2.1. Culturing Flies

1. Fly food ingredients: 51 mL water, 69 g Agar (granulated), 90 g Brewer's Yeast, 110 g molasses, 400 g malt extract, 400 g corn flour, 50 g soy flour, 31.3 mL propionic acid, 36.0 mL 20% Nipagin in 95% EtOH
2. Culture bottles (175 mL)
3. Culture vials (28.5 mL)
4. Fly stocks (*see* **Note 1**)

**2.2. Collecting Larvae and Irradiation**

1. Metal probes
2. Fly food vials
3. Apple juice plates (ingredients for 1 L): 750 mL water, 22.5 g Bacto Agar, 25.0 g sucrose (technical grade), 250 mL apple juice, 7.0 mL 20% Nipagin in 95% EtOH
4. An X ray irradiator
5. A $\gamma$-ray irradiator

**2.3. Dissecting Larvae and Fixing Larval Imaginal Discs**

1. Dumont forceps #5
2. Glass dishes (*see* **Note 2**)
3. 1× PBS buffer: 8 g NaCl, 0.2 g KCl, 1.44 g $Na_2HPO_4$, 0.24 g $KH_2PO_4$ in 800 mL of distilled $H_2O$. Adjust the pH to 7.4 with HCl. Add $H_2O$ to 1 L
4. A stereo microscope
5. 4% Paraformaldehyde in 1× PBS
6. Ice buckets

**2.4. Staining Larval Imaginal Discs**

1. *Primary antibodies*: Rabbit anti-pH3 antibody (Upstate Biotechnology, used at 1:1000 in 5% normal goat serum/PBT) or mouse monoclonal anti-pH3 antibody (6G3, Cell Signalling, used at 1:2000 in 5% normal goat serum/PBT)
2. *Secondary antibodies*: Goat anti-rabbit or goat anti-mouse Alexa 488, 546, or 660 antibody (Invitrogen, used at 1:250 in 5% normal goat serum/PBT; *see* **Note 3**)
3. PBT (0.3% Triton X-100/1× PBS)
4. Normal goat serum
5. Sodium Azide

**2.5. Dissecting and Mounting Larval Imaginal Discs**

1. Same as **Section 2.3** except using 1× PBT
2. 100 μL and 1-mL Pasteur pipettes
3. Pipettemen
4. Glass slide coverslips 22 × 22 mm #1.5 (e.g., VWR)
5. Microscope glass slides 25 × 75 mm (e.g., VWR)
6. Lens paper (e.g., VWR)
7. Tissue paper (e.g., Kimwipes)
8. Vectashield mounting medium with DAPI (Victor)
9. Transparent nail polish
10. Slide holders (e.g., VWR)

### 2.6. Imaging Larval Imaginal Discs and Measuring Mitotic Cells

1. A confocal microcope (e.g., Leica SP2/SP5)
2. A fluorescence microscope (e.g., Zeiss Axio Imager M2m)
3. Imaging software (e.g., Leica Confocal or Photoshop)

## 3. Methods

### 3.1. Culturing Flies

1. Decide on the amount of fly food needed and proportionally increase or decrease each ingredient of the food recipe. Add the appropriate amount (60%) of hot water to a kettle. Close the lid and bring the water to a boil. Then turn down the heat to stop the water from boiling. Add agar a small amount at a time using a metal whisk. Mix well after each addition and then turn up the heat so that the ingredients in the kettle boil for 2 more minutes. Then slowly add the brewer's yeast and mix well.
2. Dissolve the molasses and malt extract in 20% of the total volume of hot water. Stir into the agar/brewer's yeast/water mixture.
3. Dissolve the corn flour and soy flour in 20% of the total volume of warm water. Pour into the kettle.
4. Cook the mix for 2 h with the lid on. Stir well.
5. Then reduce the heat further and cool for 1 h. Stir well.
6. When cooled to 80°C, add the propionic acid and Nipagin. Stir very well.
7. Pump food into vials and bottles (12 mL food for vials, 45 mL food for bottles). While dispensing, keep stirring occasionally to prevent food from thickening.
8. Cover vials and bottles with cheesecloth and let food solidify overnight before putting in cotton plugs. Place plastic bags over the cheesecloth if the humidity is low.
9. Wrap trays in plastic bags and store at 18°C. Label with the date the food was made.
10. Do crosses and transfer flies into fresh vials or bottles of fly food at least every 2–3 days) to prevent larvae from overcrowding (*see* **Note 4**). Depending on the number of flies, daily transfers may be necessary.

### 3.2. Collecting Larvae and Irradiation

1. Add agar and water to a 2-L flask. Add sucrose and apple juice to another flask. Stir well and then autoclave both flasks.

2. After autoclaving, cool the flasks down to about 50°C. Add the apple juice/sucrose solution to the agar solution slowly, to avoid the creation of bubbles.
3. Add the Nipagin solution and mix well without creating bubbles. Pour into 100 mm × 15 mm petri dishes. Wait until agar solidifies. Then store the plates at 4°C. Use a metal probe to pick wandering third-instar larvae (those that have just started crawling out of the food) and put them into fresh vials or onto apple juice plates (*see* **Note 5**) immediately before irradiation.
4. Carefully calculate the proper time required for low irradiation doses (250 or 500 rads of X-rays or $\gamma$ rays), set the timer accordingly, put larvae into the irradiator, and turn it on (*see* **Note 6**).

### 3.3. Dissecting Larvae and Fixing Larval Imaginal Discs

1. Set up the light source and stereo microscope properly before dissecting larvae (**Fig. 3**; *see* **Note 7**).
2. Exactly 1 h after irradiation, dissect larvae in 1× PBS with a pair of fine Dumont #5 forceps in each hand. Hold a larva

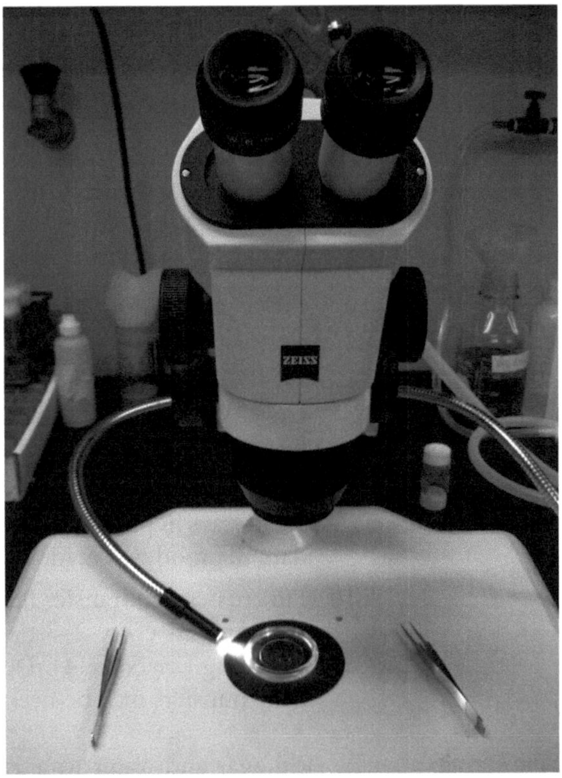


Fig. 3. Stereo microscope and light setup for larval dissections. The *glass dish* is under the microscope with the light source to the *left side*. Each hand will hold a pair of forceps during dissection.

with one pair of forceps and use the other pair to cut the larva about one-third of the way down from the mouth hook. Turn the anterior portion of the larva inside out by holding the head with one pair of forceps and pushing the mouth hook in with the other pair (*see* **Note 8**). Using the forceps, gently remove excess fat tissue, the salivary glands and the gut.

3. Immediately transfer the dissected, inside-out larval tissues into 4% paraformaldehyde in 1× PBS (in a 1.5-mL microtube) on ice. At least five larvae (ten imaginal wing discs) can be dissected and collected in the microtube. Then invert the microtube several times, making sure that all of the larvae are in the solution, and leave the tube at room temperature for 10 min, for fixation. Then remove the paraformaldehyde solution and rinse the larvae four times with 1× PBS.

### 3.4. Staining Larval Imaginal Discs

1. Incubate the fixed larvae, with wing imaginal discs attached to the body wall, with rabbit anti-phospho-histone H3 (pH3, used at 1:1000) in 5% normal goat serum/PBT overnight in a cold room at 4°C.
2. The next day, wash the larvae in PBT for 10 min three times at room temperature (*see* **Note 9**), and then incubate them with a fluorescent secondary antibody (e.g., goat anti-rabbit Alexa 546, used at 1:250) for 2 h at room temperature or overnight at 4°C (the latter is preferred).
3. Following the incubation with fluorescent antibody, wash the larvae in PBT for 10 min three times at room temperature and keep the larvae in PBT at 4°C before disc dissection and mounting.

### 3.5. Dissecting and Mounting Larval Imaginal Discs

1. Use the same setup as in **Section 3.3**, except that the dissections are done in PBT rather than in 1 × PBS.
2. Locate the wing imaginal discs, which are on either side of the body, attached to the trachea near the anterior spiracles (16). Gently remove the wing discs with forceps, without tearing or damaging them. Leave the dissected wing discs (at least 5–10) in PBT in the glass dishes used for dissection.
3. Clean glass microscope slides. Transfer 5–10 imaginal wing discs to a microscope slide using a 1-mL Pasteur pipette or a pipetteman. Carefully remove excess PBT by sucking it up with a pipette and then blotting with tissue paper.
4. Put 13 μL of Vectashield mounting medium onto the discs and cover them with a coverslip (avoid bubbles).
5. Seal the edges of the coverslip with transparent nail polish. Label the slides, put them into a slide holder, and store them at 4°C.

### 3.6. Imaging Larval Imaginal Discs and Measuring Mitotic Cells

1. Capture images of the discs using a confocal microscope (e.g., Leica SP2/SP5) or a fluorescence microscope (e.g., Zeiss Axio Imager M2m). Follow instructions in the user's manual from the manufacturers of the microscope. For confocal images, use a 20× objective lens and a 1024×1024 scan format; set the pinhole diameter at 1 Airy unit.
2. Use Leica Confocal software or Photoshop to retrieve images. To convert color images to black and white, retrieve the images in Photoshop and select "grayscale" under the "Image > mode" menu (*see* **Note 10**).
3. Count the number of pH3 positive foci.

## 4. Notes

1. Information about fly stocks is available through http://www.flybase.org.
2. Use glass dishes for dissecting larvae, since larval imaginal discs can stick to plastic. If plastic dishes are used, they must be the kind that are coated with glass on the inside.
3. Alexa Fluor 488 dye generally gives the brightest fluorescence and is therefore preferred if only a single primary antibody is used. When other fluorescence with a similar emission spectrum to Alexa Fluor 488 (e.g., GFP) is present, other fluorescent dyes (e.g., 546) can be used.
4. It is important to avoid larval overcrowding, as overcrowded larvae are developmentally delayed and smaller than uncrowded larvae. If necessary, use a spatula to transfer a small number of embryos or younger larvae (e.g., first instar larvae) to fresh fly food vials or bottles.
5. Apple juice plates are flat and can hold a rather large number of larvae. They are suitable for X-ray irradiation where equal distance from the radiation source is required (**Figs. 4** and **5d**) Normal fly food can be poured into plates for this purpose. However, transparent apple juice agar is more helpful in locating larvae for dissection. We use normal fly food vials to hold larvae for γ-ray irradiation since our γ-ray irradiator (GAMMACELL 1000) provides even penetrance of irradiation to a relatively large population of larvae, by rotating the vials. The GAMMACELL 1000 can irradiate five vials at a time (**Fig. 5a–c**).
6. Periodic calibration of the irradiators is important for consistency of the experimental data. Radiation safety training

Fig. 4. Vials, apple juice agar plates, and metal probe. Note that few larvae are in the vials and on the plates for illustration purposes; more larvae (>50) can be irradiated in a single vial or plate.

Fig. 5. Irradiation equipment and setups. (**a**) A Cs-137 $\gamma$-ray irradiator. (**b**) A metal rotating cylinder is used to hold five vials (*arrow*) simultaneously for irradiation. (**c**) The cylinder with vials (*arrow*) is put into the irradiation chamber and is ready for irradiation. A shielding metal door is closed (not shown) during irradiation. (**d**) An X-ray irradiator with larvae on an apple juice agar plate (*arrow*) ready for irradiation.

is also required for performing experiments involving irradiators.

7. To facilitate dissections, it is important to set up the microscope and light source properly. We find that it is easier to see the larvae and the tips of the forceps clearly using a 4 × objective zoom, since this setup provides adequate depth of field and high enough magnification. A 5 × zoom has higher magnification but less depth of field, which makes it more difficult to see both the larvae and forceps without constantly re-adjusting the focus. We also prefer having the light source at the left side instead of above the tissue; larval tissues are mostly transparent and can be seen more easily when lighted from the side. Light coming from above generates reflections, which make everything more difficult to see.
8. Like other skills, dissecting larvae requires practice and patience. The first few tries can be frustrating. However, once hand-eye coordination is achieved, it will take only about 10–20 s to dissect a larva.
9. The antibody solutions can be reused at least three times if sodium azide is added to a final concentration of 0.02% (w/v), to prevent microbial contamination.
10. We find that, for quantification, black and white images are easiest on the eyes and are preferred for presentation and publication if a single antibody is being used.

## Acknowledgments

We are grateful to Christine Summers in the Bohmann lab for providing recipes and protocols for making fly food and apple juice agar plates. We thank Dr L. Silver-Morse for critical reading of the manuscript. S.J.Y. is supported by a postdoctoral training grant (T32CA009363) from the National Institutes of Health. The Li lab is supported by grants from the National Institutes of Health, an American Cancer Society Research Scholar Grant, and a Leukemia & Lymphoma Society Research Scholar Grant to W.X.L.

### References

1. Yan, S. J., Lim, S. J., Shi, S., Dutta, P., and Li, W. X. (2011) Unphosphorylated STAT and heterochromatin protect genome stability, *FASEB J. 1*, 232–241.
2. Morgan, T. H. (1910) Sex limited inheritance in Drosophila, *Science 32*, 120–122.
3. Bellen, H. J., Tong, C., and Tsuda, H. (2010) 100 years of Drosophila research and

its impact on vertebrate neuroscience: a history lesson for the future, *Nat Rev Neurosci 11*, 514–522.
4. Cohen, B. (1995) Nobel committee rewards pioneers of development studies in fruitflies, *Nature 377*, 465.
5. Raju, T. N. (1999) The Nobel chronicles. 1933: Thomas Hunt Morgan (1866–1945), *Lancet 353*, 157.
6. Elledge, S. J. (1996) Cell cycle checkpoints: preventing an identity crisis, *Science 274*, 1664–1672.
7. Song, Y. H. (2005) Drosophila melanogaster: a model for the study of DNA damage checkpoint response, *Mol Cells 19*, 167–179.
8. Weatherbee, S. D., and Carroll, S. B. (1999) Selector genes and limb identity in arthropods and vertebrates, *Cell 97*, 283–286.
9. Yan, S. J., Gu, Y., Li, W. X., and Fleming, R. J. (2004) Multiple signaling pathways and a selector protein sequentially regulate Drosophila wing development, *Development 131*, 285–298.
10. Garcia-Bellido, A., and Merriam, J. R. (1971) Parameters of the wing imaginal disc development of Drosophila melanogaster, *Dev Biol 24*, 61–87.
11. Brand, A. H., and Perrimon, N. (1993) Targeted gene expression as a means of altering cell fates and generating dominant phenotypes, *Development 118*, 401–415.
12. Dietzl, G., Chen, D., Schnorrer, F., Su, K. C., Barinova, Y., Fellner, M., Gasser, B., Kinsey, K., Oppel, S., Scheiblauer, S., Couto, A., Marra, V., Keleman, K., and Dickson, B. J. (2007) A genome-wide transgenic RNAi library for conditional gene inactivation in Drosophila, *Nature 448*, 151–156.
13. Loya, C. M., Lu, C. S., Van Vactor, D., and Fulga, T. A. (2009) Transgenic microRNA inhibition with spatiotemporal specificity in intact organisms, *Nat Methods 6*, 897–903.
14. Pospisilik, J. A., Schramek, D., Schnidar, H., Cronin, S. J., Nehme, N. T., Zhang, X., Knauf, C., Cani, P. D., Aumayr, K., Todoric, J., Bayer, M., Haschemi, A., Puviindran, V., Tar, K., Orthofer, M., Neely, G. G., Dietzl, G., Manoukian, A., Funovics, M., Prager, G., Wagner, O., Ferrandon, D., Aberger, F., Hui, C. C., Esterbauer, H., and Penninger, J. M. (2010) Drosophila genome-wide obesity screen reveals hedgehog as a determinant of brown versus white adipose cell fate, *Cell 140*, 148–160.
15. Neely, G. G., Kuba, K., Cammarato, A., Isobe, K., Amann, S., Zhang, L., Murata, M., Elmen, L., Gupta, V., Arora, S., Sarangi, R., Dan, D., Fujisawa, S., Usami, T., Xia, C. P., Keene, A. C., Alayari, N. N., Yamakawa, H., Elling, U., Berger, C., Novatchkova, M., Koglgruber, R., Fukuda, K., Nishina, H., Isobe, M., Pospisilik, J. A., Imai, Y., Pfeufer, A., Hicks, A. A., Pramstaller, P. P., Subramaniam, S., Kimura, A., Ocorr, K., Bodmer, R., and Penninger, J. M. (2010) A global in vivo Drosophila RNAi screen identifies NOT3 as a conserved regulator of heart function, *Cell 141*, 142–153.
16. Bodenstein, D. (1994) The postembryonic development of Drosophila. In *Biology of Drosophila* (ed. M. Demerec), Facsimile edn., pp. 275–367, Cold Spring Harbor, NY: Cold Spring Harbor Laboratory Press.

# Chapter 9

# Screening for Radiation Sensitizers of *Drosophila* Checkpoint Mutants


## Mara Gladstone and Tin Tin Su



## Abstract

Anti-cancer therapy is largely comprised of radiation, surgery, and chemotherapy treatments. Although a single mode of therapy can be effective in treating certain types of cancer, none presents a cure. Multi-modal therapy, the use of two or more agents in combination (e.g., radiation and chemotherapy together), shows potential for a more effective treatment of cancer. The challenge then is identifying effective therapy combinations. In this chapter, we describe the use of Drosophila as a whole animal in vivo model to screen for small molecules that effectively combine with ionizing radiation to kill checkpoint mutants preferentially over wild-type. The differential use of wild-type and checkpoint mutants has the potential to identify molecules that act in a genotype-specific manner to eradicate checkpoint mutant tissues when combined with radiation, while sparing wild-type tissues.



**Key words:** Drosophila, radiation, checkpoint, molecule library, drug screen, cancer.


## 1. Introduction

Cancer remains a leading cause of death in developed countries. Tumor growth and survival relies on two inputs: "intrinsic" uncontrolled proliferative conditions within cancer cells and "extrinsic" growth and survival signals from surrounding non-cancerous tissues. The latter includes extracellular growth and survival factors, cell–cell and cell–matrix interaction and angiogenesis. An effective treatment for cancer would target both intrinsic and extrinsic conditions. Yet, most of the current therapies against cancer were developed in cell-based assays that fail to recapitulate the extrinsic growth and survival conditions found in vivo (1).


W.X. Li (ed.), *Cell Cycle Checkpoints*, Methods in Molecular Biology 782,
DOI 10.1007/978-1-61779-273-1_9,

A second shortcoming of most of the current anti-cancer agents is their side effects. Most therapies, such as radiation and chemotherapy, effectively eliminate rapidly dividing cancer cells. Unfortunately, these treatments also have deleterious effects on the surrounding tissue. Multi-modal therapy, in which two or more anti-cancer agents are applied in combination, has the potential to reduce side effects because each agent may be applied at a reduced dose (2–4). Simultaneous application of multiple agents can also prevent the selection of target cells or pathogens with resistance to a single agent applied alone. This is well documented in the treatment for Tuberculosis and AIDS where combination therapies have been effectively used to eradicate pathogenic microorganisms that can become resistant to single agents alone (5, 6). For these reasons, multi-modal therapy is being assessed in multiple clinical trials. A major challenge, however, is in identifying effective therapy combinations. Current approach is to combine, in an empirical manner, agents that were approved for use as single agents and test these in human clinical trials (www.clinicaltrials.gov). Such an approach is costly and ineffective.

For these reasons, it would be extremely valuable to be able to identify, *from the onset,* combination of therapeutic agents that can inhibit both intrinsic and extrinsic conditions that contribute to cancer growth and survival.

In this chapter, we describe a novel screening method that uses *Drosophila* as a model to identify *from the onset* small molecules that synergize with radiation. The screen is derived from our knowledge of Drosophila mutants in Checkpoint Kinase 1 (Chk1), encoded by the *grapes* gene in Drosophila (7, 8). We found that *grp* mutant larvae survive radiation exposure as well as wild-type under normal conditions. More cells may die in *grp* mutants due to defective checkpoints, but as long as surviving cells can undergo compensatory proliferation to regenerate tissues, the animal can survive. We found that mutations and growth conditions that interfered with compensatory proliferation also reduced the survival of *grp* mutants when combined with radiation. Furthermore, as expected, the combined effect of radiation with treatments that can interfere with compensatory proliferation is greater on *grp* mutants than on wild-type. We have leveraged these results to design a screen for molecules that when applied in conjunction with radiation, kill *grp* mutants preferentially over wild-type. Such molecules could target compensatory proliferation, for example, and thus sensitize tissues and organisms to radiation, i.e., “radiation sensitizers.”

A pilot screen through a small molecule library using the screen described here identified small molecules, some of which are known radiation sensitizers of human cancers. Subsequent screens in our labs have identified novel radiation sensitizers that we are now testing in mammalian models. The differential use of wild-type and checkpoint mutants has the potential to identify molecules that act in a genotype-specific manner to eradicate checkpoint mutant tissues when combined with radiation, while sparing wild-type tissues. This would reduce side effects. The use of whole animals in the screen may allow the discovery of small molecules that inhibit not only on intrinsic pathways, but also act on extrinsic pathways. We believe that this screening method could be adapted to find radiation sensitizers as well as "chemo-sensitizers," small molecules that increase the killing effect of known chemotherapy agents.

## 2. Materials

### 2.1. Drosophila Husbandry

1. Standard Cornmeal-Agar-Molasses Drosophila food (1487 g cornmeal, 2344 g molasses, 164 g agar, 938 g yeast, 375 mL Tegosept mold inhibitor [10% methyl-4-hydroxy benzoate in 95% EtOH], 94 mL proprionic acid, and 21,100 mL water).
2. Culture bottles, 6 oz.
3. Yeast pellets (Fleischmann's yeast, division of AB Mauri Food Inc.).
4. Population cages.
   These are 30 cm × 30 cm × 37 cm boxes made of 0.5-cm-thick clear acrylic. At least one circular hole (21 cm diameter) should be present on one of the sides to provide access. This hole is covered with nylon netting to contain flies during use. Pantyhose work well for this purpose!
5. Blotting paper (Whatman 3MM Chr paper).

### 2.2. Embryo Collections

1. Yeast paste (*see* **Note 1**).
2. Grape juice agar plates: for 1 L: 720 mL $dH_2O$, 12 g sucrose, 250 mL grape juice, 27.2 g agar. Heat mixture in a microwave for 10–15 min, in 3–5-min intervals, stirring in between. Once the mixture is cool enough to touch the beaker, add 20 mL ethanol and 10 mL glacial acetic acid. Mix. Pour about 100 mL into an 11 in. by 6 in. Styrofoam plate. 1 L makes 10 plates. Flame plates using a Bunsen burner to remove air bubbles on their surface before the agar solidifies.
3. An incubator set at 25°C, 60% humidity, and 12 h light–12 h dark cycle.

4. Egg collection baskets. These are short cylinders (2.5 cm inner diameter × 2.5 cm length) made of 0.3-cm-thick acrylic. One end is covered with glued-on nylon mesh (e.g., Nitex screen).
5. Paintbrushes and spatulas.

### 2.3. Irradiation

1. TORREX X-ray generator 120 D (EG & G Astrophysics Research).
2. Spatulas and plastic containers (Tupperware is convenient).
3. Metal sieves (VWR scientific USA standard testing sieves): 400, 600, and 800 μM.
4. Dish for irradiating. We use a plastic container the size of a standard 100 mm × 20 mm Petri dish, with a large hole cut in the lid for air exchange. We place larvae in the dish with a little water to prevent drying, place a large piece of Nitex screen and close the lid to prevent the larvae from escaping during irradiation.

### 2.4. Feeding Small Molecules to Larvae

1. Minimal-nutrient Corn meal agar food. For 1 L: 12 g agar, 77 g cornmeal, and 960 mL dd$H_2O$. Autoclave at 121°C on the liquid cycle for 20 min to dissolve. Mix immediately after autoclaving to prevent cornmeal from clumping. Cool with mixing before use.
2. –20°C freezer for small molecule storage.
3. Narrow plastic vials (Genesee Scientific 32–116).
4. Cotton plugs "Flugs" (Genesee Scientific 49–102).
5. DMSO.
6. Microwave, dd$H_2O$.
7. Dissecting microscope with GFP illumination (Leica M2125, with ebq 100 GFP light).

### 2.5. Quantification of Survival

1. 95% ethanol
2. Light box (Laboratory Supplies Co, Inc. Model G129)
3. Permanent marker

### 2.6. Data Collection and Analysis

Microsoft Excel (or similar)

### 2.7. Small Molecule Libraries

In our pilot screen we used the Diversity Set of molecules (~2000) from the National Cancer Institute's Developmental Therapeutics Program. These are supplied as 10-mM solutions in DMSO in 96-well plates, containing 10–20 μL per well.

## 3. Methods

Briefly, the screen involves irradiating correctly sized larvae en masse, aliquoting larvae into food vials supplemented with drug, and quantifying eclosion 10 days after the irradiation (**Fig. 1**).

### 3.1. *Drosophila* Husbandry

1. To have enough adult flies for embryo collections, maintain about 70 bottles of flies for each genotype at any given time. We use standard 6 oz bottles with standard cornmeal-agar-molasses food to culture flies.
2. Day 1: Transfer newly eclosed adults to fresh, room temperature bottles with a sprinkle of dry Baker's yeast. Aim for approximately 300 adults per bottle.
3. Days 1–3: Allow flies to lay eggs.
4. Day 3: Transfer adults to a population cage. Place in the cage a 15 cm × 57 cm piece of 3MM paper, accordion-folded to

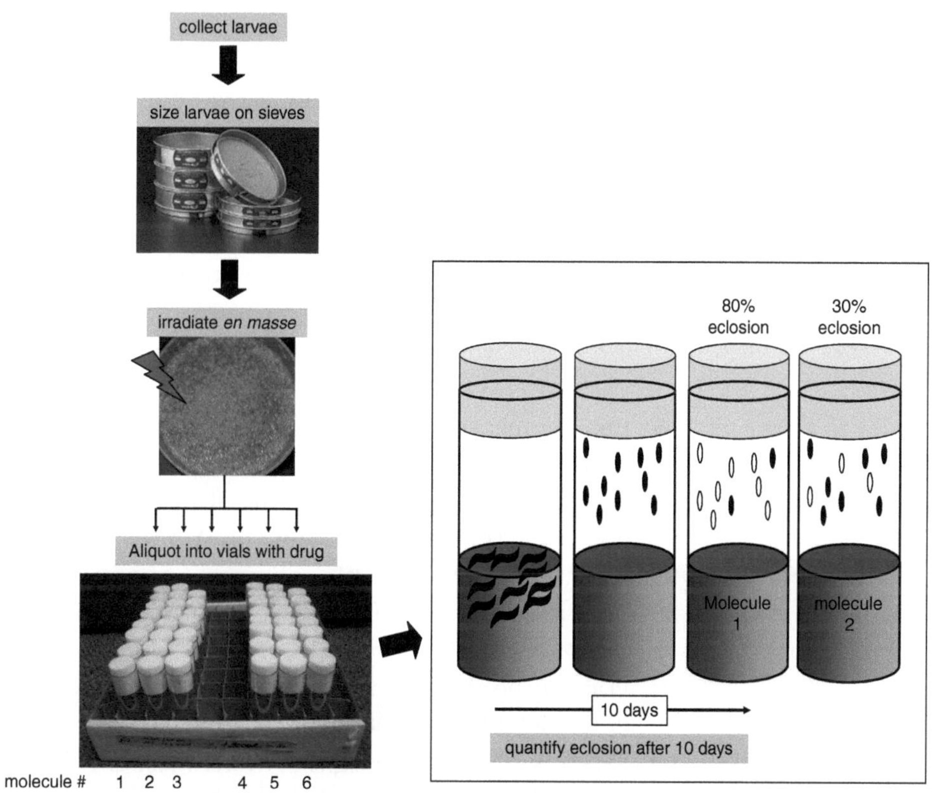


Fig. 1. A schematic depiction of the screen. The major steps are illustrated. The larvae that died following radiation/drug treatment remain in the pupa case (*black ovals* on the side of the vial). These are easily distinguished from the larvae that survived radiation/drug treatment to eclose from the pupa case (*empty ovals*). The survival results for two molecules that produce 30 and 80% eclosion are depicted.

stand up vertically; this is to increase the vertical surface area for flies to rest on and to reduce moisture (*see* **Note 2**).

5. Days 4–5: Check the bottles for larvae. If the bottles are too crowded, split larvae into fresh bottles with yeast sprinkles. Ideally, you want about 300 adults to eclose from each bottle so that you can repeat Step 2 above in the next generation. Keep about 70 bottles with larvae and discard the rest. Allow larvae to develop into adulthood (about 2 weeks). These will be used for the next cage.
6. Feed the population cage daily. Use a fresh 11 in. by 6 in. grape juice-agar Styrofoam plate (*see* **Section 2**) with a generous tablespoonful of yeast paste (*see* **Note 1**) spread in a wide line along its length.
7. Change the 3MM paper frequently as necessary.

### 3.2. Embryo Collections

1. Use a fresh grape juice-agar Styrofoam plate (*see* **Section 2**) for collecting embryos. Spread a small amount of yeast (approximately 1 teaspoon; *see* **Note 1**) thinly in a line down the length of the plate before placing the plate in the cage.
2. Allow flies to lay embryos for 4 h.
3. While the flies are laying eggs, allow fresh food bottles to equilibrate to room temperature. Do NOT put yeast pellets in the bottles. We find that adding yeast pellets can lead to bottle-to-bottle variations in growth rate.
4. Remove egg collection plate from the cage and use a paintbrush to remove stray flies that are stuck to the plate.
5. To collect embryos, squirt the plate with distilled water and use a paintbrush to free embryos from the agar while dissolving the yeast paste.
6. Pass the mix of embryos, yeast and water through an egg collection basket.
7. Repeat Steps 5 and 6 until most of the embryos on the plate are in the basket. Rinse the embryos well with distilled water to remove any yeast.
8. Place approximately 300 embryos into each culture bottle. We use a paintbrush with stiff bristles for this.
9. Place the bottles in the 25°C incubator. Allow the embryos to develop into third-instar larvae (4–5 days).
10. Check bottles daily for signs of overcrowding. Overcrowding will be apparent when "pits" in the surface of the food are visible. This occurs because larvae prefer to feed as a monolayer; "pits" increase the surface area to accommodate a large number of feeding larvae. Overcrowded

bottles should be split immediately into two or more fresh bottles.

11. Embryo collections should be taken every day of the week to maximize the number of drugs screened per cage. We find that embryo production declines after about 10 days for most genotypes.

### 3.3. Irradiation

1. Irradiate third-instar larvae at room temperature. We use a TORREX X-ray generator (Astrophysics Research) set at 5 mA and 115 kV. We use shelf 7, placing the larvae to be irradiated directly in the center in order to allow even irradiation (this may be different for other X-ray machines).
2. Turn on the X-ray machine and warm-up according to the manufacturer's instructions.
3. Using a spatula, scoop larvae-food mixture out of the bottles and place in a large plastic container (*see* **Note 3**).
4. Rinse off the food and collect larvae of uniform size using metal sieves (*see* **Section 2**). This is best done by filling the plastic container that contains the larvae-food mixture with tap water (*see* **Note 4**), mixing, and pouring the mixture of larvae, food and water onto a stack of three sieves (800-μM sieve on top, 600-μM sieve in the middle, and 400-μM sieve on the bottom). Washing the larvae-food mixture separates the larvae and the food allowing the larvae to pass through the sieves to be sorted by size. Avoid pouring blocks of food onto the sieves. Excess food in the sieves interferes with proper third-instar selection and may require additional rinsing. Vigorous washing can stress and even kill larvae; so it is important to be gentle and to work quickly.
5. Discard chunks of food and large larvae that collect on the 800-μM sieve. Discard small larvae that go through the top two sieves to collect on the 400-μM sieve. Use the larvae that collect in the middle sieve. We usually freeze to kill unwanted larvae before discarding.
6. Use a spatula to collect larvae gently from the middle, 600-μM sieve and place in a closable plastic dish for irradiation (*see* **Section 2**). Prevent crowding. Ideally, the larvae will be in a single monolayer. If you have too many larvae, perform multiple irradiations.
7. Determine the number of vials that will be needed by estimating the number of third-instar larvae collected. *See* **Section 3.4**, Step 7, for number of larvae per vial for each genotype.
8. Irradiate (*see* **Note 5**).

9. The dose of radiation should be chosen to cause 60–70% lethality in the genotype you are using. This should be determined in a dose-response curve prior to the initiation of the screen. Be sure to use the same culture conditions for dose-response determination and in the screen. These include culture conditions, sizing with sieves, and the use of minimal cornmeal-agar food after irradiation. We use 4000 rads (40 Grays) of X-rays for wild-type Sevelin strain and for *grp*$^{1}$ mutants, and 3500 rads for *p53*$^{5A\text{-}1\text{-}4}$ mutants, for example. *p53*$^{5A\text{-}1\text{-}4}$ is a null allele that results from a partial deletion of the gene while the *grp*$^{1}$ is a strong hypomorph that results from transposon insertion (9–12).

### 3.4. Feeding Small Molecules to Larvae

1. Minimal-nutrient corn-meal agar food (*see* **Section 2**) should be prepared either earlier in the day or the day before irradiation (*see* **Note 6**). Minimal-nutrient corn-meal agar can be stored at 4°C and simply reheated in the microwave when needed (*see* **Note 6**).
2. If the small molecules are stored in a 96-well liquid format, the plate should be removed from the freezer while the larvae are being irradiated, and placed in a dark area (*see* **Note** 7).
3. Before this step, the number of vials needed should have already been determined. Take out the appropriate number of vials (generally around 80 if the cages are set up properly) and label each with the plate and well number. Be sure to include at least three control vials with the solvent, which for our screens has been DMSO.
4. Add 3 mL of cornmeal-agar food to each vial (*see* **Note 8**). The food should be cool enough to touch because the small molecules may be sensitive to heat.
5. Pipette 3 μL of a small molecule per vial (or 3 μL of DMSO per control vial). Deliver the drug/DMSO directly into the food and use the pipette tip to immediately stir the food/drug mix. We have found that DMSO levels in the food that are greater than 0.1% can reduce larval survival.
6. Repeat Steps 4 and 5 until all of the vials are filled. We typically prepare drug vials in batches of 10. That is, we pour food into 10 vials, add drug to each, and repeat the process.
7. Allow the food to cool and solidify (*see* **Note 9**). Add ~50 larvae to each vial. Close the vial using a cotton plug (*see* **Section 2**).
8. Place the vials in the 25°C incubator.
9. Starting with the 5th day after irradiation, identify homozygous *grp* mutants. *grp* mutants are maintained as heterozygotes over a balancer chromosome carrying a GFP

transgene. Homozygous mutants, of interest in the screen, lack GFP. These are best distinguished from GFP+ siblings in early pupa stages before body pigments obscure the GFP signal. Using a dissecting microscope with GFP illumination, use a permanent marker to circle each non-GFP pupa (*see* **Note 10**). Continue daily for an additional 5 days or until no new pupae form.

### 3.5. Quantification of Survival

1. At 10 days after irradiation, discard live flies from the vials. This can be done by shaking flies through a funnel into a flask containing 95% ethanol. We find that most flies that were going to eclose have done so by this time.
2. Place each vial on a light box and determine which pupa cases still contain flies and which do not. Empty pupa cases are translucent, while those containing flies are opaque. Keep track of which cases have been counted by crossing them off with a permanent marker as you count.
3. Record how many pupa cases are empty and how many are full.
4. Determine the survival rate (% eclosion) by dividing the number of empty pupa cases by the total number of pupae for each small-molecule drug vial.

### 3.6. Data Collection and Analysis

1. Analyze data for each batch of irradiated larvae of a given genotype separately. Plot % eclosion against the number of molecules that produced that level of eclosion. It may be helpful to bin the data (e.g., group all molecules that produce 50–55% eclosion). The histogram should appear Gaussian (**Fig. 2**).

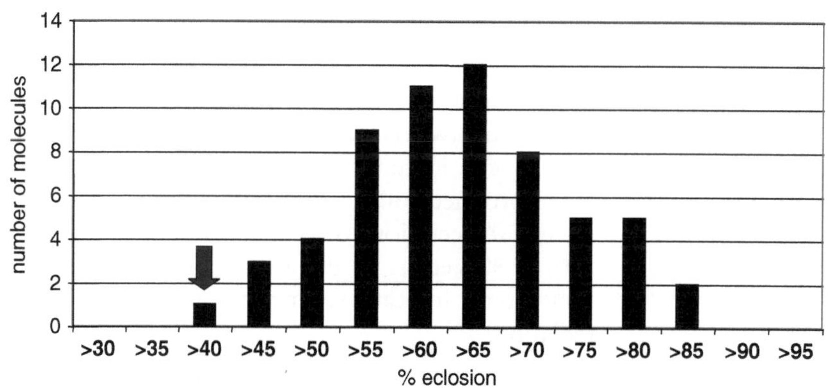


Fig. 2. A sample of screening data. The effect of a sample of 80 small molecules on survival of *grp* larvae irradiated with 4000 R of X-rays. The number of molecules is plotted against the interval of percentage of eclosion as shown. *Arrows* indicate molecules that produced eclosion rates that are two standard deviations below the average.

2. Calculate the average % eclosion and its standard deviation for the batch. Average % eclosion in solvent control vials should be similar to the average of the batch.
3. Identify molecules that decreased survival by greater than two standard deviations from the average. Also calculate the $X^2$ value for each drug by using the average for the batch to calculate expected eclosed/expected non-eclosed numbers. Any drugs that are two standard deviations or more below the average and also have a $X^2$ with a $p < 0.05$ are potential radiation sensitizers that should be tested further.

### 3.7. Testing for Reproducibility

1. Repeat Steps 3.3 and 3.4 for potential hits from the primary screen.
2. Include ten vials of solvent control. During data analysis, use this population to calculate the average and standard deviation and to calculate expected eclosed/non-eclosed numbers for the $X^2$ test.
3. Determine whether % eclosion of potential hits is two or more standard deviations lower than the average, and whether the $X^2$ test produces a $p < 0.05$. If so, the potential hit is a reproducible hit and should be tested further.

### 3.8. Secondary Screen

1. Repeat Steps 3.3 and 3.4 for reproducible hits, but include wild-type −/+ IR and mutant −/+ IR samples.
2. Determine the survival rate of each of the vials.
3. Reproducibility in radiation resistance should be assessed by calculating the average eclosion from at least three experiments for each dose of drug, with and without irradiation. Standard deviation of the population should be calculated for each sample. Student's *t*-test should be used to determine if the differences between samples are significant. This analysis will determine if a particular concentration of drug affected a particular genotype specifically.
4. Using the eclosion rate for drug alone and IR alone, determine what the eclosion rate would be if the effect of combining the two is additive. Any eclosion rate below this can be considered synergistic. For example, if the drug alone produces 60% eclosion and the radiation alone produces 50% eclosion, the expected additive effect is 60% × 50% or 30% eclosion. Eclosion rates significantly lower than 30% would indicate synergistic effects between drug and radiation.
5. For further testing in human cell lines, we choose drugs that show the best synergy and are genotype specific such that mutants are killed to a greater extent than wild-type.

## 4. Notes

1. Make up live yeast paste by mixing baker's yeast (Fleischmann's yeast, division of AB Mauri Food Inc.) with double-distilled water and make a smooth paste about the consistency of peanut butter. This can be covered and stored at 4°C.
2. The population cages we use have one smaller hole on the top and a larger hole on the front. To cover the smaller hole, we place a nylon hose over the hole and tie a knot. To cover the larger hole, using medium/normal size nylon hoses, we cut the legs off of the nylons right at the top of the thigh. We place the elastic band on the larger hole and then tie off the nylons.
3. Scrape off only the top layer of food where the larvae are. This is also the layer with liquefied food that is easier to rinse off. If you dig too deep, you will start to collect chunks of food that are harder to get rid of.
4. It is important that the water is lukewarm, not hot or cold. It is also important that the flow of the water is not too heavy. To control the flow of water, we place a hand under the faucet to spread the stream out as much as possible.
5. Your X-ray machine should be calibrated on a regular basis and certified for safety as required by the radiation safety authority of your institution.
6. Autoclave mixture on liquid cycle for 25 min. Remove cornmeal agar from the autoclave and mix well, immediately. This should help prevent chunks from forming. If large chunks do form, you must start over. If cornmeal agar is made before the day of irradiating or if it solidifies while making the vials, it can be liquefied again in the microwave. Microwave on high for 3–5 min while keeping an eye on it. You do not want the mixture to bubble over or become too dry. You may have to add $ddH_2O$ as necessary to restore the volume if the food is drying out.
7. We find that in order to prolong the life of the small molecules it is best to avoid freeze–thaw cycles. Thaw the original plate once when performing the *Drosophila* experiments and make 1–2 replicates to be used for future experiments. If you do not use all of the drugs in one plate on the first day, keep the plate at room temperature for 1–2 days rather than re-freezing.
8. The best way to add 3 ml of food is to have the volume of 3 mL marked off on an empty vial. Hold the marked

standard vial next to an empty vial and pour warm food into the latter until it reaches the mark in the standard vial. Trying to pipette corn-meal agar food can lead to a mess.

9. It is important that the food solidifies completely before the larvae are added. If not, the larvae can become trapped in the food once it solidifies and die.
10. We find that red permanent markers work best for visibility under GFP illumination. It is also easier to see GFP when a piece of yellow paper (e.g., a Post-it note) is placed under the vial.

## Acknowledgments

The authors thank Petros Yoon for critical reading of the manuscript and for his contribution to the screen. The work in the Su lab is supported by an NIH grant (GM087276) to T. T. Su.

### References

1. Alberts, B., Johnson, A., Lewis, J., Raff, M., Roberts, K., and Walter, P. (2002) *Molecular Biology of the Cell*, Fourth ed., Garland, New York.
2. Ma, B. B., Bristow, R. G., Kim, J., and Siu, L. L. (2003) Combined-modality treatment of solid tumors using radiotherapy and molecular targeted agents, *J Clin Oncol 21*, 2760–2776.
3. Parchment, R. E., Gordon, M., Grieshaber, C. K., Sessa, C., Volpe, D., and Ghielmini, M. (1998) Predicting hematological toxicity (myelosuppression) of cytotoxic drug therapy from in vitro tests, *Ann Oncol 9*, 357–364.
4. Smith, R. S., Chen, Q., Hudson, M. M., Link, M. P., Kun, L., Weinstein, H., Billett, A., Marcus, K. J., Tarbell, N. J., and Donaldson, S. S. (2003) Prognostic factors for children with Hodgkin's disease treated with combined-modality therapy, *J Clin Oncol 21*, 2026–2033.
5. David, H. L. (1970) Probability distribution of drug-resistant mutants in unselected populations of Mycobacterium tuberculosis, *Appl Microbiol 20*, 810–814.
6. Sharma, P. L., Nurpeisov, V., Hernandez-Santiago, B., Beltran, T., and Schinazi, R. F. (2004) Nucleoside inhibitors of human immunodeficiency virus type 1 reverse transcriptase, *Curr Top Med Chem 4*, 895–919.
7. Jaklevic, B., Uyetake, L., Lemstra, W., Chang, J., Leary, W., Edwards, A., Vidwans, S., Sibon, O., and Tin Su, T. (2006) Contribution of growth and cell cycle checkpoints to radiation survival in Drosophila, *Genetics 174*, 1963–1972.
8. Jaklevic, B. R., and Su, T. T. (2004) Relative contribution of DNA repair, cell cycle checkpoints, and cell death to survival after DNA damage in Drosophila larvae, *Curr Biol 14*, 23–32.
9. Fogarty, P., Kalpin, R. F., and Sullivan, W. (1994) The Drosophila maternal-effect mutation grapes causes a metaphase arrest at nuclear cycle 13, *Development 120*, 2131–2142.
10. Fogarty, P., Campbell, S. D., Abu-Shumays, R., Phalle, B. S., Yu, K. R., Uy, G. L., Goldberg, M. L., and Sullivan, W. (1997) The Drosophila grapes gene is related to checkpoint gene chk1/rad27 and is required for late syncytial division fidelity, *Curr Biol 7*, 418–426.
11. Purdy, A., Uyetake, L., Cordeiro, M. G., and Su, T. T. (2005) Regulation of mito-

sis in response to damaged or incompletely replicated DNA require different levels of Grapes (Drosophila Chk1), *J Cell Sci 118*, 3305–3315.

12. Rong, Y. S., Titen, S. W., Xie, H. B., Golic, M. M., Bastiani, M., Bandyopadhyay, P., Olivera, B. M., Brodsky, M., Rubin, G. M., and Golic, K. G. (2002) Targeted mutagenesis by homologous recombination in D. melanogaster, *Genes Dev 16*, 1568–1581.

# Chapter 10

# Study of Cell Cycle Checkpoints Using *Xenopus* Cell-Free Extracts

**Seetha V. Srinivasan and Jean Gautier**

## Abstract

Cell cycle checkpoints are involved in the coordinated response to DNA damage and thus play a key role in maintaining genome integrity. Several model systems have been developed to study the mechanisms and complexity of checkpoint function. Here we describe the application of cell-free extracts derived from *Xenopus* eggs as a model system to investigate DNA replication, damage, and checkpoint activation. We outline the preparation of cell-free extracts, DNA substrates and their subsequent use in assays aimed at understanding cell cycle checkpoints, and related processes. Several advances made over the years have enabled the continued use of the *Xenopus* system to answer a variety of questions in DNA replication, repair and checkpoint signaling. It is anticipated that the versatile *Xenopus* system is amenable to future modification as well to continue studies attempting to understand these important physiological processes.

**Key words:** *Xenopus*, cell-free extracts, DNA replication, DNA damage, ATM/ATR kinases, phosphorylation, chromatin binding, cell cycle checkpoint.

## 1. Introduction

Cells subjected to DNA damage activate response pathways that cause a transient cell cycle arrest to facilitate DNA repair and/or programmed cell death (1). Such surveillance mechanisms can also monitor genome integrity and typically operate through signal transduction pathways called checkpoints. The net effect of checkpoints is to delay cell cycle progression in order to aid the cell in coping with genotoxic insults (2, 3).

Diminishing checkpoint function or increasing the induction of damage can compromise the integrity of the genome, the consequences of which are severe. Classically, yeast and mammalian

W.X. Li (ed.), *Cell Cycle Checkpoints*, Methods in Molecular Biology 782,
DOI 10.1007/978-1-61779-273-1_10, © Springer Science+Business Media, LLC 2011

cell systems have been used to study this important phenomenon and have resulted in the identification of radiation-sensitive and checkpoint genes (4, 5). However, the use of these systems in certain situations can be limiting. For instance, standard genetic screens may not be very useful in studying essential genes. Also, the DNA damage response is more complex in vertebrates than in yeast and therefore, certain critical regulators of the damage response, such as p53 and BRCA1, can only be found in vertebrates (6). Mammalian cell lines have been used to circumvent the above drawbacks. However, cell-based model systems do not allow the use of specific biochemical readouts because they are frequently based on cell growth, survival, or other phenotypes that derive from complex outputs (7). One solution to overcome the above limitations is to use the cell-free, *Xenopus* model system.

The *Xenopus* cell-free extract has found several applications in the study of DNA replication, repair and checkpoint signaling. This system has certain key advantages (8, 9): First, the extracts contain cytoplasmic and nuclear proteins that can support up to 12 cell divisions in the absence of transcription. Second, the protein concentration in the extracts is high and sufficient to carry out complete rounds of semi-conservative, cell cycle regulated DNA replication. Third, the extracts allow the study of essential proteins by immunodepletion/neutralization coupled to rescue with the recombinant protein of interest. The depletion of endogenous proteins from the extract using specific antibodies (10) makes it possible to study the role of the depleted protein in DNA replication, repair, and checkpoint signaling (11–13).

DNA replication requires the sequential loading of replication initiation factors such as ORC, Cdc6, Cdt1, and MCM2-7 that assemble on chromatin to form the pre-replication complex. The subsequent initiation of DNA replication requires the activity of Cdk2/cycE, Cdc7/Dbf4, Cdc45, GINS, MCM10, and possibly other components of the replication machinery (14). Classic DNA replication experiments with *Xenopus* extracts involve the addition of sperm chromatin to egg extracts made from unfertilized *Xenopus* eggs. This leads to nuclear assembly capable of undergoing one complete round of semi-conservative DNA replication (8). By varying the method used to prepare the extract, different types of extracts representing specific phases of the cell cycle can be obtained. These extracts have been particularly useful in understanding the regulation of entry into S phase and mitosis (15–20). More recently, a “nucleus-free” system has been developed (13) that enables the study of plasmid replication and also alleviates other challenges faced in extracts involving nuclear assembly. Both types of *Xenopus* extracts (extracts involving “nuclear assembly” and those that are “nucleus-free”) can be used to study checkpoint signaling by experimentally interfering with DNA replication and inducing a checkpoint.

In this chapter, we describe techniques that use the *Xenopus* model system to recapitulate several aspects of the DNA damage response, including cell cycle checkpoints. The protocols used in the preparation of various extracts have been described in **Section 3.1**. The intact and damaged DNA templates used in conjunction with the extracts are described in **Section 3.2**. The subsequent section (**Section 3.3**) covers the various assays that involve the use of the *Xenopus* system to investigate cell cycle checkpoints.

## 2. Materials

The animals and hormones described below are materials common to the protocols used for the preparation of extracts.

1. Animals: *Xenopus laevis* (the African clawed frog), females and males (Nasco). Females are used to obtain eggs for extract preparation while males are used in the preparation of demembranated chromatin.
2. Hormones to prime frogs: pregnant mare serum gonadotropin (Calbiochem/EMD chemicals) and human chorionic gonadotropin (Sigma).

### 2.1. CSF Extract

1. MMR(Marc's Modified Ringers) buffer: 5 mM HEPES (pH 7.7–7.8), 0.1 mM EDTA, 100 mM NaCl, 2 mM KCl, 1 mM $MgCl_2$, 2 mM $CaCl_2$.
2. Dejellying buffer: 2% L-cysteine HCl monohydrate in water (pH 7.8 is critical).
3. XB Buffer: 10 mM HEPES (pH 7.7–7.8), 1 mM $MgCl_2$, 0.1 mM $CaCl_2$, 100 mM KCl, 50 mM sucrose.
4. Cytostatic factor (CSF)-XB Buffer: 10 mM HEPES (pH 7.7–7.8), 2 mM $MgCl_2$, 0.1 mM $CaCl_2$, 100 mM KCl, 5 mM EGTA, 50 mM sucrose.
5. 10 mg/mL cytochalasin B.
6. Energy Mix: 150 mM creatine phosphate, 20 mM ATP, 20 mM $MgCl_2$ (Store at –20°C).
7. Protease inhibitors: leupeptin, chymostatin, pepstatin; stock: 10 mg/mL each in DMSO.
8. 2 M sucrose.
9. For 6-DMAP chromatin: 6-dimethylaminopurine; 1 M $CaCl_2$.
10. Becton Dickinson 14-mL Falcon tube cat # 352059.

### 2.2. Activated CSF Extract/Interphase Extract I

In addition to the materials required for CSF extract (*see* **Section 2.1**), the activated extract requires:

1. 10 mg/mL cycloheximide.
2. 1 M $CaCl_2$.

### 2.3. ELB Extract/LSS Extract

The materials required for the preparation of this extract are the same as those needed for CSF extract (*see* **Section 2.1**) except that ELB buffer is used instead of CSF buffer.

1. Egg Lysis Buffer (ELB): 250 mM sucrose, 1× ELB Salts, 1 mM DTT, 50 μg/mL cycloheximide.
2. 10× ELB Salts (for ELB buffer): 25 mM $MgCl_2$, 0.5 M KCl, 100 mM HEPES, pH 7.7–7.8, with KOH; filter sterilize. Store at 4°C.

### 2.4. Interphase Extract II

1. Dejellying buffer: 20 mM Tris, pH 8.5, 110 mM NaCl. Add 500 μL of 1 M DTT per 100 mL buffer just before use.
2. 5× MMR: 100 mM HEPES–KOH, pH 7.5, 2 M NaCl, 10 mM KCl, 5 mM $MgSO_4$, 10 mM $CaCl_2$, 0.5 mM EDTA. Dilute to 1× or 0.25× as needed.
3. 5× S buffer: 250 mM HEPES–KOH, pH 7.5, 250 mM KCl, 12.5 mM $MgCl_2$, 1.25 M sucrose. Keep at 4°C, add 7 μL of 14.3 M β-mercaptoethanol/ml for a final concentration of 2 mM; and protease inhibitors (optional).

*Store the following reagents in small aliquots at –20°C:*

1. Ca Ionophore A23187. Stock: 10 mg/mL in DMSO.
2. 10 mg/mL Cytochalasin B in DMSO or ethanol.
3. 1 M creatine phosphate stock in milliQ water.
4. 10 mg/mL creatine phosphokinase in milliQ water.
5. 15 mg/mL leupeptin in DMSO.
6. 10 mg/mL cyclohexamide in milliQ water.
7. Other materials: Beckman polyallomer tube cat # 347287, 21-G needle, polypropylene culture tubes cat # 352059, 100% glycerol, styrofoam box or wooden platform, liquid nitrogen, cryotubes (1.5- or 2-mL tubes), forceps, Becton Dickinson 14-mL Falcon tube cat # 352059.

### 2.5. PEG-Fractionated Membrane-Free Egg Cytosol

The starting materials for this extract are the same as those for Interphase Extract I (*see* **Section 2.2**). In addition, the following reagents are required:

1. LFB Buffer: 50 mM KCl, 40 mM HEPES KOH (pH 8.0), 20 mM $K_2HPO_4/KH_2PO_4$ (pH 8.0), 2 mM DTT, 2 mM $MgCl_2$, 1 mM EGTA, 10% sucrose, 1 μg/μL Aprotinin, pepstatin, leupeptin (made fresh before use).
2. Polyethylene glycol: PEG 6000.
3. Mg-ATP.

***2.6. High Speed Supernatant (HSS)***

1. Marc's Modified Ringer's Buffer (MMR): 100 mM NaCl, 2 mM KCl, 0.5 mM $MgSO_4$, 2.5 mM $CaCl_2$, 0.1 mM EDTA, 5 mM HEPES–KOH, pH 7.7–7.8.
2. Dejellying solution: 2% L-cysteine HCl monohydrate in water (pH 7.8 is critical).
3. Egg Lysis Buffer (ELB): 1× ELB salts, 1 mM DTT, 50 μg/mL cycloheximide, 0.25 M sucrose.
   The 10× salts (stock) for Egg Lysis Buffer: 25 mM $MgCl_2$, 500 mM KCl, 100 mM HEPES–KOH, pH 7.7, filter sterilized and stored at 4°C.
4. 5 mg/mL cytochalasin B in DMSO; stored at –20°C.
5. 10 mg/mL cycloheximide in water; stored at –20°C.
6. 1 M Dithiothreitol (DTT) in water; stored at –20°C.
7. 10 mg/mL Aprotinin/leupeptin (1000×) solution in water; store at –80°C.
8. Energy Mix: 10 mM Creatine phosphate, 10 μg/mL Creatine kinase, 2 mM ATP, 2 mM $MgCl_2$, 5 mM HEPES, pH 7.5, 1 mM DTT.
9. Polypropylene tubes – Falcon tubes BD biosciences # 352059.
10. Beckman polyallomer tubes # 349623 (if thin tubes are used, cat # is 326819).

***2.7. Nucleoplasmic Extract (NPE)***

1. Marc's Modified Ringer's Buffer (MMR): 100 mM NaCl, 2 mM KCl, 0.5 mM $MgSO_4$, 2.5 mM $CaCl_2$, 0.1 mM EDTA, 5 mM HEPES–KOH, pH 7.7–7.8.
2. Dejellying solution: 2% L-cysteine HCl monohydrate in water (pH7.8 is critical).
3. Egg lysis buffer: 250 mM sucrose, 2.5 mM $MgCl_2$, 50 mM KCl, 10 mM HEPES, pH 7.7, 50 μg/ml cycloheximide, 1 mM DTT.
4. Protease inhibitor cocktail (1000×): Leupeptin, chymostatin, pepstatin. Dissolve to a concentration of 10 mg/mL in DMSO. Store in 100-μL aliquots in –20°C.
5. 10 mg/mL aprotinin.
6. 5 mg/mL cytochalasin B (Sigma) stock; stored at –20°C.
7. 10 mg/mL cycloheximide in water and stored at –20°C.
8. Nocodazole.
9. 1 M Dithiothreitol (DTT) in water; stored at –20°C.
10. ATP regeneration mix: Prepare stock of 1 M phosphocreatine, 0.2 M ATP and 5 mg/mL creatine phosphokinase.
11. Demembranated Sperm Chromatin (*See* **Sections 2.9** and **3.2.1**).

12. Falcon Tube (Becton Dickinson cat # 352063).
13. Becton Dickinson tube cat # 352059.
14. Beckman tube cat # P60720.

### 2.8. Immuno-depletion of Extracts to Study Protein Function

1. Protein-A sepharose CL-4B beads
2. Compact Reaction Columns (CRC columns USB #13929 and 13912)
3. Phosphate-buffered saline (PBS buffer), pH 7–7.5

### 2.9. Demembranated Sperm Chromatin

1. Benzocaine. Dissolve 1 g in 10 ml of 100% ethanol. Dispense drop-wise into 2 L of water with continuous stirring. This gives a 0.05% solution of benzocaine. The solution can be re-used for up to 6 months if stored at 4°C.
2. Marc's Modified Ringer's Buffer (MMR): 100 mM NaCl, 2 mM KCl, 0.5 mM $MgSO_4$, 2.5 mM $CaCl_2$, 0.1 mM EDTA, 5 mM HEPES–KOH, pH 7.7–7.8.
3. Nuclei Preparation Buffer (NPB): 250 mM sucrose, 15 mM HEPES–KOH, pH 7.7, 1 mM EDTA, 0.5 mM Spermidine, 0.2 mM Spermine, 1 mM DTT. A 2× solution of NPB is also prepared for use.
4. 10% Triton X-100 in water; store at 4°C.
5. 10% Bovine Serum A 5lbumin and store at –20°C or make fresh.
6. Glycerol.
7. 0.3 M PMSF made in 100% ethanol. Make fresh before use.
8. 10 mg/mL leupeptin.
9. 10 mM Spermidine.
10. 10 mM Spermine.
11. 1.5 M sucrose solution; store at 4°C.

### 2.10. Generation of Double-Strand Breaks (DSB) in Chromatin

1. *Xenopus* egg Extract.
2. Demembranated sperm nuclei (*see* **Section 3.2.1**).
3. PflMI restriction enzyme.
4. Chromatin isolation buffer II: 50 mM HEPES, pH 7.8, 2.5 mM $MgCl_2$, 10 mM KCl, supplement with 0.125% Triton X-100.
5. Sucrose cushion: 30% sucrose in Chromatin Isolation buffer II without Triton X-100.

### 2.11. Preparation of 6-DMAP Chromatin

1. Demembranated Sperm chromatin (*see* **Section 3.2.1**).
2. 6-DMAP Extract (*see* **Section 3.1.1**, last paragraph).
3. Chromatin isolation buffer I: 50 mM KCl, 5 mM $MgCl_2$, 2 mM DTT, 50 mM HEPES, pH 7.5–7.8, 0.5 mM

Spermine, 0.15 mM spermidine, 1 μg/μL Aprotinin, pepstatin, leupeptin, 0.125% Triton X-100 (made fresh before use).
Sucrose Cushion: chromatin isolation buffer I without Triton X-100, containing 30% sucrose).

4. Mg-ATP.

#### 2.12. Preparation of Plasmid with a Single Interstrand Crosslink (ICL)

1. pBS KS-vector.
2. Linker sequence containing PflMI restriction site: 5′CACCCATGGAATGGACGC3′ and 5′GGCCGCGTCCATTCCATGGGTGAGCT3′.
3. PflMI and DraIII restriction enzymes.
4. Gel purification kit.
5. T4 DNA ligase.
6. Exonuclease-It kit (Bayou Biolabs cat # X-101).
7. PCR purification kit.

#### 2.13. Preparation of DNA with Double-Strand Breaks

1. pBR322 plasmid.
2. HaeIII restriction enzyme (NEB).
3. Phenol/chloroform to extract digested DNA.

#### 2.14. Preparation of Biotinylated Substrates

1. DNA fragment of interest.
2. Gel purification kit.
3. T4 polymerase.
4. dATP, dGTP, dTTP, biotin-dCTP.
5. 0.5 M EDTA, pH 8.0.
6. PCR purification kit.

#### 2.15. DNA Damage Checkpoint Induced by Single-Strand DNA Gaps in Chromatin

1. Activated CSF extract.
2. Demembranated sperm nuclei.
3. Etoposide or Exonuclease III.
4. 1 M Tris–HCl, pH 8.0.
5. 1 M $MgCl_2$.
6. Nuclei Preparation Buffer (NPB): 250 mM sucrose, 15 mM HEPES–KOH, pH 7.7, 1 mM EDTA, 0.5 mM Spermidine, 0.2 mM Spermine, 1 mM DTT, 10 μg/mL leupeptin, 0.3 mM PMSF.
7. 100 mM caffeine in 10 mM PIPES at pH 8.0.
8. $^{32}P$-dATP.
9. Stop Solution: 0.5% SDS, 80 mM Tris (pH 8.0), 8 mM EDTA.

10. Proteinase K.
11. Phenol/chloroform to extract DNA.
12. Trichloroacetic acid.
13. Whatmann 3 MM blotting paper.

#### 2.16. DNA Damage Checkpoint Induced by DSB in Chromatin

1. Activated CSF extract.
2. Circular control plasmid or plasmids containing double-strand breaks (DSB).
3. 100 mM caffeine in PIPES, pH 8.0.
4. 100 μM Wortmannin in DMSO.
5. Anti *Xenopus* ATM antibodies.
6. All the reagents required for PEG fractionated membrane-free cytosol (**Section 3.1.5**).
7. 6-DMAP chromatin (*see* **Section 3.2.3**).
8. 6-DMAP extract (*see* **Section 3.1.1**, last paragraph).
9. $^{32}$P-dATP.
10. Stop Solution: 0.5% SDS, 80 mM Tris (pH 8.0), 8 mM EDTA.
11. Proteinase K.
12. Phenol/chloroform to extract DNA.
13. Trichloroacetic acid.
14. Whatmann 3 MM blotting paper.

#### 2.17. Assay to Study Checkpoint Induced by ICL

1. *Xenopus* extracts (HSS/NPE, **Sections 3.1.6** and **3.1.7**).
2. Control and ICL DNA (**Section 3.2.4**).
3. Energy Mix: 10 mM Creatine phosphate, 10 μg/mL Creatine kinase, 2 mM ATP, 2 mM $MgCl_2$, 5 mM HEPES, pH 7.5, 1 mM DTT.
4. 100 mM caffeine in PIPES, pH 8.0.
5. $^{32}$P-dCTP.
6. Stop Solution: 0.5% SDS, 80 mM Tris (pH 8.0), 8 mM EDTA.
7. Proteinase K.
8. Phenol:chloroform:isoamylalcohol (25:24:1).
9. 3 M Sodium acetate.
10. Ethanol.
11. DNA loading buffer with bromophenol blue and cyan blue dyes.
12. Trichloroacetic acid.

#### 2.18. Study of the Checkpoint Induced by DNA Damage In Trans on DNA Replication

1. *Xenopus* Extracts (HSS/NPE; *see* **Sections 3.1.6** and **3.1.7**).
2. Energy Mix: 10 mM creatine phosphate, 10 μg/mL creatine kinase, 2 mM ATP, 2 mM $MgCl_2$, 5 mM HEPES, pH 7.5, 1 mM DTT.
3. Control or ICL DNA (*see* **Section 3.2.4**).
4. $^{32}P$ dCTP.

#### 2.19. Phosphorylated Histone H2AX Detection (Endogenous)

1. Activated CSF extract or Interphase Extract I (*see* **Section 3.1.2**).
2. Demembranated sperm chromatin (*see* **Section 3.2.1**).
3. Chromatin isolation buffer I: 50 mM KCl, 5 mM $MgCl_2$, 2 mM DTT, 50 mM HEPES, 0.5 mM Spermine 3HCl, 0.15 mM spermidine 4HCl, 1 μg/μL Aprotinin, pepstatin, leupeptin, 0.125% Triton X-100 (made fresh before use); this buffer is supplemented with 1 mM NaF, 1 mM Na vanadate, and 0.125% Triton X-100.
4. Chromatin isolation buffer I (without Triton X-100) supplemented with 30% sucrose.

#### 2.20. Histone H2AX Phosphorylation Assay (Exogenous Substrate)

1. *Xenopus* extracts (*See* **Sections 3.1.1**, **3.1.2**, and **3.1.3**).
2. Biotinylated DNA fragments to induce DNA damage (M280 Dynal Biotechnology).
3. ELB Buffer: 250 mM sucrose, 1× ELB Salts, 1 mM DTT, 50 μg/mL cycloheximide
   (10× ELB Salts – for ELB buffer: 25 mM $MgCl_2$, 0.5 M KCl, 100 mM HEPES, pH 7.7–7.8 with KOH); filter sterilize. Store at 4°C.
4. Wild-type H2AX peptide: AVGKKASQASQEY.
5. Mutant H2AX peptide: AVGKKAAQAAQEY.
6. EB Buffer: 20 mM HEPES, pH 7.5, 50 mM NaCl, 10 mM $MgCl_2$, 1 mM DTT.
7. ATP.
8. $\gamma$-$^{32}P$ ATP.
9. 0.5 M EDTA, pH 8.0.
10. p81 phospho-cellulose filter papers (Upstate Biotechnology).
11. Acetic acid.
12. 100 mM caffeine in PIPES, pH 8.0.
13. EB Kinase buffer: 20 mM HEPES, pH 7.5, 50 mM NaCl, 10 mM $MgCl_2$, 1 mM DTT, 1 mM NaF, 1 mM $Na_3VO_4$, 10 mM $MnCl_2$.

#### 2.21. Phosphorylation of ATM/ATR Target Proteins

1. *Xenopus* Extracts (*see* **Section 3.1**).
2. Energy Mix: 10 mM Creatine phosphate, 10 μg/mL Creatine kinase, 2 mM ATP, 2 mM $MgCl_2$, 5 mM HEPES, pH 7.5, 1 mM DTT.
3. Control and damaged DNA (either chromatin or small DNA templates can be used; *see* **Section 3.2**).
4. 100 mM caffeine.
5. Phospho Chk1 antibody (Cell Signaling).
6. Phospho ATM antibody (Rockland Immunochemicals, PA).
7. Geminin/Roscovitine.
8. Curcumin dissolved in ethanol.

#### 2.22. Chromatin Binding Assay

1. *Xenopus* extracts (*see* **Section 3.1**).
2. Demembranated sperm chromatin (*see* **Section 3.2.1**).
3. Chromatin isolation buffer I: 50 mM KCl, 5 mM $MgCl_2$, 2 mM DTT, 50 mM HEPES, 0.5 mM Spermine, 0.15 mM spermidine, 1 μg/μL Aprotinin, pepstatin, leupeptin, 0.125% Triton X-100 (made fresh before use).
4. Fractionated Extracts (*see* **Section 3.1.5**).
5. 6-DMAP Chromatin (*see* **Section 3.2.3**).
6. Laemmli loading buffer.

#### 2.23. Binding Assay Using HSS/NPE and ICL Plasmid

1. *Xenopus* extracts (HSS/NPE *see* **Sections 3.1.6** and **3.1.7**).
2. Protein-A sepharose CL-4B beads.
3. Compact Reaction Columns (CRC columns USB # 13929 and 13912).
4. Phosphate-buffered saline (PBS buffer), pH 7–7.5.
5. Control or ICL plasmid DNA (*see* **Section 3.2.4**).
6. Chromatin isolation buffer II: 50 mM HEPES, pH 7.8, 2.5 mM $MgCl_2$, 100 mM KCl, supplement with 0.125% Triton X-100.
7. Sucrose cushion: 30% sucrose in chromatin isolation buffer II without Triton X-100.
8. Low retention 1.5-mL Eppendorf tubes (Fisherbrand #02-681-341).
9. 2× Laemmli Buffer.
10. For Immunodepletions (also *see* **Section 3.1.8**).

## 3. Methods

The methods described below include (**a**) the preparation of different types of extracts (**Sections 3.1.1**, **3.1.2**, **3.1.3**, **3.1.4**, **3.1.5**, **3.1.6**, **3.1.7** and **3.1.8**), (**b**) the preparation of DNA substrates – both chromosomal (**Sections 3.2.1**, **3.2.2** and **3.2.3**) and small template DNA (**Sections 3.2.4**, **3.2.5** and **3.2.6**) and (**c**) assays used to study checkpoints – (i) DNA replication assays (**Sections 3.3.1**, **3.3.2**, **3.3.3** and **3.3.4**), (ii) assays to study protein modifications such as phosphorylation to detect checkpoint activation (**Sections 3.3.5**, **3.3.6**, **3.3.7** and **3.3.8**), and (iii) Chromatin/DNA binding assays (**Sections 3.3.9** and **3.3.10**).

### 3.1. Preparation of Extracts

Checkpoints studies using *Xenopus* involve at least four different types of extracts. Three of the extracts (Cytostatic Factor Extract, Low Speed Interphase Extract, and High Speed Extract) are derived from *Xenopus* eggs while the fourth type (Nucleoplasmic Extract) requires *Xenopus* eggs and sperm chromatin. Depending on the type of extract used in the experiment, the addition of sperm chromatin can support DNA replication in either "nuclear assembly" or "nucleus-free" contexts: while sperm chromatin added to activated CSF or LSS extracts leads to nuclear assembly, the HSS/NPE system is a nucleus-free system wherein the DNA template is initially incubated with HSS to form pre-replication complexes and DNA replication is initiated by the addition of NPE (13).

The description of extract preparation in this chapter includes Cytostatic-arrested (CSF) extracts arrested in M phase and activated CSF extracts. Since CSF and activated CSF extracts are obtained by relatively low speed centrifugation, they are also referred to as Low Speed Supernatant (LSS). These extracts can be further fractionated for checkpoint studies to obtain fractionated membrane-free cytosol. A modification of the LSS extract protocol has made it possible to store the extract, without appreciable loss of activity, for later use. We have referred to such an extract as the Interphase Extract II, which is also described in this chapter. Furthermore, we describe the preparation of High Speed Supernatant (HSS), which is a membrane-free extract that supports the assembly of the pre-replication complexes on chromatin (13, 21). The HSS extract used with Nucleoplasmic Extract (NPE) can bypass the nuclear envelop formation step, yet allows normal DNA replication of chromatin and plasmid (13, 21, 22). The NPE system is particularly useful since it allows for the modification of the nuclear environment and supports plasmid

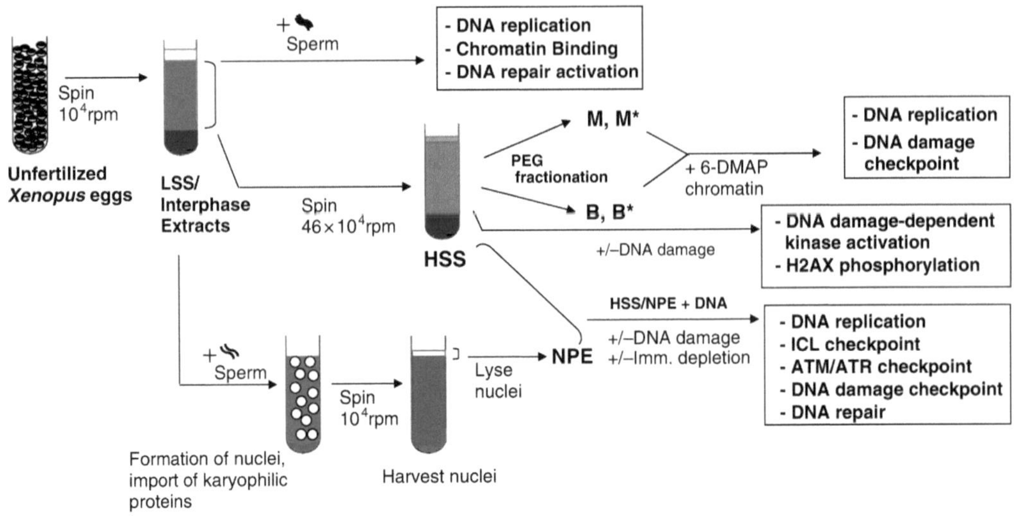


Fig. 1. Schematic representation of *Xenopus* extracts used to study cell cycle checkpoints.

replication very efficiently (10, 13). The preparation of demembranated sperm chromatin and NPE are also outlined in this chapter. Energy Mix is added to extracts before use.

**Fig. 1** summarizes the various extracts and their uses.

#### 3.1.1. CSF Extract

The preparation of fresh CSF-arrested extracts has been previously described (7, 23). The protocol described below is used to obtain extracts in the M phase of the cell cycle. To obtain such extracts, we omit the activation of the eggs by Ca ionophore and instead supplement the buffer with 5 mM EGTA.

1. Prime 4–6 female frogs with 500 μL/frog of PMSG (PMSG stock: 100 IU/μL). This step is generally performed 3–7 days before the frogs are induced with hCG.
2. Induce females to lay eggs with 800 IU of hCG/frog injected the night prior to egg collection. Place each frog in a container with 1 L of 1× MMR (pH 7.7–7.8) after HCG injection. Eggs are laid overnight and collected next morning (*see* **Notes 1** and **2**).
3. Wash eggs in 1× MMR and remove as much buffer as possible.
4. Dejelly the eggs in 2% cysteine in water (pH 7.8 is critical) (*see* **Note 3**).
5. Remove the cysteine solution by pouring off the excess liquid.
6. Rapidly wash eggs three times with XB and then remove all XB.
7. Wash eggs three times in CSF-XB and remove as much buffer as possible.

8. Wash eggs two times in XB containing protease inhibitors (10 μg/mL leupeptin, pepstatin, and aprotinin).
9. Transfer eggs into 1 mL of CSF-XB with protease inhibitors and 100 μg/mL cytochalasin B in 1.5 mL Eppendorf tubes (*see* **Note 4**). Alternatively, if there is a large volume of eggs, 14-mL Falcon tubes can be used.
10. Pack the eggs by spinning for 1 min at 700 rpm and remove excess buffer (*see* **Note 5**).
11. Combine eggs from different tubes and fill each eppendorf tube as much as possible. Crush the eggs at 4–8°C for 15 min at 10–12,000 rpm.
12. Collect the extract with an 18-gauge needle by puncturing the side of the tube and gently aspirate out the cloudy intermediate cytoplasmic layer. This layer is located between the top opaque lipid layer (yellow) and the solid pellet of pigments and egg debris (dark brown).
13. Supplement the cytosolic extract with 1/1,000 volume of cytochalasin B (10 mg/mL stock), 1/15 volume of energy mix, 1:1000 protease inhibitors (leupeptin, chymostatin, pepstatin 10 mg/mL each in DMSO), and 1/40 volume of 2 M sucrose.
14. Clarify the cytosolic extract by centrifugation for 30 min at 13,000×*g* in an eppendorf centrifuge at 4°C. The clear cytoplasmic layer is collected for immediate use (*see* **Notes 6** and 7).

For double-stranded break (DSB) – induced checkpoint assay, the extract needs to be treated with 6-DMAP. In such a case, the CSF extract is incubated for 30 min with 2 mM 6-DMAP at 23°C. The extract is then supplemented with 0.4 mM $CaCl_2$ and incubated for 15 min at 23°C.

#### 3.1.2. Activated CSF Extract/Interphase Extract I

Upon activation by calcium, CSF extracts exit M phase and enter interphase. Such activated extracts are used in DNA replication, chromatin binding, and checkpoint assays.

1. Supplement freshly prepared CSF extracts (*see* **Section 3.1.1**) with 50 μg/mL of cycloheximide and 0.4 mM $CaCl_2$.
2. Incubate for 15 min at 21°C. This treatment mimics the Calcium flux that normally takes place after fertilization of eggs and triggers degradation of mitotic cyclins, inactivation of Cdc2/Cyclin B, and exit from mitosis (*see* **Note** 7).

#### 3.1.3. ELB Extract/LSS Extract

A variation of the above activated extract is the preparation of Crude S-phase extract/Low Speed Supernatant Extract where the MMR buffer does not contain EGTA. Therefore the eggs are already activated and the buffer does not need to be supplemented with $CaCl_2$. The protocol in this case is exactly the same as for CSF extract except that the eggs are washed with ELB (Egg Lysis Buffer) instead of CSF-XB buffer. Since the ELB buffer already contains sucrose, it is also not necessary to supplement the cytosolic extract with 1/40 volume of 2 M sucrose (during Step 13 of CSF extract **Section 3.1.1**). Supplement extract with energy mix before use.

#### *3.1.4. Interphase Extract II*

The chief advantage of this type of extract is that it can be stored as frozen drops at –80°C and thawed for use when necessary (24). The frozen extract can keep for about 6 months without appreciable loss of quality.

1. Induce 4–6 frogs to lay eggs in 0.25× Interphase Extract II MMR (the frogs are induced to lay eggs as described in **Section 3.1.1** above) (*see* **Notes 1** and **2**).
2. Collect and pool good eggs (those that have even coloration of dark and light surfaces, symmetrical, and dense), discard bad ones (fluffy, disfigured, discolored or floating eggs) by aspirating them out with a plastic transfer pipette.
3. De-jelly with de-jelly buffer (100 mL buffer per frog), for 5–10 min. If there are more eggs, change de-jelly buffer every 2–3 min, but extending the time of dejellying may be detrimental to quality of the extract (*see* **Note 3**).
4. Wash three times with 0.25× MMR buffer.
5. Add 1/10,000 of 10 mg/mL Calcium ionophore A23187 and incubate for 5 min to activate eggs.
6. Wash three times with 0.25× MMR buffer.
7. Wash three times with S-buffer. Prepare 50 mL + 7 μL β-mercaptoethanol per frog and in the last wash, add 1/1000 leupeptin.
8. Transfer eggs to 14-ml Falcon tube (*see* **Notes 4** and **5**).
9. Spin in Sorvall swing-bucket HB-6 rotor for 1 min at 1000 rpm. This is the packing spin. Remove excess buffer.
10. Spin at 10,000 rpm in Sorvall for 15 min at 4°C. This is the crushing spin.
11. Transfer cytoplasm to 1.5-ml Beckman polyallomer tube (cat # 347287) using 21-G needle/syringe.
12. Add cytochalasin B (1/500).
13. Seal tubes tightly with parafilm and rotate in the cold room for 5 min to completely homogenize the extract.
14. Centrifuge in Beckman table top centrifuge with TLS-55 swing-bucket rotor for 55,000 rpm for 15 min at 4°C.
15. Take only the cytoplasmic and membrane layers. Leave out the mitochondrial layer, the oily layer and the debris

(typically, the debris accumulates at the bottom-most part of the tube; immediately above the debris is the oily layer above which is the mitochondrial layer). Use a cut pipette tip (P1000 or P200) to recover the desired fractions.

16. Supplement with the following:
    30 mM Creatine phosphate.
    150 μg/mL Phospho creatine kinase.
    20 μg/mL Cycloheximide.
    Invert tube several times to make sure all components are adequately mixed.
    Use extract immediately or freeze as follows:
    Freezing the Interphase Extract II:
    (i) Add glycerol (3% final concentration) to extract and mix well by inverting tube.
    (ii) On a Styrofoam platform place P60 plastic petri dishes filled with liquid nitrogen. Fill labeled cryotubes with liquid nitrogen and place on Styrofoam rack.
    (iii) With a cut P200 tip, drop beads of 22 μL of extract into P60 dish containing liquid nitrogen. Allow extract to flash-freeze in the liquid nitrogen (*see* **Note 8**).
    (iv) Store the beads of frozen extract in –80°C.

*3.1.5. PEG-Fractionated Membrane-Free Egg Cytosol*

This procedure was earlier described (23, 25) to study DNA replication initiation. Interphase extracts are subjected to fractionation by stepwise PEG precipitation to yield fractions referred to as M and B (*see* **Note 9**).

1. Dilute interphase extracts fourfold with cold LFB buffer.
2. Subject diluted extracts to an ultracentrifugation step at 80,000×*g* for 40 min at 4°C in a Beckmann TL100 tabletop ultracentrifuge, using a TLS55 rotor.
3. Carefully pipette supernatants and transfer to a fresh tube (avoid pellet contamination). Supplement with 0.075 volume of a 50% PEG to give a final concentration of 3.5%.
4. Incubate samples on ice for 30 min and spin for 10 min at 10,000×*g* at 4°C.
5. Resuspend pellets in a volume of LFB containing 2.5 mM Mg-ATP corresponding to 1/5th of the starting volume of extract. This yields fraction B.
6. Adjust the corresponding supernatant to 9% PEG, incubate on ice for 30 min and spin for 10 min at 10,000×*g* at 4°C to yield fraction M.

*3.1.6. High Speed Supernatant (HSS)*

1. Prime 4–6 female frogs with 500 μL/frog of PMSG (PMSG stock: 100 IU/μL).

2. Induce females to lay eggs with 800 IU of hCG/frog injected the night prior to egg collection. Place each frog in a container with 1 L of 1× MMR (pH 7.7–7.8) after HCG injection. Eggs are laid overnight and collected the next morning (**Notes 1** and **2**).
3. Wash eggs with 1× MMR. Remove bad eggs/floaters/debris with a transfer pipette.
4. Combine eggs in a glass beaker and dejelly them using 2% cysteine solution (pH 7.8) for about 10 min (*see* **Note 3**).
5. Wash eggs three times with 0.5× MMR, pH 7.8 (*see* **Note 10**).
6. Wash eggs three times with ELB, pH 7.8. At the last wash, add 1 mL of 10 mg/mL cycloheximide per 100 mL ELB (*see* **Note 4**).
7. Transfer eggs into 14-mL polypropylene tubes (Falcon tube #352059) without caps and centrifuge them at 1000 rpm for 1 min at 4°C (Sorvall HB-6 rotor, use adapters to enable centrifugation of the tubes). This is the packing spin. Remove excess buffer after this spin and add cytochalasin B to a final concentration of 5 μg/mL (stock 10 mg/mL).
8. Centrifuge eggs at 10,000 rpm for 20 min at 4°C. This is the crushing spin and will result in formation of different layers.
9. Transfer the cytoplasmic layer (central layer) to a 50-mL tube. Remove the layer by using a 5-cc syringe with a 25-gauge needle. Add 100 μg/ml cycloheximide, 5 μg/mL cytochalasin B, 1 mM DTT, and protease inhibitors.
10. Place cytoplasmic layer in Ultracentrifuge tubes (polyallomer tubes Beckman #349623; make sure to have at least 3.75 mL per tube (*see* **Note 11**).

    Centrifuge tubes at 46,000 rpm for 2 h 15 min at 4°C (SW50.1 rotor).
11. Collect clear cytoplasm with pipette.
12. Aliquot into tubes (100 μL/tube), freeze with liquid nitrogen and store at –80°C.

*3.1.7. Nucleoplasmic Extract (NPE)*

The preparation of nucleoplasmic extracts was initially reported by Walter et al. (13). Nucleoplasmic Extract when used in combination with HSS has several advantages and is especially useful in replicating both chromatin and plasmids (13, 26). Therefore, this type of extract finds application in assays where restriction enzymes and crosslinks are used to simulate DNA damage and the ensuing checkpoints are investigated (11, 13, 26).

1. Inject 8–14 female frogs with PMSG (500 μL/frog). Leave them for 3–5 days.
2. Induce egg-laying by injecting each frog with 1 mL of HCG and placing them in individual containers with 1 L 1× MMR. It must be noted that the final yield of NPE will be a small fraction of the Crude S extract obtained from the eggs (*see* **Notes 1** and **2**).
3. Combine good eggs, remove and discard buffer. Wash eggs with 1× MMR.
4. Dejelly the eggs with 2% L-cysteine (made up in water) pH 7.7. (The pH is extremely important since acidic pH will not dejelly eggs.) Gentle swirling to ensure complete dejellying is permissible, but care should be taken to avoid damaging the eggs (**Note 3**).
5. Wash the eggs rapidly three times with 0.5× MMR and three times with 1× ELB, pH 7.7.
   Transfer eggs to 14-mL Falcon Tube (cat # 352059).
6. Pack eggs by centrifugation for 1 min at 1000 rpm. Remove excess buffer, add Protease inhibitor cocktail, Aprotinin 5 μg/mL (1:2000), and 2.5 μg/mL cytochalasin B.
7. Crush eggs by centrifugation at 10,000 rpm in a Sorvall Centrifuge with a HB6 rotor for 15 min at 4°C.
8. The crude cytoplasmic layer is withdrawn by puncturing the side of the tube with a 21-gauge needle; the crude S (the middle layer) is supplemented with:
   Protease inhibitor cocktail, 10 μg/mL Aprotinin, 5 μg/mL Cytochalasin B, 50 μg/mL cycloheximide, 3.3 μg/mL nocodazole (stock 1000× in DMSO), and 1 mM DTT.
9. The crude S extract thus obtained is re-centrifuged at 10,000 rpm for 10 min at 4°C in a Falcon tube of appropriate size.
10. All of the residual lipids and the viscous dark brown material located just below it are completely aspirated off. About 1 mL of extract could be lost at this step; but it is critical that all of the upper layers are removed. The remaining cytoplasm is decanted into a fresh tube and supplemented with:
    - ATP regeneration system (must be prepared fresh before use): Per mL of extract, add 10 μL ATP (0.2 M stock), 10 μL phosphocreatine (1 M stock), and 1 μL creatine phosphokinase (5 mg/mL stock).
    - Demembranated sperm chromatin (*see* **Section 3.2.1**).

This nuclear assembly reaction is carried out in volumes of 4.5 mL per tube. The assembly reaction is made by thoroughly mixing 1 mL of extract with 90 μL of 200,000/μL sperm. Pipetting several times with a cut blue tip is recommended before transferring the mix into a 5-mL Falcon tube (Becton Dickinson cat # 352063). To this mix, 3.5 mL of extract is added and mixed by inverting multiple times. It is essential to have at least 4 mL of extract in each tube to ascertain that the layer of nuclei formed on the top after centrifugation will be thick enough to harvest.

11. The nuclear assembly reaction is incubated at room temperature and mixed by inversion every 10 min. A few microliters of the reaction are observed under the microscope. It is recommended to wait approximately 30 min after the time when round nuclei first become visible. The total time after sperm addition is around 70–80 min. Large nuclei are essential to ensure a good layer of nuclei in the subsequent step.
12. To collect the nuclei, the reaction is centrifuged for 2 min at 10,000 rpm at 4°C in a HB-4 rotor in a Sorvall centrifuge. The Falcon tube (Becton Dickinson # 352063) with the extract is placed in a tube (Becton Dickinson # 352059) containing 2 mL of water to help support it during the spin.
    After this step, the nuclei should be visible as a clear layer about 2 mm thick on the top of the extract. If the layer is thinner than expected, let the nuclei grow more next time and make sure that the sperm concentration used is correct.
13. Hold the tube to the light to better distinguish the clear nuclear layer from the underlying cytoplasm and remove the nuclei with a cut-off 200 μL pipette tip. Transfer all of the nuclei to a Beckman 5 × 21 mm ultracentrifuge tube (cat # P60720).
14. Supplement the nuclei with ELB to a final volume of 225 μL (assuming a 4.5-mL nuclear assembly reaction); mix thoroughly with a cut-off pipette tip and centrifuge for 30 min at 55,000 rpm at 2°C. Centrifugation is carried out in a Beckman TL100 table top ultracentrifuge using a TL55 swinging bucket rotor furnished with Teflon adapters (Beckman) to accommodate the 5 × 21 mm tubes.
    If there is more than 750 μL of the NPE+ELB mixture, a thick-walled 11 × 34 mm tube (Beckman cat # 343778) can be used with a centrifugation time of 40 min.
15. Carefully aspirate any lipids at the top of the sample after the spin and harvest the clear nucleoplasm taking care

to avoid the pellet consisting of nuclear envelopes and chromatin (*see* **Notes 12** and **13**).

16. The expected volume of NPE from each 4.5 mL nuclear assembly reaction is around 180 μL. The NPE is quick-frozen (small aliquots) in liquid nitrogen and stored at −80°C.

*3.1.8. Immunodepletion of Extracts to Study Protein Function*

Immunodepletion allows the specific removal of an endogenous protein from the extract by using antibodies against the protein of interest. The removal of protein/proteins from the extract is a valuable tool to investigate protein function; however, care should be taken to retain the overall activity and functionality of the extract. Therefore, the general protocol may need modification based on the antibody used and the subsequent assay that employs the immunodepleted extract. The following protocol is for rabbit polyclonal antibodies that bind avidly to Protein A beads.

1. Wash Protein-A sepharose CL-4B beads at least three times with large volumes of 1× PBS (beads to buffer ratio is about 1:10) to remove the Na-azide present in the storage buffer. After each wash, the beads are pelleted by centrifuging them at 2500 rpm for 1 min.
2. Add 1–4 volumes of sera to one volume of beads. If affinity purified antibodies are used, the volume of antibody added can be lower. The bead-antibody suspension is placed in compact reaction columns (CRC columns USB) and mixed thoroughly by incubating overnight at 4°C on a rotator.
3. Wash the bead-antibody mixture in the same buffer that was used to make the extract. If the beads are washed in the CRC column itself, then at least five washes are recommended. After each wash, the columns are placed in eppendorf tubes and spun at 2000 rpm for 1 min to remove residual liquid. Care should be taken to prevent drying of the beads during these spins. If necessary, the time/speed of the spin should be reduced.
4. Resuspend the beads with the extract that needs to be depleted. Typically, for one volume of beads, 3–5 volumes of extract are used, but this can vary depending on the protein being depleted. The column containing the bead/extract mixture is incubated at 4°C on a rotator for 30–40 min.
5. To collect the extract, place the column in a fresh eppendorf tube and spin at 2000 rpm for 1 min. The beads are retained in the column and the extract flows through into the eppendorf tube. For additional rounds of depletion, the depleted extract is incubated in a fresh column containing a second batch of (antibody-treated) beads and Steps 4 and 5 are repeated. Complete depletion of most proteins requires two to three rounds of depletion.

6. The depleted extract must be compared to a control extract by Western Blotting to monitor the efficiency of depletion (the control extract is one which is subjected to the same procedure with control IgG instead of antibodies). This is strongly recommended before using the depleted extract in assays designed to investigate protein function.

### 3.2. Preparation of DNA Templates

To study DNA replication and related processes, *Xenopus* extracts are used in conjunction with a variety of DNA templates including sperm chromatin and plasmid DNA. In this section we describe the preparation of both chromosomal as well as small DNA templates typically used to investigate DNA replication and cell cycle checkpoints.

While intact templates are widely used to understand normal DNA replication, the DNA is sometimes subjected to damage in order to investigate damage-related physiological processes such as checkpoint and repair assays. Therefore, the preparation of damaged DNA templates is also discussed.

Chromatin templates are used in several assays to monitor DNA replication and binding of DNA-associated proteins. In some instances, the DNA is appropriately damaged to investigate DNA damage-related phenomena. The protocols to make both intact and damaged chromatin are described below (**Sections 3.2.1**, **3.2.2** and **3.2.3**).

#### 3.2.1. Demembranated Sperm Chromatin

Apart from its use in studying DNA replication, the demembranated sperm chromatin described below can also be used as a starting material to prepare NPE, a nucleus-free system used in replication, repair and checkpoint assays.

1. Prime male frogs with 25U PMSG (per frog) 3–5 days before use and inject with 125U HCG (per frog) the day before Sperm Chromatin Preparation (*see* **Note 1**).
   It is suggested that 4–5 male frogs be used to obtain adequate sperm of high concentration (500,000 nuclei/μL).
2. Put frogs in 0.05% benzocaine for 30 min.
3. Dissect frogs (incision on the ventral side), cut out testes and immediately place in cold 1× MMR in a petri dish on ice.
4. Remove fat tissue and blood vessels and rinse three times in MMR. The final testes should be as free from fat and blood tissues as possible. If necessary, use a dissecting microscope to rapidly tease out unwanted tissue.
5. Wash three times in 1× NPB.
6. Remove buffer and macerate extensively with a clean, sharp razor blade. Thorough maceration is critical to obtain

maximum recovery. Place the testes in a 60-mm plastic petri dish on ice and macerate for about 20 min.

7. Add 10 mL cold NPB and aspirate a few times with a plastic Pasteur pipette. Filter sperm suspension through two layers of cheesecloth into a 50-mL Falcon Tube. Squeeze excess liquid and any remaining sperm from the cheesecloth. Wash cheesecloth in 2–4 mL NPB and squeeze out any remaining sperm into the tube. Split the contents of the 50-mL tube into two 14-mL Falcon tube (Becton Dickinson # 352059).
8. Centrifuge using a HB-6 rotor at 6000 rpm for 15 min at 4°C to recover sperm and remove debris.
9. Remove supernatant and wash pellet with 10 mL NPB, trying to remove any blood vessels and debris. Repeat the wash process 2–3 times and spin for 10 min between washes. If blood vessels/debris still remains, centrifuge at low speed (800 rpm) and transfer the liquid containing the sperm to a new tube. The debris should remain at the bottom of the tube. Re-centrifuge at 6000 rpm for 15 min.
10. Remove supernatant from the last wash and resuspend pellet in 1 mL NPB + 0.2% Triton X-100. Incubate for 15 min at room temperature with gentle mixing.
11. Add 10 mL of cold NPB + 3% BSA and mix gently to obtain a homogenous suspension. Centrifuge for 10 min at 6000 rpm.
12. Wash 2 times with 10 mL NPB + 0.3% BSA.
13. Resuspend in 1 mL (or less) NPB + 0.3% BSA and 30% glycerol.
14. Count sperm using a hemocytometer and DAPI staining. A range of dilutions should be counted since the sperm sometimes clump together leading to inaccurate counts.
15. Freeze nuclei using liquid nitrogen and store at –80°C in aliquots (*see* **Note 14**).

#### *3.2.2. Generation of Double-Strand Breaks (DSB) in Chromatin*

The *Xenopus* system can be used to assess the physiological consequences of DNA damage and identify the protein complexes that assemble at the site of DNA damage. The use of S-phase and M-phase extracts (**Section 3.1**) further enables the identification of cell cycle "phase dependent" composition of the DNA repair complexes. To introduce DNA damage in chromatin, we use PflMI restriction enzyme in our laboratory, although other suitable enzymes can also be used.

1. To the extract, add demembranated sperm nuclei/chromatin (5000 nuclei/μL). Incubate for 10 min at 21°C.

2. Remove 15 μL of extract into a separate eppendorf tube. This is the control.
3. To the remaining extract–sperm mixture, add 0.05 U/μL of PflMI restriction enzyme. Incubate at 21°C.
4. At desired time intervals, remove 15 μL of extract and stop the reaction with 750 μL of ice-cold chromatin isolation buffer (supplemented with 0.125% Triton X-100). Keep on ice for 10 min.

This protocol is used to perform both the chromatin binding assay to detect proteins involved in various stages of the cell cycle and in other checkpoint assays.

The Interphase Extract II can also be used for these types of studies.

*3.2.3. Preparation of 6-DMAP Chromatin*

To investigate DNA damage induced checkpoints, 6-DMAP chromatin is used in certain replication assays. It is prepared as follows:

1. Incubate demembranated sperm chromatin (**Section 3.2.1**) at 40,000 nuclei/μL for 20 min at 23°C in 6-DMAP extract (50 μL) (**Section 3.1.1**, last paragraph).
2. Dilute the above extract-nuclei mixture (50 μL) tenfold using chromatin isolation buffer I.
3. Layer carefully on 1.2 mL of sucrose cushion (chromatin isolation buffer I containing 30% sucrose).
4. Pellet the 6-DMAP chromatin at 6000×*g* for 15 min at 4°C and resuspend in chromatin isolation buffer I supplemented with 2.5 mM Mg-ATP.

The 6-DMAP chromatin cannot be stored and must be used immediately.

Small DNA templates: While chromosomal DNA is used in various assays to study checkpoints in *Xenopus*, the use of small DNA templates has found increasing applications due to certain unique advantages. For instance, the generation of site-specific damage on DNA can easily be accomplished by using plasmids with known sequences. Furthermore, the induced damage can be designed to mimic the damage caused by UV lesions or chemotherapeutic agents such as mitomycin C, and cisplatin. Such templates, in combination with *Xenopus* extracts, notably the HSS/NPE system, are useful to study checkpoint signaling. The protocols to obtain two types of small DNA molecules are described in **Sections 3.2.4** and **3.2.5** below.

*3.2.4. Preparation of Plasmid with a Single Interstrand Crosslink (ICL)*

DNA interstrand crosslinks (ICLs) can be induced by chemotherapeutic, environmental or endogenous agents. They represent a class of damaged DNA where the strands are covalently linked, usually leading to a distorted DNA helix (11). Interstrand crosslinks are highly toxic since they can block DNA replication and transcription. Their removal or bypass is accomplished by the concerted effort of multiple repair pathways and repair failure can result in mutations, rearrangements, tumors, or cell death (11, 12).

Checkpoint signaling induced by ICLs and their repair is not completely understood, but recent studies have begun to unravel the mechanisms involved (11, 12). In our laboratory, we use *Xenopus* cell-free extracts to investigate checkpoint signaling induced by interstrand crosslinks and their subsequent repair. The protocol to obtain DNA with interstrand crosslinks is outlined below:

1. The pBS KS-vector (Stratagene) is used as the backbone and a linker sequence containing a PflMI restriction site was inserted between SacI and NotI sites in the multiple cloning site of the vector.
   Linker sequence: 5′CACCCATGGAATGGACGC3′and 5′GGCCGCGTCCATTCCATGGGTGAGCT3′.
   The resulting plasmid is referred to as pBS-PflMI.
2. The pBS-PflMI plasmid is digested with PflMI and DraIII restriction enzyme for at least 2 h at 37°C.
3. The digest is run on a 0.8–1% agarose gel and the plasmid band is cut out of the gel.
4. The gel slice containing the plasmid is purified from the agarose gel slice using the Qiagen gel extraction kit by following the protocol provided by the manufacturers (Qiagen).
5. The purified plasmid is ligated to the crosslinked oligonucleotide duplex using T4 DNA ligase (New England Biolabs) overnight at 14°C. (5′ GTGTAC AAG*CTGACCATGGA 3′ and 5′ ATGGTCAG*CTT GTACACGTA 3′. The crosslink is located between the G*s). The generation of the crosslinked duplex is a somewhat elaborate process and has been described elsewhere (27, 28).
6. After ligation, remove noncircular and nicked molecules using Exonuclease-It (Bayou Biolabs; Harahan, LA cat # X-101).
7. Purify using the PCR purification kit (Qiagen).
   As an alternative to Steps 6 and 7, the circular plasmid can be purified by two rounds of cesium chloride centrifugation.
8. Confirm the presence of ICL in the final plasmid by qRT-PCR using primers flanking the ICL. A control duplex without the crosslink is also generated as described in the above steps.

#### 3.2.5. Preparation of DNA with Double-Strand Breaks

DNA molecules with DSBs have been generated in our laboratory by using either restriction enzymes or PCR. To obtain DNA fragments with DSBs, we use the circular pBR322 plasmid as template and digest it to completion with restriction endonucleases. We tested different enzymes generating different types of DNA ends (blunt, 3′ overhang, or 5′ overhang) and did not observe differences in the DNA-damage response (29). Such DNA can be used in assays that investigate the effect of damage on DNA replication and also in chromatin binding experiments.

1. Digest 0.5 mg of pBR322 with HaeIII (NEB). HaeIII cuts pBR322 plasmid 25 times, thus generating 26 fragments containing 2 DSBs each.
2. Digest DNA and extract twice in phenol/chloroform, then precipitate in ethanol and sodium acetate.
3. Resuspend DBS-containing DNA in water at a concentration of 1 mg/mL.
4. Dilute the DSBs stock solution into the extracts to the desired concentration.

Alternatively, we have used λ-DNA that was digested with a series of restriction enzymes giving rise to different numbers of restriction fragments. λ-DNA is digested with Xbe I, Nco I, Hind III, and BstE I enzymes that generate 2, 5, 7, and 14 fragments, respectively. This approach enables us to increase the concentration of DSBs in the extracts while keeping the mass of added DNA constant.

To obtain 1 kb DNA fragments by PCR, we use the M13 ssDNA template using 22nt primers complementary to positions 5570 and 6584 as described earlier (30).

#### 3.2.6. Preparation of Biotinylated Substrates

Biotinylated substrates are extensively used in replication, checkpoint and repair assays. We have established a system in which we use biotinylated DNA substrates to understand the initiation of DNA replication and DNA damage checkpoint signaling (31).

1. One microgram of the gel purified DNA fragment is end-labeled with 1 unit T4 polymerase in the presence of 33 μM each of dATP, dGTP, dTTP, and biotin-dCTP for 15 min at 12°C.
2. Stop reaction by addition of 50 mM EDTA and incubate at 76°C.
3. Purify the labeled DNA using the PCR purification kit (Qiagen) as per the manufacturers' instructions.
4. Quantify the purified DNA by photospectrometry.

In our study, we observed that DNA damage response and checkpoint activation were induced by the relatively short substrates (31), thus increasing the potential use of such substrates in checkpoint studies.

### 3.3. Use of the *Xenopus* System to Study Cell Cycle Checkpoints/ Checkpoint Assays

The maintenance of genome integrity in cells is constantly challenged by DNA damaging agents. The response to such DNA lesions involves the prompt detection, signaling and repair of the damaged DNA. Two proteins with established roles in damage signaling are the ATM and ATR protein kinases (6).

The ATM- and ATR-dependent checkpoints can be classified based on whether or not they depend on active DNA replication and fork progression. During G1 or S phases, double-strand breaks (DSB) and ssDNA can induce replication-independent checkpoints, namely the G1/S checkpoint and the Intra-S checkpoint. However, there also exists replication-dependent checkpoints such as the Replication checkpoint and the S/M checkpoint (32).

*Replication-independent checkpoints*: The G1/S checkpoint can result from IR or radiomimetic agents and functions to prevent origin firing until the damage is repaired. This pathway can be either p53 dependent or independent. The two p53-independent pathways involve ATM and ATR: Firstly, the presence of DSBs activates an ATM-dependent checkpoint resulting in the phosphorylation and inhibition of Cdk2 activity. This inhibition of Cdk2 prevents the loading of Cdc45 onto chromatin and blocks origin firing (32). The second p53-independent G1/S checkpoint ensues upon the generation of aberrant DNA structures comprising of ssDNA-RPA intermediates generated in G1. Such intermediates can be generated experimentally by treatment with exonuclease III or by addition of etoposide, an inhibitor of DNA topoisomerase II (33). This RPA-dependent signaling results in ATR activation, which in turn inhibits origin firing (34).

The Intra-S checkpoint inhibits late origin firing. The proper installation of this checkpoint requires the phosphorylation of serine residues 278 and 343 of Nbs1 by ATM. After sensing the damage, ATM and ATR activation is followed by phosphorylation of their targets- Chk2 and Chk1 respectively. The ultimate result of the ensuing cascade is to downregulate Cdk2 and Cdc7 protein kinases which in turn prevent cdc45 loading and origin activation.

*S-phase checkpoint*: The S-phase checkpoint is activated in response to UV, IR and HU treatments and usually requires that active DNA replication be elicited. This checkpoint results in ATR activation and phosphorylation of its target Chk1 on serine residues 317 and 345 (35). The phosphorylation of Chk1 in turn culminates in the phosphorylation-dependent degradation of Cdc25A and inhibition of Cdk2-Cyclin E (36). Other regulators of the S-phase checkpoints are Claspin/Mrc1, MRN complex, BRCA1 and FANC proteins. The phosphorylation of Nbs1 by ATM, is another event crucial for the S-phase checkpoint (37).

*Replication-dependent checkpoints*: The Replication checkpoint and the S/M checkpoint are other replication-dependent checkpoints that are active during the S phase. The Replication checkpoint is initiated by stalled forks in response to genotoxic stresses, aberrant DNA structures or DNA damage and inhibits origins through the activity of ATR and Chk1. The S/M checkpoint ensures that DNA replication is complete before the cell enters mitosis. This checkpoint prevents mitotic entry by enabling Chk2 phosphorylation and activation which in turn inhibits Cdc25 phosphatase activity and prevents the activation of the Cyclin B-cdk1 complexes. The S/M checkpoint also involves ATM substrates such as Nbs1 and BRCA1 (38, 39).

*G2/M checkpoint*: The G2/M checkpoint targets Cdk1-Cyclin B kinase and prevents mitotic entry in response to DNA damage or incomplete S phase. Cell cycle arrest in this case involves both the ATM and the ATR pathways (40). The G2 checkpoint is abrogated by caffeine and is sometimes used as an assay to test the involvement of the ATM pathway.

In addition to the above functions, the ATM/ATR kinases also assist in DNA repair by inducing repair proteins and activating them by post-translational modifications such as phosphorylation. The activation of ATM is in turn an indicator of checkpoint activation and can be assayed by assessing the phosphorylation of its substrates such as Smc1, Chk2, or H2AX.

The above paragraphs (in **Section 3.3**) are a few examples of the proteins involved in and the mechanisms that drive various cell cycle checkpoints activated as a result of DNA damage. To assess whether a checkpoint has been activated or to further investigate checkpoint-related phenomenon; the proteins involved, their levels, phosphorylation status, and other modifications are frequently used as readouts. For instance, phosphorylations of Chk1, Chk2, or H2AX indicate that the ATR, ATM, or the ATM/ATR pathways have (respectively) been activated.

The *Xenopus* cell-free system can be used to investigate cell cycle checkpoints in a variety of ways such as monitoring DNA replication, studying the phosphorylation or activation of target proteins, and the assembly or localization of repair proteins on the site of DNA damage. The following sections describe the assays used to study checkpoints.

*DNA replication assay to study checkpoint activation:* Extracts derived from *Xenopus* eggs can support semi-conservative DNA replication of genomic DNA upon the addition of DNA templates (8). However, in the presence of DNA damage, checkpoints are activated. These checkpoints prevent the initiation or progression of DNA replication upon DNA damage (41, 42) and coordinate DNA replication, recombination and repair processes (6, 43). This inhibition occurs both when the damage is present on the template during replication in "cis" as well as when DNA damage signaling is induced in "trans" by DNA containing exogenous DSBs. Both types of DNA damage can activate a checkpoint.

The use of the *Xenopus* system to investigate such checkpoints is described below.

#### *3.3.1. DNA Damage Checkpoint Induced by Single-Strand DNA Gaps in Chromatin*

We have developed a cell-free system that recapitulates the inhibition of DNA replication in the presence of single-strand DNA gaps (44, 45).

Single-strand DNA gaps are generated by incubating chromatin in cell-free extracts in the presence of etoposide, an inhibitor of topoisomerase II, or by in vitro treatment of chromatin by DNA exonuclease III. Etoposide generates lesions in the chromatin templates that are undergoing DNA replication (44) by blocking the activity of DNA topoisomerase covalently linked to DNA 5′ termini (46).

In addition to studying checkpoints induced by ssDNA gaps in chromatin, the protocol described below (the portion that describes the monitoring of DNA replication) can also be used to assess the replication of chromatin and/or small DNA templates in various extracts.

1. Incubate 20 μL of activated extract with 5000 demembranated sperm nuclei/μL in the presence of etoposide at 23°C for 90 min (*see* **Note 15**).
2. Concentrations of etoposide ranging from 10 to 50 μM are effective in inducing a checkpoint response, as seen by the inhibition of genomic DNA replication (*see* **Note 16**).
3. Etoposide-induced inhibition of DNA replication is rescued by the addition of 5 mM caffeine, an inhibitor of checkpoint signaling kinases, including ATM and ATR (47).
4. Monitor DNA replication by incorporation of $^{32}$P-dATP into the chromatin. Add 0.2 μCi of a-$^{32}$P-dATP to each replication reaction.
5. Stop DNA replication reactions by diluting the samples in 200 μL of stop solution.
6. Incubate the samples with 1 mg/mL of proteinase K for 30 min at 37°C.
7. Extract DNA with 1 volume of phenol/chloroform.
8. Centrifuge the samples for 10 min at room temperature.
9. Recover the aqueous phase and precipitate with 2 volumes of ethanol and 1/10 volume of sodium acetate (3 M stock).
10. Resuspend the pellet in DNA loading buffer and run on a 0.8% agarose gel in TBE.
11. Fix the gel in 7% TCA (some protocols require higher percentage of TCA). Position the gel between two layers of Whatmann 3 MM paper and stacks of filter paper. Dry the gel overnight on the bench.
12. Expose the dried gel for autoradiography.

*3.3.2. DNA Damage Checkpoint Induced by DSB in Chromatin*

To investigate the cell cycle response to DNA damage at the onset of S phase, we use the *Xenopus* cell-free system designed to study initiation of DNA replication (48). Activated extracts are treated with either circular plasmid DNA, plasmid DNA containing DSBs, or λ-DNA containing DSBs. The treatment of the cytosolic extracts with DSBs-containing DNA activates a checkpoint in *trans*. In this protocol, the damaged DNA is only used to trigger the checkpoint and is subsequently removed, following which the extract is tested for its ability to replicate intact DNA templates. The damaged template is removed to avoid any interference with genomic DNA replication, such as titration of essential factors required in the elongation step of genomic DNA replication.

1. Incubate 100 μL of activated extract/Interphase I extract at 23°C in the presence of 50 ng/μL of circular plasmid DNA or digested plasmid (DSB) for 15 min to activate the checkpoint (*see* **Note 15**).
2. For rescue experiments, pre-treat extracts with 5 mM caffeine, 200 nM wortmannin, or affinity-purified anti-X-ATM antibodies for 15 min at 23°C, prior to incubation with the damaged DNA.

M and B fractions are prepared to study the initiation of DNA replication. The M and B fractions are prepared from cytosolic extracts as described earlier (**Section 3.1.5**), except that the fractions are prepared from extracts in which the checkpoint response has been activated by DSBs. Fractions derived from extract treated with DSBs are called M* and B*.

1. For replication assays, mix 0.5 μL of 6-DMAP chromatin (10,000 nuclei/μL) (*see* **Section 3.2.3**) with 1 μL each of either M and B or M* and B* fraction obtained from extract treated with different types of DNA molecules and/or caffeine, wortmannin, and ATM-neutralizing antibodies.
2. Incubate the reactions for 15 min at 23°C. Add 10 μL of 6-DMAP extract (*see* **Section 3.1.1**, last paragraph, for 6-DMAP extract).
3. Monitor DNA synthesis by the incorporation of α-$^{32}$P-dATP for 90 min at 23°C, followed by agarose gel electrophoresis (as described in **Section 3.3.1**; Steps 4–13).

*3.3.3. Assay to Study Checkpoint Induced by ICL*

To investigate whether the damaged DNA activates an ATM/ATR checkpoint, replication assays are conducted in the presence of caffeine which is an inhibitor of checkpoint kinases. The experiment described below uses DNA containing an interstand crosslink, however the protocol can be modified to investigate whether other damaged DNA templates similarly activate the ATM/ATR checkpoints.

The protocol for the ICL induced damage assay is described below:

1. *Xenopus* extracts (HSS and NPE) are prepared as described above (**Sections 3.1.6** and **3.1.7**).
2. Take two sets of two tubes. To each set of tubes, add 2.3 μL of HSS, 0.15 μL of Energy Mix, and 5 ng control DNA (tube 1) or 5 ng ICL DNA (tube 2) (Control and ICL DNA as described in **Section 3.2.4**). To the first set of tubes, 5 mM caffeine (made up in buffer) is also added. The second set serves as a "no-caffeine control" and only buffer is added instead. Mix. Incubate for 30 min at 21°C.
3. Add 4.6 μL NPE and 0.2 μL $^{32}$P-dCTP (Perkin Elmer # BLU513H250UC), mix by pipetting and incubate for 2 h at 21°C (*see* **Note 17**).
4. Stop reaction by adding 200 μL of Stop Solution. Mix well.
5. Add 1/20 volumes (10 μL) of Proteinase K and incubate at 55°C for 1–2 h.
6. Cool samples to room temperature and spin down for 10 s.
7. Add 200 μL of phenol:chloroform:isoamylalcohol (25:24:1 Sigma) and mix well by inverting several times.
8. Centrifuge samples at 13,000 rpm for 10 min. Transfer the aqueous phase to a fresh tube. Add 1/10 volumes (20 μL) of sodium acetate (3M Stock) and 2.5 volumes (500 μL) of cold 100% ethanol. Incubate at –20°C overnight.
9. Spin samples for 10 min at 13,000 rpm in a microcentrifuge.
10. Discard supernatant, briefly dry pellet and dissolve it in 15-μL DNA loading buffer with bromophenol blue and cyan blue dyes.
11. Load samples and electrophorese them on a 0.8% agarose gel.
12. Treat gel with trichloroacetic acid (Sigma) 50% made up in water for 20–30 min.
13. Dry gel overnight between 3 M blotting paper and several layers of absorbent paper towels. The next day, the gel may be dried for an additional hour at 70°C using a vacuum dryer. Subject dried gel to autoradiography; multiple exposure times may be required, ranging from 15 min to overnight.

#### *3.3.4. Study of the Checkpoint Induced by DNA Damage* In Trans *on DNA Replication*

Checkpoint activation can be assessed by monitoring the effect of the ICL DNA (or other forms of damaged DNA) on the replication of an intact plasmid. The experiment described below uses ICL DNA; however, a similar protocol can be used to study the impact of other types of damaged DNA on the replication of intact DNA templates.

1. Prepare *Xenopus* extracts (HSS and NPE) as described (**Sections 3.1.6** and **3.1.7**).
2. Add 2.3 μL of HSS, 0.15 μL of Energy Mix, and 5 ng Control or ICL DNA. Mix. Incubate for 30 min at 21°C.
3. Add 4.6 μL NPE; mix by pipetting and incubate for 1 h at 21°C (*see* **Note 17**).
4. Add the plasmid that is to be replicated to each of the tubes and also add 0.2 μL $^{32}P$ dCTP. Mix well and incubate at 21°C for 20 min.
5. Analyze the replication of the plasmid DNA by agarose gel electrophoresis as described earlier (**Section 3.3.1**).

Since the two plasmids (control /ICL plasmid and the plasmid to be replicated) are of different sizes, they can be distinguished by gel electrophoresis.

*3.3.5. Study of Phosphorylation to Assess Checkpoint Activation*

It is known that DNA damage blocks DNA polymerases, allows DNA unwinding by MCM helicases and triggers an ATM or ATR-dependent checkpoint (49, 50). The ATR checkpoint pathway is activated during S phase to sense and coordinate cellular responses to DNA damage. Several checkpoint proteins are recruited to the site of damaged DNA and play a role in the assembly of the ATR kinase signaling complex (51, 52). The activated ATR phosphorylates its target proteins, including Chk1, to initiate the cellular response to the DNA damage. This phosphorylation of Chk1 is commonly used as a readout for ATR activation. In *Xenopus*, Chk1 phosphorylation at the S344 site is observed in response to DNA damage induced by UV rays, methylmethanesulfonate (MMS), 4-nitroquinoline1-oxide (4-NQO), interstrand crosslinks (ICL), and aphidicolin (11, 53, 54). This is best detected by a phospho-specific antibody capable of detecting *Xenopus* Chk1 phosphorylated at S344.

The phosphorylation of Histone H2AX at the Serine 139 residue is another frequently used indicator of DNA damage. Histone H2AX is a well-characterized substrate for activated protein kinases and is phosphorylated by both ATM (in response to IR) and ATR kinases (in response to UV) or DNA replication blocks (55, 56) at Serine 139 (57, 58). In combination with 53BP1, histone H2AX is also thought to play a role in G2-M checkpoint activation (59). In our laboratory, we have monitored the phosphorylation of both endogenous histone H2AX as well as exogenous H2AX peptide (60, 61).

Several other proteins get phosphorylated upon DNA damage. For instance, MRE11, a protein that is involved in the repair of double-strand chromosome breaks (62) and Nbs1 that forms a complex with MRE11 and Rad50 (63) are both phosphorylated in response to DNA damage. The above examples collectively suggest that the phosphorylation of various proteins can be used to assess the presence of DNA damage as well as, in some instances, determine the activation of cell cycle checkpoints. The use of phosphorylation assays and *Xenopus* cell-free extracts in checkpoint studies is outlined in the following sections (**Sections 3.3.5**, **3.3.6**, **3.3.7** and **3.3.8**).

#### 3.3.6. Phosphorylated Histone H2AX Detection (Endogenous Substrate)

The detection of phosphorylated Histone H2AX is a sensitive assay to monitor DNA damage induced checkpoint signaling.

1. Incubate 50 μL of interphase extract (or extract in which the occurrence of DNA damage will be assessed) with 10,000 nuclei/μL for 90 min at 23°C.
2. Isolate postreplicative chromatin by diluting the extracts in chromatin isolation buffer I containing 1 mM NaF, 1 mM sodium vanadate, and 0.125% Triton X-100.
3. Layer samples onto chromatin isolation buffer containing 30% sucrose and lacking Triton X-100, then spin at 6000×*g* for 20 min at 4°C.
4. Prepare a positive control by incubating sperm nuclei for 30 min in interphase extract to decondense chromatin.
5. Isolate the chromatin and digest for 4 h with *Not*I.
6. Reisolate the digested chromatin through a sucrose cushion and incubate in interphase extract for 60 min.
7. Boil chromatin in Laemmli buffer and process for SDS-PAGE electrophoresis.
8. Use antiphosphorylated H2AX antibody for Western blotting at 1/6000 dilution.

#### 3.3.7. Histone H2AX Phosphorylation Assay (Exogenous Substrate)

Another assay to monitor checkpoint signaling is the measurement of H2AX peptide phosphorylation (29, 61). Histone H2AX is a well-characterized substrate for protein kinases that are activated by DNA damage and is phosphorylated in vivo at serine 139 by ATM and ATR (29). We have reported earlier the use of the C-terminal peptide of mouse H2AX as a reporter substrate to monitor the response to damage (29). The actual assay involves incubation of interphase extracts (LSS/ Interphase extracts I/II or Activated CSF) with fragmented DNA and either wild-type (AVGKKASQASQEY) or control/mutant (AVGKKAAQAAQEY) H2AX peptides. The presence of fragmented/damaged DNA results in a rapid phosphorylation of the exogenous peptide.

1. The extract (LSS or Activated CSF extracts) is incubated with DNA fragments to elicit a DSB response. The DNA fragments we used are biotinylated at one end and immobilized on Streptavidin beads (M280 Dynal Biotechnology) at concentrations of 20 and 60 ng/μL (corresponds to $1.2 \times 10^{11}$ and $3.6 \times 10^{11}$ ends per microliter, which in turn simulate irradiation doses of 70 Gy and 210 Gy respectively for a human lymphocyte).
2. Separate the DNA-bound beads from the extracts (according to the manufacturer's instructions – Dynal Biotechnology) and wash DNA six times with ELB buffer supplemented with 0.1% (v/v) Triton. The resulting DNA-bound and soluble fractions of the extracts can be evaluated for various parameters.
3. Mix 2 μl of the above extract with 20 μL of EB buffer supplemented with 50 μM ATP, 0.4 μL $\gamma$-$^{32}$P ATP (stock – 10 mCi/mL), and 1 mg/mL of either wild-type or mutant H2AX peptide (Sigma).
4. Incubate samples for 15 min at 30°C.
5. Stop reactions by adding 2 μL of 0.5 M EDTA.
6. Spot reactions on p81 phosphocellulose filter papers (Upstate Biotechnology).
7. Wash filters three times in 5% (v/v) acetic acid and twice with water.
8. Air dry filters and quantify radioactivity in a scintillation counter.

It should be noted that the cpm incorporated into the control mutant H2AX peptide is subtracted from the cpm of the wild-type H2AX peptide and the values are normalized to those from control extracts treated with streptavidin beads in the absence of DNA (no DSB). Second, the extracts may be depleted of a specific protein by using antibodies (*see* **Section 3.1.8**) and the depleted extract can be analyzed as above to assess the function of a particular protein. Third, the extract can be pre-incubated with 5 mM caffeine to serve as a control in the H2AX peptide phosphorylation assay.

*An alternate protocol for H2AX peptide phosphorylation is as follows:*

1. Take 2 μL of LSS Extract (that has been preincubated with undamaged or damaged DNA as above) and mix with 20 μL of EB Kinase Buffer supplemented with 0.5 mg/mL histone H2AX peptide (wild-type or mutant), 50 μM ATP and 1 μL of $\gamma^{32}$P ATP (10 mCi/μL, greater than 3000 Ci/mMol).
2. Incubate samples at 30°C for 20 min.

3. Stop reaction by adding 20 μL of 50% acetic acid.
4. Spot samples on p81 phospho-cellulose filter papers.
5. Wash filters three times with 10% (v/v) acetic acid and twice with water. Air dry filters.
6. Quantify radioactivity using a scintillation counter.
7. Subtract the number of cpm incorporated into the control mutant H2AX peptide from the number incorporated into the wild-type H2AX peptide; normalize the values to those from control extracts treated with streptavidin beads in the absence of DNA (negative control).

#### 3.3.8. Phosphorylation of ATM/ATR Target Proteins

Response to DNA damage involves the sensing of the damage, transduction of the signal and activating the response. Each component of the damage response is carried out by specific proteins. The ATM and ATR proteins are master regulators of the pathway. The levels and phosphorylation status of these proteins and their downstream targets are commonly used readouts to investigate the activation of a checkpoint.

For instance, to distinguish between ATR and ATM activation, phosphorylation of Chk1 protein at S344 and phosphorylation of ATM at S1981 can be investigated as follows:

1. Prepare *Xenopus* extracts (HSS and NPE) as described earlier (**Sections 3.1.6** and **3.1.7**).
2. Take two sets of two tubes. To each set of tubes, add 2.3 μL of HSS, 0.15 μL of Energy Mix, and 5 ng control DNA (tube 1) or damaged DNA (tube 2). To the first set of tubes, 5 mM caffeine (made up in buffer) is also added. The second set serves as a "no-caffeine control" and only buffer is added instead. Mix. Incubate for 30 min at 21 °C.
3. Add 4.6 μL NPE and incubate for 90 min at 21 °C (*see* **Note 17**).
4. Analyze the soluble extracts by Western blotting with phospho Chk1 antibody (Cell Signaling) and phospho ATM antibody (Rockland Immunochemicals, PA).

To differentiate between ICL and canonical DNA replication checkpoints, the above experiment can be carried out with Geminin (50 ng/mL stock, added at 1/50 final concentration) and Roscovitine (500 μM in DMSO) instead of caffeine. We have observed that ICL induced Chk1 phosphorylation is resistant to geminin/roscovitine treatment indicating the existence of a replication-independent pathway (11).

To evaluate the role of the Fanconi Anemia (FA) pathway in the ICL-induced checkpoint, the above experiment is carried out with curcumin – an inhibitor of the FA pathway (100 μM in ethanol; Sigma) – instead of caffeine.

Binding assays to study localization of damage-dependent proteins: The association or binding of proteins on to chromatin or small DNA templates is frequently used to understand the composition of complexes that bind DNA and to investigate the function of a specific protein. Furthermore, understanding the binding patterns of proteins onto intact or damaged DNA can provide insights into DNA damage, signaling, and repair. The protocols to identify the proteins bound to either intact or damaged DNA (either chromatin or plasmid DNA) are described below.

#### 3.3.9. Chromatin Binding Assay

A critical aspect of the DNA-damage response is the damage-dependent localization of proteins to the DNA. This is exemplified by the formation of damage-induced foci within the nuclei of mammalian cells. Cell-free systems allow the rapid mixing and subsequent separation of chromatin, nuclear, and cytoplasmic fractions. To analyze the status of the proteins that bind on the chromatin in a replication or checkpoint-dependent manner, chromatin binding assays can be performed (*see* **Notes 18** and **19**).

1. Perform chromatin-binding assays in activated extracts or in fractionated extracts. In the case of activated extracts, assemble chromatin in 50 μL of interphase extracts in which a checkpoint has or has not been activated.
2. Incubate 10,000 nuclei/μL for 60–120 min, and dilute the extract with 800 μL of chromatin isolation buffer I.
3. In the case of fractionated extracts, assemble replication reactions as above with the following modifications:
   Scale up reactions tenfold. Incubate 10 μL of M and B fractions (*see* **Section 3.1.5**) for 15 min with 5 μL of 6-DMAP chromatin (10,000 nuclei/μL) (*see* **Section 3.2.3**).
4. Following incubation, dilute each reaction in 200 μL of chromatin isolation buffer I supplemented with 0.1% Triton X-100.
5. Layer the chromatin onto the same buffer containing 30% sucrose (omit the Triton X-100 in this sucrose containing buffer).
6. Centrifuge the chromatin at 6000×*g* for 15 min at 4°C.
7. Resuspend the pellet in Laemmli loading buffer.
8. Run the samples on 10% SDS-PAGE and analyze by Western blotting with specific antibodies.

Either intact or damaged chromosomal DNA can serve as templates in the above assay.

A similar protocol can also be conducted with HSS/NPE (1:2 ratio) and demembranated sperm chromatin as the DNA template on which proteins are bound.

*3.3.10. Binding Assay Using HSS/NPE and ICL Plasmid*

To investigate the role of specific proteins in checkpoint signaling, we deplete the specific protein and determine the activation of downstream pathways and the phosphorylation of target proteins (11). Although the assay described below involves ICL plasmid, it must be noted that the protocol can easily be adapted to use chromatin or other small DNA templates instead of the ICL plasmid.

Alternatively, this assay can also be modified to carry out rescue experiments where recombinant wild-type or mutant protein is added back to the extract.

This assay is extensively used to study the levels, phosphorylation status and identity of the proteins that are bound to DNA under intact or damaged conditions.

We have used this assay in our laboratory to investigate checkpoint signaling from and the repair of a single, site-specific interstand crosslink (11).

1. *Xenopus* HSS, NPE extracts were prepared as described earlier (64) (*see* **Note 18**).
2. Take 2.3 μL of HSS, 0.15 μL of Energy Mix, and 5 ng plasmid DNA (control DNA or the ICL DNA) in an eppendorf tube. Mix. Incubate for 30 min at 21 °C. Add 4.6 μL NPE, mix well, and incubate for 1 h at 21 °C.
3. Dilute the reaction with 800 μL of Chromatin isolation buffer II supplemented with Triton X-100.
4. Place on ice for 5 min.
5. Layer the entire mixture on 320 μL of 30% sucrose in chromatin isolation buffer II (without Triton X-100). Low retention 1.5-mL Eppendorf tubes (Fisherbrand # 02-681-341) are used for this step.
6. Centrifuge at 6000 rpm (8000 rpm for plasmid DNA template) in HB-6 rotor for 30 min at 4 °C.
7. Carefully remove most of the liquid from the tubes after the spin but leave about 10 μL in the tube. This should contain the DNA/chromatin-bound protein. Freeze the tube in liquid nitrogen and store at −20 °C overnight.
8. Thaw out tubes with DNA bound protein and resuspend the pellets with 10 μL of 2× Laemmli buffer. Denature samples by boiling for 3 min and analyze by SDS-PAGE and Western Blotting.

A control sample without DNA must also be analyzed simultaneously to ascertain that there is no contamination from non-chromatin bound proteins.

An undepleted extract is also used as a control in these assays.

## 4. Notes

1. Frogs are generally not fed after this initial prime in order to reduce the possibility of fecal matter contaminating the eggs. However, the experiments must be scheduled so that the initial priming and egg collection are done within a week so that the frogs are not deprived from eating for extended periods.
2. Depending on age and size, a female *Xenopus* will lay between 5 and 10 mL of dejellied eggs. This yields between 2 and 4 mL of egg cytosol.
3. Over-dejellying of eggs while preparing any of the extracts is detrimental and will cause eggs to burst/pop and render them unusable. Typically dejellying takes about 10 min. However, each batch of eggs is unique and therefore dejellying times must be adjusted accordingly.
4. If any floating or totally white egg/shells are seen, quickly remove and discard them.
5. If Falcon tubes and HB-6 rotors are used, then the caps of the tubes should be removed before centrifugation to allow proper closing of the rotor's lid.
6. The cytoplasmic layer can also be pipetted by directly sliding a pipette tip against the wall of the tube through the top lipid layer. Alternatively, the lipid layer can be removed first using a cotton swab.
7. Cytosolic extracts (CSF or activated) must be used immediately after preparation.
   Freezing and thawing the extract triggers apoptosis.
8. While freezing the Interphase extracts II, to avoid clumping of drops, it must be ascertained that the previous drop is completely frozen before adding the next drop. Frozen drops generally sink to the bottom of the liquid Nitrogen containing P60 dish. The frozen drops can be picked with a clean forceps and dropped into a frozen cryotube. Several drops can be placed in a single cryotube. To avoid explosion, the cryotube must not be closed unless all the liquid nitrogen in it has evaporated.
9. For M/B fractionation, the quality of the eggs and the timing of the preparation are critical to get functional fractions. We perform the fractionation as soon as possible after the preparation of the extract. The complete procedure should not take more than 3 h to recover functional fractions. The quality of M and B fractions can also be tested in pilot experiments. M or B fractions do not support DNA

replication by themselves, but only in combination. If background replication is observed with either M or B fraction alone, the concentration of PEG used for fractionation can be modified with a 1% window: 3.5± 0.5% for B and 9 ±0.5% for M.

10. For the preparation of HSS, the MMR buffer can be made as a 10× stock, stored at 4°C and diluted to 1× MMR. It is recommended to check that the pH of the diluted MMR is 7.7–7.8 before use.
11. For the ultracentrifugation step, if thin tubes are used, the cat# is Beckman 326819.
12. If the NPE is contaminated by insoluble material that appears cloudy or opaque, the extract is re-centrifuged in the TL55 rotor with Teflon adapters for 15 min.
13. It is recommended that NPE is flash-frozen in liquid Nitrogen and stored in –80°C as small aliquots. Repeated freeze–thaw cycles can result in loss of activity and must be avoided.
14. Demembranated sperm chromatin is stored at –80°C in small aliquots (10 μL/tube) for replication assays and in larger aliquots (200 μL/tube) for making Nucleoplasmic extract. Repeated freeze–thawing should be avoided.
15. Pipetting the extracts gently but thoroughly is critical to the success of all procedures to achieve a homogenous extract. The formation of aggregates or incomplete resuspension can be detrimental to the experiment.
16. Alternatively, supplement activated extract with ExoIII chromatin at the same concentration of nuclei.

    Exo III chromatin preparation: Incubate $10^6$ sperm nuclei with 100. U DNA exonuclease III (Roche) for 10 min at 37°C (in 60 mM Tris–HCl, pH 8.0, and 0.6 mM $MgCl_2$ buffer); stop the reaction by addition of 1 mL NPB buffer
17. The volume of HSS and NPE can be altered, but it is recommended to keep the ratio of HSS to NPE at 1:2.
18. In chromatin binding assays, depleted extracts can be used to investigate the function of a specific protein in coordinating the binding specificities of other proteins.
19. In chromatin binding assays, there is the possibility of contaminating the chromatin-bound proteins with cytosolic proteins. Therefore, it is recommended to always include a "no DNA" control (Extract is processed exactly as other samples except that the DNA is omitted from the reaction tube).

## Acknowledgments

This work was supported by grants from the National Institute of Health (CA92245 and GM77495) to J.G. and Department of Defense Grant (W81XWI1-09-1-0504) to S.V.S.

### References

1. Jackson, S. P. and Bartek, J. (2009) The DNA-damage response in human biology and disease. Nature **461**, 1071–1078.
2. Clarke, P. R. and Allan, L. A. (2009) Cell-cycle control in the face of damage – a matter of life or death. Trends Cell Biol **19**, 89–98.
3. Hensey, C. and Gautier, J. (1995) Regulation of cell cycle progression following DNA damage. Prog Cell Cycle Res **1**, 149–162.
4. Murray, A. W. (1995) The genetics of cell cycle checkpoints. Curr Opin Genet Dev **5**, 5–11.
5. Thompson, S. L., Bakhoum, S. F. and Compton, D. A. Mechanisms of chromosomal instability. Curr Biol **20**, R285–R295.
6. Zhou, B. B. and Elledge, S. J. (2000) The DNA damage response: putting checkpoints in perspective. Nature **408**, 433–439.
7. Costanzo, V. and Gautier, J. (2004) Xenopus cell-free extracts to study DNA damage checkpoints. Methods Mol Biol **241**, 255–267.
8. Blow, J. J. and Laskey, R. A. (1986) Initiation of DNA replication in nuclei and purified DNA by a cell-free extract of Xenopus eggs. Cell **47**, 577–587.
9. Smythe, C. and Newport, J. W. (1991) Systems for the study of nuclear assembly, DNA replication, and nuclear breakdown in Xenopus laevis egg extracts. Methods Cell Biol **35**, 449–468.
10. Tutter, A. V. and Walter, J. C. (2006) Chromosomal DNA replication in a soluble cell-free system derived from Xenopus eggs. Methods Mol Biol **322**, 121–137.
11. Ben-Yehoyada, M., Wang, L. C., Kozekov, I. D., Rizzo, C. J., Gottesman, M. E. and Gautier, J. (2009) Checkpoint signaling from a single DNA interstrand crosslink. Mol Cell **35**, 704–715.
12. Raschle, M., Knipscheer, P., Enoiu, M., Angelov, T., Sun, J., Griffith, J. D., Ellenberger, T. E., Scharer, O. D. and Walter, J. C. (2008) Mechanism of replication-coupled DNA interstrand crosslink repair. Cell **134**, 969–980.
13. Walter, J., Sun, L. and Newport, J. (1998) Regulated chromosomal DNA replication in the absence of a nucleus. Mol Cell **1**, 519–529.
14. Machida, Y. J., Hamlin, J. L. and Dutta, A. (2005) Right place, right time, and only once: replication initiation in metazoans. Cell **123**, 13–24.
15. Gautier, J., Minshull, J., Lohka, M., Glotzer, M., Hunt, T. and Maller, J. L. (1990) Cyclin is a component of maturation-promoting factor from Xenopus. Cell **60**, 487–494.
16. Gautier, J., Norbury, C., Lohka, M., Nurse, P. and Maller, J. (1988) Purified maturation-promoting factor contains the product of a Xenopus homolog of the fission yeast cell cycle control gene cdc2+. Cell **54**, 433–439.
17. Gautier, J., Solomon, M. J., Booher, R. N., Bazan, J. F. and Kirschner, M. W. (1991) cdc25 is a specific tyrosine phosphatase that directly activates p34cdc2. Cell **67**, 197–211.
18. Lohka, M. J. and Masui, Y. (1983) Formation in vitro of sperm pronuclei and mitotic chromosomes induced by amphibian ooplasmic components. Science **220**, 719–721.
19. Murray, A. W. and Kirschner, M. W. (1989) Cyclin synthesis drives the early embryonic cell cycle. Nature **339**, 275–280.
20. Murray, A. W., Solomon, M. J. and Kirschner, M. W. (1989) The role of cyclin synthesis and degradation in the control of maturation promoting factor activity. Nature **339**, 280–286.
21. Lebofsky, R., Takahashi, T. and Walter, J. C. (2009) DNA replication in nucleus-free Xenopus egg extracts. Methods Mol Biol **521**, 229–252.
22. Lupardus, P. J., Van, C. and Cimprich, K. A. (2007) Analyzing the ATR-mediated checkpoint using Xenopus egg extracts. Methods **41**, 222–231.
23. Costanzo, V., Robertson, K. and Gautier, J. (2004) Xenopus cell-free extracts to study the DNA damage response. Methods Mol Biol **280**, 213–227.
24. Trenz, K., Errico, A. and Costanzo, V. (2008) Plx1 is required for chromosomal

DNA replication under stressful conditions. Embo J **27**, 876–885.
25. Costanzo, V., Robertson, K., Ying, C. Y., Kim, E., Avvedimento, E., Gottesman, M., Grieco, D. and Gautier, J. (2000) Reconstitution of an ATM-dependent checkpoint that inhibits chromosomal DNA replication following DNA damage. Mol Cell **6**, 649–659.
26. Walter, J. and Newport, J. (2000) Initiation of eukaryotic DNA replication: origin unwinding and sequential chromatin association of Cdc45, RPA, and DNA polymerase alpha. Mol Cell **5**, 617–627.
27. Dooley, P. A., Tsarouhtsis, D., Korbel, G. A., Nechev, L. V., Shearer, J., Zegar, I. S., Harris, C. M., Stone, M. P. and Harris, T. M. (2001) Structural studies of an oligodeoxynucleotide containing a trimethylene interstrand cross-link in a 5′-(CpG) motif: model of a malondialdehyde cross-link. J Am Chem Soc **123**, 1730–1739.
28. Dooley, P. A., Zhang, M., Korbel, G. A., Nechev, L. V., Harris, C. M., Stone, M. P. and Harris, T. M. (2003) NMR determination of the conformation of a trimethylene interstrand cross-link in an oligodeoxynucleotide duplex containing a 5′-d(GpC) motif. J Am Chem Soc **125**, 62–72.
29. Costanzo, V., Paull, T., Gottesman, M. and Gautier, J. (2004) Mre11 assembles linear DNA fragments into DNA damage signaling complexes. PLoS Biol **2**, E110.
30. de Jager, M., van Noort, J., van Gent, D. C., Dekker, C., Kanaar, R. and Wyman, C. (2001) Human Rad50/Mre11 is a flexible complex that can tether DNA ends. Mol Cell **8**, 1129–1135.
31. Kurth, I. and Gautier, J. (2010) Origin-dependent initiation of DNA replication within telomeric sequences. Nucleic Acids Res **38**, 467–476.
32. Kastan, M. B. and Bartek, J. (2004) Cell-cycle checkpoints and cancer. Nature **432**, 316–323.
33. Costanzo, V. and Gautier, J. (2003) Single-strand DNA gaps trigger an ATR- and Cdc7-dependent checkpoint. Cell Cycle **2**, 17.
34. Kim, J. M., Yamada, M. and Masai, H. (2003) Functions of mammalian Cdc7 kinase in initiation/monitoring of DNA replication and development. Mutat Res **532**, 29–40.
35. Liu, Q., Guntuku, S., Cui, X. S., Matsuoka, S., Cortez, D., Tamai, K., Luo, G., Carattini-Rivera, S., DeMayo, F., Bradley, A., Donehower, L. A. and Elledge, S. J. (2000) Chk1 is an essential kinase that is regulated by Atr and required for the G(2)/M DNA damage checkpoint. Genes Dev **14**, 1448–1459.
36. Busino, L., Chiesa, M., Draetta, G. F. and Donzelli, M. (2004) Cdc25A phosphatase: combinatorial phosphorylation, ubiquitylation and proteolysis. Oncogene **23**, 2050–2056.
37. Zhao, S., Weng, Y. C., Yuan, S. S., Lin, Y. T., Hsu, H. C., Lin, S. C., Gerbino, E., Song, M. H., Zdzienicka, M. Z., Gatti, R. A., Shay, J. W., Ziv, Y., Shiloh, Y. and Lee, E. Y. (2000) Functional link between ataxia-telangiectasia and Nijmegen breakage syndrome gene products. Nature **405**, 473–477.
38. Buscemi, G., Savio, C., Zannini, L., Micciche, F., Masnada, D., Nakanishi, M., Tauchi, H., Komatsu, K., Mizutani, S., Khanna, K., Chen, P., Concannon, P., Chessa, L. and Delia, D. (2001) Chk2 activation dependence on Nbs1 after DNA damage. Mol Cell Biol **21**, 5214–5222.
39. Yarden, R. I., Pardo-Reoyo, S., Sgagias, M., Cowan, K. H. and Brody, L. C. (2002) BRCA1 regulates the G2/M checkpoint by activating Chk1 kinase upon DNA damage. Nat Genet **30**, 285–289.
40. O'Connell, M. J., Walworth, N. C. and Carr, A. M. (2000) The G2-phase DNA-damage checkpoint. Trends Cell Biol **10**, 296–303.
41. Garner, E. and Costanzo, V. (2009) Studying the DNA damage response using in vitro model systems. DNA Repair (Amst) **8**, 1025–1037.
42. Smith, E. and Costanzo, V. (2009) Responding to chromosomal breakage during M-phase: insights from a cell-free system. Cell Div **4**, 15.
43. Cimprich, K. A. (2003) Fragile sites: breaking up over a slowdown. Curr Biol **13**, R231-R233.
44. Costanzo, V., Shechter, D., Lupardus, P. J., Cimprich, K. A., Gottesman, M. and Gautier, J. (2003) An ATR- and Cdc7-dependent DNA damage checkpoint that inhibits initiation of DNA replication. Mol Cell **11**, 203–213.
45. MacDougall, C. A., Byun, T. S., Van, C., Yee, M. C. and Cimprich, K. A. (2007) The structural determinants of checkpoint activation. Genes Dev **21**, 898–903.
46. Burden, D. A. and Osheroff, N. (1998) Mechanism of action of eukaryotic topoisomerase II and drugs targeted to the enzyme. Biochim Biophys Acta **1400**, 139–154.
47. Sarkaria, J. N., Busby, E. C., Tibbetts, R. S., Roos, P., Taya, Y., Karnitz, L. M. and Abraham, R. T. (1999) Inhibition of ATM and ATR kinase activities by the radiosensitizing agent, caffeine. Cancer Res **59**, 4375–4382.

48. Chong, J. P., Mahbubani, H. M., Khoo, C. Y. and Blow, J. J. (1995) Purification of an MCM-containing complex as a component of the DNA replication licensing system. Nature **375**, 418–421.
49. Pichierri, P. and Rosselli, F. (2004) Fanconi anemia proteins and the s phase checkpoint. Cell Cycle **3**, 698–700.
50. Pichierri, P. and Rosselli, F. (2004) The DNA crosslink-induced S-phase checkpoint depends on ATR-CHK1 and ATR-NBS1-FANCD2 pathways. Embo J **23**, 1178–1187.
51. Cook, J. G. (2009) Replication licensing and the DNA damage checkpoint. Front Biosci **14**, 5013–5030.
52. Melo, J. and Toczyski, D. (2002) A unified view of the DNA-damage checkpoint. Curr Opin Cell Biol **14**, 237–245.
53. Kumagai, A., Guo, Z., Emami, K. H., Wang, S. X. and Dunphy, W. G. (1998) The Xenopus Chk1 protein kinase mediates a caffeine-sensitive pathway of checkpoint control in cell-free extracts. J Cell Biol **142**, 1559–1569.
54. Lupardus, P. J., Byun, T., Yee, M. C., Hekmat-Nejad, M. and Cimprich, K. A. (2002) A requirement for replication in activation of the ATR-dependent DNA damage checkpoint. Genes Dev **16**, 2327–2332.
55. Fernandez-Capetillo, O., Celeste, A. and Nussenzweig, A. (2003) Focusing on foci: H2AX and the recruitment of DNA-damage response factors. Cell Cycle **2**, 426–427.
56. Paull, T. T., Rogakou, E. P., Yamazaki, V., Kirchgessner, C. U., Gellert, M. and Bonner, W. M. (2000) A critical role for histone H2AX in recruitment of repair factors to nuclear foci after DNA damage. Curr Biol **10**, 886–895.
57. Burma, S., Chen, B. P., Murphy, M., Kurimasa, A. and Chen, D. J. (2001) ATM phosphorylates histone H2AX in response to DNA double-strand breaks. J Biol Chem **276**, 42462–42467.
58. Rogakou, E. P., Pilch, D. R., Orr, A. H., Ivanova, V. S. and Bonner, W. M. (1998) DNA double-stranded breaks induce histone H2AX phosphorylation on serine 139. J Biol Chem **273**, 5858–5868.
59. Fernandez-Capetillo, O., Chen, H. T., Celeste, A., Ward, I., Romanienko, P. J., Morales, J. C., Naka, K., Xia, Z., Camerini-Otero, R. D., Motoyama, N., Carpenter, P. B., Bonner, W. M., Chen, J. and Nussenzweig, A. (2002) DNA damage-induced G2-M checkpoint activation by histone H2AX and 53BP1. Nat Cell Biol **4**, 993–997.
60. Costanzo, V., Robertson, K., Bibikova, M., Kim, E., Grieco, D., Gottesman, M., Carroll, D. and Gautier, J. (2001) Mre11 protein complex prevents double-strand break accumulation during chromosomal DNA replication. Mol Cell **8**, 137–147.
61. Dupre, A., Boyer-Chatenet, L. and Gautier, J. (2006) Two-step activation of ATM by DNA and the Mre11-Rad50-Nbs1 complex. Nat Struct Mol Biol **13**, 451–457.
62. Ogawa, H., Johzuka, K., Nakagawa, T., Leem, S. H. and Hagihara, A. H. (1995) Functions of the yeast meiotic recombination genes, MRE11 and MRE2. Adv Biophys **31**, 67–76.
63. Haber, J. E. (1998) The many interfaces of Mre11. Cell **95**, 583–586.
64. Shechter, D., Costanzo, V. and Gautier, J. (2004) ATR and ATM regulate the timing of DNA replication origin firing. Nat Cell Biol **6**, 648–655.

# Chapter 11

# Analyzing DNA Replication Dynamics of Genotoxin-Treated Cells Using Velocity Sedimentation

**Tovah A. Day, Chris Sproul, Marila Cordeiro-Stone, and Cyrus Vaziri**

## Abstract

Following acquisition of DNA damage S-phase progression may potentially be affected via multiple mechanisms. For example DNA damage-activated signal transduction pathways negatively regulate the initiation of DNA synthesis at unfired origins of replication, a process termed the 'S-phase checkpoint' or the 'intra-S-phase checkpoint'. Additionally, many DNA lesions pose physical barriers to replication forks and therefore inhibit DNA synthesis directly by blocking the elongation of active replicons. Inhibition of DNA synthesis in response to DNA damage is commonly assayed by measuring incorporation of radiolabeled or halogenated nucleotides into bulk genomic DNA. However, these techniques do not distinguish between effects of DNA damage on initiation and elongation phases of DNA synthesis. The velocity sedimentation protocol described here allows investigators to determine the effects of DNA damage on initiation and elongation events. This technique involves labeling replicating DNA with $^{3}$H-thymidine, then analyzing the size distribution of labeled ssDNAs based on their differential density sedimentation profiles after centrifugation through alkaline sucrose gradients. Determining the relative abundance and growth rates of small nascent ssDNAs provides an index of initiation and elongation events, respectively. Therefore, analysis of replication dynamics using velocity sedimentation provides a potentially valuable tool for assaying S-phase checkpoints as well as other aspects of DNA replication.

**Key words:** DNA damage, DNA replication, S-phase checkpoint, ATR, Chk1, initiation of DNA replication.

## 1. Introduction

In recent years there has been considerable interest in elucidating biological responses to various forms of DNA damage. S-phase 'checkpoints' are of particular interest, having emerged as potentially important tumor-suppressive mechanisms that integrate

W.X. Li (ed.), *Cell Cycle Checkpoints*, Methods in Molecular Biology 782,
DOI 10.1007/978-1-61779-273-1_11, © Springer Science+Business Media, LLC 2011

DNA replication with DNA repair to maintain genome stability (1, 2).

Historically, the S-phase checkpoint has been viewed as a signaling mechanism that inhibits the initiation stage of DNA synthesis in response to acquisition of DNA damage. It is now clear that different forms of DNA damage trigger distinct S-phase checkpoint signaling pathways. For example, bulky DNA lesions including Benzo[a]pyrene-Dihydrodiol Epoxide (BPDE)-N2-dG adducts and ultraviolet radiation-induced lesions (such as Cyclobutane Pyrimidine Dimers or CPD) cause uncoupling of replicative helicase and polymerase activities, triggering an S-phase checkpoint that is mediated via the ATR and Chk1 protein kinases (2, 3). In contrast, the S-phase checkpoint induced by DNA Double Stranded Breaks (DSB) is ATM and Chk2-mediated (2, 3).

Effects of DNA damage on S-phase cells are commonly studied using the 'radioresistant DNA synthesis' (RDS) assay, which measures rates of total DNA synthesis. Potentially, DNA damage can affect both initiation and elongation phases of DNA replication yet the RDS assay provides no specific information on the effects of DNA damage on replicon initiation and DNA chain elongation. Therefore, to obtain a more complete understanding of the effects of DNA damage on S-phase dynamics, many investigators have studied the effects of DNA damage on elongation rates of bulk and newly synthesized DNA using velocity sedimentation to separate replication intermediates by size (4–7).

Collectively, such studies have determined that different doses of genotoxins inhibit initiation and elongation phases of DNA synthesis via distinct mechanisms. Thus, low concentrations of the carcinogen BPDE (<100 nM) or low fluences of UV radiation (0.5–1 J/m$^2$) transiently inhibit the formation of small nascent DNAs (representing initiation events) but do not affect global elongation of large DNA chains representing ongoing replicons (5, 7). Experimentally interfering with ATR/Chk1 checkpoint signaling can ablate the effects of low doses of BPDE and UV on initiation of replication (8, 9). Therefore, inhibition of the initiation of DNA synthesis induced by low doses of BPDE or UV is mediated by the S-phase checkpoint. In contrast, higher concentrations of BPDE (200–600 nM) or UV (>1 J/m$^2$) inhibit both initiation and elongation steps of DNA replication resulting in a global block to DNA synthesis which is not ATR/Chk1-mediated and instead results from polymerase stalling by DNA lesions (9). This chapter describes a protocol that may be used for distinguishing effects of various experimental manipulations (including but not limited to genotoxin treatments) on global initiation and elongation events.

McGrath and Williams pioneered techniques for velocity sedimentation of DNA in alkaline gradients (10, 11). Essentially the

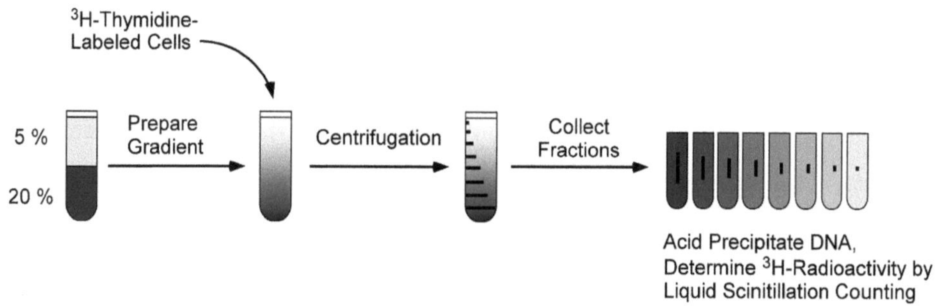


Fig. 1. Separation of different lengths of unwound nascent ssDNAs based on their differential sedimentation in alkaline sucrose density gradients. See text for details.

technique allows unwinding of DNA and analysis of the molecular weights of the resulting single-stranded DNA (ssDNA) by sedimentation in sucrose gradients. Numerous investigators have used variations of this basic technique to investigate replication and repair dynamics of viruses, bacteria, and mammalian genomes. The most commonly employed protocols involve incubating cells with radiolabeled thymidine to label genomic DNA. The labeled cells may be subject to various experimental manipulations such as genotoxin treatments or changes in genetic background. After harvest, the cells are placed directly onto an alkaline solution leading to immediate lysis and unwinding of DNA. Unwound ssDNAs are centrifuged through a sucrose gradient and separated on the basis of their length due to differential density sedimentation (**Fig. 1**). Analysis of size distribution of radiolabeled ssDNAs has provided considerable insight into mechanisms by which initiation and elongation phases of DNA are affected by genotoxins and genetic background.

The protocol described below is one we use routinely to examine the effects of genotoxins on initiation and elongation events in cultured mammalian cells. This protocol was adapted from methods originally published by the Kaufmann and Cordeiro-Stone laboratories (7).

## 2. Materials

### 2.1. Cell Culture and Lysis

1. Thymidine, [Methyl-14C], 50–62 mCi/mmol (1.85–2.29 GBq/mmol), concentration: 0.05 mCi/mL (1.85 MBq/mL), Perkin Elmer Product number NEC568D050UC.
2. Thymidine, [Methyl-3H], 23–27 Ci/mmol (851–999 MBq/mmol), concentration: 1.0 mCi/mL

(37 MBq/mL), Perkin Elmer Product number Product number: NET027L001MC.

3. Gradient solution: 0.4 M sodium hydroxide (NaOH), 2.0 M sodium chloride (NaCl), 0.01 M ethylenediamine tetraacetic acid (EDTA).
4. Scrape solution: 0.1 M sodium chloride, 0.01 M EDTA (pH 8.0).
5. Lysis buffer: 0.5 M NaOH, 0.1 M EDTA.
6. Carrier DNA solution: 500 μg/mL salmon sperm DNA, 0.4 M NaOH.
7. 1.5 M hydrochloric acid (HCl), 6% anhydrous sodium pyrophosphate ($Na_4P_2O_7$).
8. 0.5 M HCl, 6% $Na_4P_2O_7$.
9. 70% ethanol.
10. 95% ethanol.

### 2.2. Equipment

1. Cell scrapers
2. Ultracentrifuge
3. SW28 Rotor or equivalent
4. Beckman polyallomer centrifuge tubes (Fisher cat. # 326823 or equivalent)
5. Puncture apparatus, pump, fraction collector
6. Collection tubes (Fisher cat. # 14-958-E)
7. Vacuum manifold
8. GF/C Microfibre filters (Fisher cat. # 09-87432A)
9. Scintillation vials and fluid
10. Scintillation counter

## 3. Methods

In the experiment described here, replicate plates of H1299 cells are incubated with $^{14}$C-thymidine to label bulk genomic DNA (day 1). After a 'chase' period in $^{14}$C-thymidine-free medium the cells are treated with genotoxin (or left untreated for controls) and then pulse-labeled with $^{3}$H-thymidine (day 3). In cells that do not receive genotoxin, the $^{3}$H-thymidine is incorporated into both newly initiated DNAs and elongating replicons. The relative incorporation of labeled thymidine into small nascent ssDNAs (initiation events) and larger elongating strands will be evident by the size distribution of $^{3}$H-labeled DNAs within the sucrose gradients (day 4). Activation of the S-phase checkpoint (using 50 nM BPDE) reduces the incorporation of $^{3}$H-thymidine into nascent DNAs without affecting incorporation into larger (elongating) DNAs. Higher doses of genotoxins (e.g., 600 nM BPDE) inhibit incorporation of $^{3}$H-thymidine into larger DNA fragments (indicating reduced elongation).

### 3.1. Day 1

1. H1299 cells are plated into 60-mm plates (two plates for each condition) with 4 mL of media containing 10 nCi/mL $^{14}$C-thymidine.
2. Cells are incubated with $^{14}$C-thymidine for 30–40 h (approximately two generations) to label the genomic DNA. The total $^{14}$C radioactivity recovered from the gradients will eventually be used as a measure of the number of cells applied to the gradients. The appropriate cell density on day 1 will depend on the cell line. Fast-growing cancer cells should be plated at lower density to avoid over-confluence at the time of genotoxin treatment (day 3). Recommended plating densities for representative cancer and primary cells are shown in the table below.

| Cells | Cells/60 mm plate on day 1 | Approx. confluence on day 3 |
|---|---|---|
| HeLa, H1299, etc. | 50,000 | 30–40% |
| HDF | 250,000 | 30–40% |

### 3.2. Day 2

Radioactive medium is removed and replaced with 4 mL unlabeled medium. (The cold chase must be performed for at least 3 h prior to genotoxin treatment and cell harvesting).

Meanwhile, one sucrose gradient is prepared for each sample:

1. Centrifuge tubes are labeled before pouring to avoid disturbing the gradient.
2. 17.5 mL of 20% sucrose gradient solution is placed in the bottom of each Beckman centrifuge tube.
3. 17.5 mL of a 5% sucrose gradient solution is gently layered dropwise on top of the 20% sucrose layer to create two distinct phases. An interface line will be evident at the boundary of the two phases.
4. The tubes are sealed with rubber stoppers and placed in racks (we routinely use the disposable styrofoam package supplied with 50-mL falcon tubes). The racks/tubes are gently tilted until they are horizontal and allowed to equilibrate overnight.

### 3.3. Day 3

1. The volume of growth medium for each plate of cells is reduced to 1.5 mL. It is best to avoid adding new media since this may affect replication parameters.
2. Cells are treated with the genotoxin (two plates per condition) and returned to the incubator for 1 h.
3. $^3$H-thymidine is added to each plate of cells. We typically use 10 μCi/mL $^3$H-thymidine for fast-growing cells (e.g., H1299 and other cancer cell lines) or 20 μCi/mL $^3$H-thymidine for slower growing cells (e.g., primary human dermal fibroblasts). Plates are returned to the incubator for 15 min.
4. Radioactive medium is removed from each plate and discarded. Monolayers are washed twice with PBS.
5. 500-μL Scrape solution is added to each plate and cells are scraped with a rubber policeman. The solution containing scraped cells is transferred to a 1.5-mL microcentrifuge tube and stored on ice.
6. The tubes containing the equilibrated gradients are gently placed upright and the rubber stoppers are removed carefully (to avoid disrupting the gradients).
7. 500 μL of lysis buffer is gently layered on top of each gradient. 500 μL of cells in Scrape solution are layered on top of the lysis buffer.
8. The loaded gradient is stored either at room temperature for 3 h under fluorescent lighting or at 4°C for 12–15 h under fluorescent lighting. As noted in 'Troubleshooting' below, the time of exposure to fluorescent light must be determined empirically since the lighting in every lab will be different. Exposure to fluorescent light is a critical step that helps nick longer parental strands of DNA (yet does not significantly affect shorter growing $^3$H-labeled strands). As an alternative to fluorescent light, some investigators perform X-irradiation to fragment parental DNA (12). Unless fragmented, the parental DNA strands become entangled with the shorter growing DNA strands of interest, precluding efficient separation in the alkaline sucrose gradients (*see* **Note 1** below for examples of poor separation of nascent and parental DNAs).

### 3.4. Day 3 or 4

1. The loaded gradients (day 3, Step 9 above) are centrifuged at 25,000 RPM in SW28 rotor (or equivalent) for 4 h at 20°C. After centrifugation approximately twenty-nine 1-mL fractions are collected from each gradient. In our lab we puncture the bottom of the tube and collect fractions using a BR-184 Density Gradient Fractionator device purchased from Brandel. The needle used to puncture the bottom of

the tube is connected in sequence to a peristaltic pump and fraction collector with tubing). Fractions collected may be stored at 4°C in the dark for 2–3 days before proceeding to acid precipitation in the next step.

2. For acid precipitation, tubes containing individual fractions are kept on ice. 600 μL of carrier DNA solution and 2.0 mL of 1.5 M HCl/6% sodium pyrophosphate are added to each fraction.
3. Acid precipitation of the fractions is allowed to continue for 20 min on ice.
4. Precipitated DNA is collected on GF/C microfibre filters using a vacuum manifold. For each fraction, one filter is placed on the manifold and wet with 0.5 M HCl/6% sodium pyrophosphate. The contents of each fraction (containing a precipitation reaction) are poured onto the wet filter. The filter is rinsed once with 95% ethanol to dry. Each filter is then placed in a scintillation vial and dried for at least 1 h before adding scintillation fluid (~4 mL) and scintillation counting.

### 3.5. Data Analysis

The $^{3}H$ radioactivity in each fraction is normalized to the number of cells by dividing by the total number of CPM of $^{14}C$ radioactivity in all the fractions. An alternative to this normalization strategy that is often more convenient is to have an extra plate that is not treated with any radioactive nucleotides for each condition. Cells from this plate are similarly harvested into Scrape solution and the total number of cells is determined by counting with a hemacytometer or coulter counter. The total number of cells determined in this manner can also be used to normalize the $^{3}H$ radioactivity. Normalized $^{3}H$ CPM readings of each fraction are plotted on the *y*-axis versus fraction number on the *x*-axis. By normalizing $^{3}H$ radioactivity to a constant number of cells applied to gradients it is possible to determine changes in quantity of $^{3}H$ incorporated during the pulse, and the sizes of nascent DNA into which the precursor is incorporated. Additionally, the sizes of DNA within the sedimented fractions can be calibrated using radioactively labeled DNAs of known size such as SV 40 or λ DNA (5).

**Figure 2** shows velocity sedimentation profiles from a representative experiment in which H1299 cells were treated with 50 nM BPDE, or left untreated for controls. The results show that 50 nM BPDE specifically reduces the incorporation of $^{3}H$ thymidine into small nascent DNAs (fractions 16–26), demonstrating inhibition of initiation but not elongation under these experimental conditions.

To determine the effect of the genotoxin treatment or genetic manipulation on the elongation of DNA replicons, a chase with

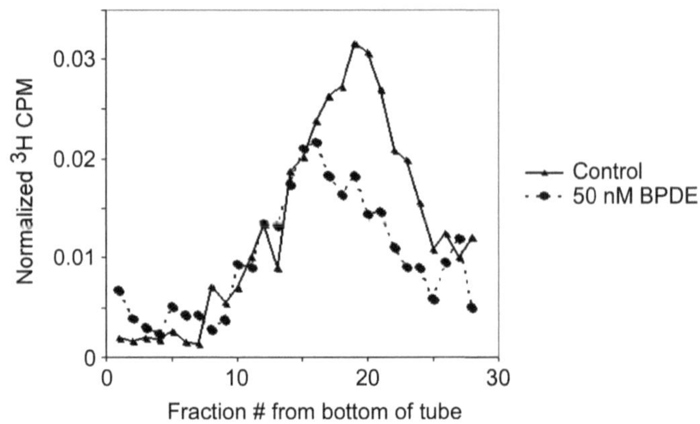


Fig. 2. Velocity sedimentation profiles of nascent $^3$H-labeled DNA from control and BPDE-treated H1299 cells. See text for details.

non-radioactive media may be added in between the $^3$H pulse and the harvest (Steps 3 and 4 of day 3). The chase period can last between 1 and 2 h.

**Figure 3** shows results of an experiment in which H1299 cells were pulse-labeled with $^3$H-thymidine to label nascent DNAs, exactly as described above (**Fig. 2**). Immediately after the incubation with $^3$H-thymidine, the cells were washed with complete to remove unincorporated label. The washed cells were then placed in medium containing 50 nM BPDE, 600 nM BPDE, or no BPDE (for control). Two hours after BPDE treatment, cells were harvested and velocity sedimentation was performed exactly as described above.

As shown in **Fig. 2**, treatment with 50 nM BPDE (a dose of genotoxin that activates the S-phase checkpoint and inhibits DNA synthesis in a Chk1-dependent manner) (9) does not detectably

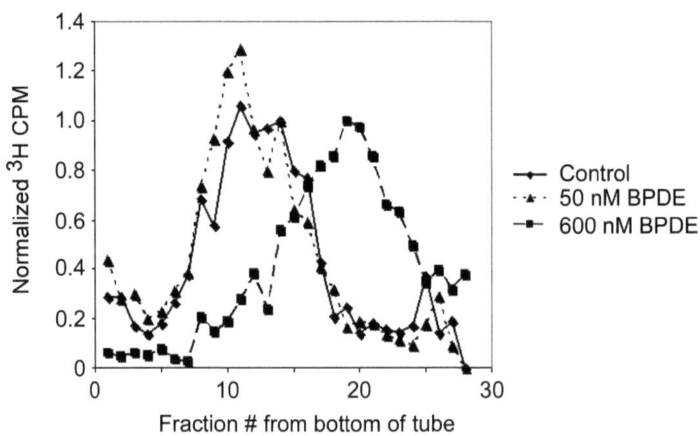


Fig. 3. Velocity sedimentation profiles demonstrating effects of different BPDE concentrations on elongation of nascent $^3$H-labeled DNA. See text for details.

affect the size increase of $^3$H-labeled DNA during the chase period. In contrast, 600 nM BPDE (a dose that inhibits DNA synthesis in a Chk1-independent manner) (9) clearly retards the size increase of $^3$H-labeled DNAs, indicative of elongation defects.

As a result of similar analyses, we and other investigators have concluded that BPDE doses not exceeding 100 nM inhibit DNA synthesis via a Chk1-mediated mechanism that involves inhibition of initiation events. Under our experimental conditions, the inhibition of DNA synthesis elicited by 50 nM BPDE is transient and this dose of genotoxin does not affect cell viability. In contrast BPDE doses exceeding 100 nM inhibit DNA synthesis due to inhibition of initiation and elongation events. H1299 cells fail to resume DNA synthesis following treatment with 600 nM BPDE. Although Chk1 phosphorylation is induced robustly in response to 600 nM

BPDE, the inhibition of DNA synthesis under these conditions is not Chk1-dependent and occurs mainly due to elongation blocks.

In our experience the velocity sedimentation assays described above have been reproducible and relatively problem-free. However, aberrant sedimentation profiles are occasionally obtained, and typically result from poor sucrose gradient quality, inadequate cell lysis, and denaturation of DNA, and poor separation of nascent replicating DNA strands from bulk genomic DNA (*see* **Note 1**).

An alternative technique that is being used increasingly for studying DNA replication dynamics involves measuring tracts of newly replicated DNA (evident due to incorporation of halogenated nucleotide analogues) in combed DNA fibers (13, 14). Therefore the advantages and disadvantages of velocity sedimentation analysis relative to DNA fiber assays are considered below (*see* **Notes 2** and **3**, respectively).

## 4. Notes

1. The quality of the sucrose gradient can be examined by collecting equal fractions from the bottom of the tube to the top and measuring the refractive index of each fraction with a refractometer. The refractive index should decrease linearly from the bottom of the tube to the top. **Figure 4** shows the refractive index of gradients formed after layering 18 mL 5% alkaline sucrose on top of 18 mL 20% alkaline sucrose, placing in a horizontal position, and waiting 4, 6, 8, or 15 h before placing upright and collecting equal fractions from the bottom of the tube. As shown in **Fig. 4**, after 4–6 h, the

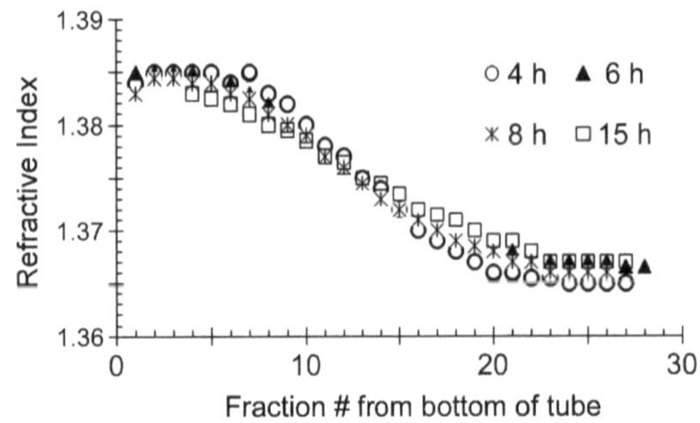


Fig. 4. Refractive index of alkaline sucrose gradient fractions taken from the bottom of the centrifugation tube to the top. Density gradient should be approximately linear.

gradient is linear from fractions 5 to 20, while the fractions on the bottom (1–5) or the top (20–28) have not mixed and do not form a linear gradient. By allowing the gradient to form for 12–15 h, a more uniform linear gradient forms across the length of the entire centrifuge tube.

For successful velocity sedimentation analyses, the DNA must be completely denatured, and it is important that the smaller nascent daughter DNA is completely separated from the full length parental DNA. To facilitate this, the gradient tubes with cells in lysis solution suspended on top must be left under fluorescent lighting to induce DNA breaks and facilitate complete denaturation (as noted above in Step 9 of day 3). The length of time and distance from the lamp(s) required to accomplish this must be determined empirically since the fluorescent lighting in every lab will vary.

If not well lysed and denatured, the $^{14}C$ radioactivity will sediment to the bottom of the tube and carry $^{3}H$-labeled nascent DNA. **Figure 5a** shows the distribution of $^{14}C$ in a sample that was not adequately lysed. The majority of the $^{14}C$-labeled DNA has sedimented to the bottom of the tube in this sample. **Figure 5b** shows distribution of DNA in a sample that lysed adequately, with the large parental genomic DNA sedimenting near the bottom of the tube, but not reaching the end of the gradient.

2. Advantages of Velocity Sedimentation Relative to DNA fiber assays.
   (i) The equipment required for velocity sedimentation is routine and available in most laboratories. In contrast, DNA combing assays require sophisticated microscopes and analytical software to visualize and measure lengths of individual DNA fibers.
   (ii) Velocity sedimentation is very reproducible whereas the quality of DNA fibers obtained by combing is influenced by cell lines as well environmental conditions

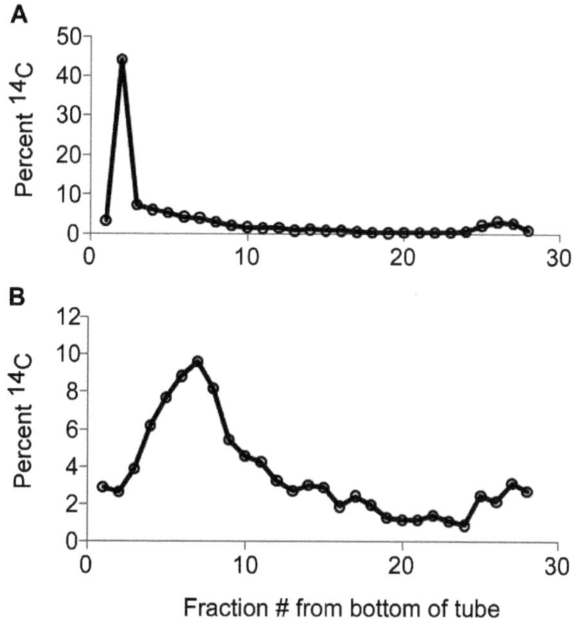


Fig. 5. Effect of inadequate DNA fragmentation on velocity sedimentation of bulk parental DNA. (**a**) Distribution of $^{14}$C-labeled DNA in a sample that was not adequately lysed/denatured. In this example the majority of the $^{14}$C DNA sedimented at the bottom of the tube. (**b**) Distribution of $^{14}$C-labeled genomic DNA in a sample that has been appropriately denatured. In this example the larger $^{14}$C-labeled genomic DNA species sediment near the bottom, but do not reach the end of the gradient.

including humidity and temperature. The quality of DNA fibers on the same slide can be very heterogeneous and experience is needed to identify fibers that are appropriate for analysis.

(iii) The quality of commercial anti-BrdU and CldU antibodies varies tremendously between lots and considerable optimization is needed to avoid cross-reactivity of antibodies between the different halogenated nucleotide analogs.

3. Disadvantages of Velocity Sedimentation Relative to DNA fiber assays.

(i) Velocity sedimentation is labor-intensive, and only a limited number of samples may be analyzed in each experiment (due to limited rotor capacity). In contrast, DNA fibers from numerous samples may be prepared simultaneously and analyzed at later times.

(ii) DNA fiber techniques allow analysis of DNAs at specific loci (which may be identified by FISH with appropriate probes of interest).

(iii) DNA combing allows more sensitive measurements of elongation rates relative to velocity sedimentation.

Thus, using DNA fiber analyses, it has been demonstrated that rates of elongation are sensitive to checkpoint signaling. The effect of checkpoint signaling on elongation rates is below the limit of sensitivity for velocity sedimentation analyses.

(iv) In contrast with velocity sedimentation analyses (which only provide information on initiation and elongation), DNA fiber analyses allow visualization and enumeration of termination events as well as measurements of inter-origin distances.

Despite the pros and cons described above, velocity sedimentation and DNA combing are best viewed as highly complementary approaches for analysis of replication dynamics. Results of studies using both techniques have generally provided consistent and important information regarding mechanisms that regulate DNA replication in response to genotoxins.

## References


1. Zhou, B. B., and Elledge, S. J. (2000) *Nature* **408,** 433–9.
2. Bartek, J., Lukas, C., and Lukas, J. (2004) *Nat Rev Mol Cell Biol* **5,** 792–804.
3. Sancar, A., Lindsey-Boltz, L. A., Unsal-Kacmaz, K., and Linn, S. (2004) *Annu Rev Biochem* **73,** 39–85.
4. Lehmann, A. R. (1972) *Eur J Biochem* **31,** 438–45.
5. Kaufmann, W. K., and Cleaver, J. E. (1981) *J Mol Biol* **149,** 171–87.
6. Cordeiro-Stone, M., Boyer, J. C., Smith, B. A., and Kaufmann, W. K. (1986) *Carcinogenesis* **7,** 1783–6.
7. Cordeiro-Stone, M., Boyer, J. C., Smith, B. A., and Kaufmann, W. K. (1986) *Carcinogenesis* **7,** 1775–81.
8. Heffernan, T. P., Simpson, D. A., Frank, A. R., Heinloth, A. N., Paules, R. S., Cordeiro-Stone, M., and Kaufmann, W. K. (2002) *Mol Cell Biol* **22,** 8552–61.
9. Liu, P., Barkley, L. R., Day, T., Bi, X., Slater, D. M., Alexandrow, M. G., Nasheuer, H. P., and Vaziri, C. (2006) *J Biol Chem.*
10. McGrath, R. A., and Williams, R. W. (1966) *Nature* **212,** 534–5.
11. McGrath, R. A., and Williams, R. W. (1967) *Biophys J* **7,** 309–17.
12. Lehmann, A. R. (1972) *J Mol Biol* **66,** 319–37.
13. Merrick, C. J., Jackson, D., and Diffley, J. F. (2004) *J Biol Chem* **279,** 20067–75.
14. Chastain, P. D., 2nd, Heffernan, T. P., Nevis, K. R., Lin, L., Kaufmann, W. K., Kaufman, D. G., and Cordeiro-Stone, M. (2006) *Cell Cycle* **5,** 2160–7.

# Chapter 12

## Methods for Studying Checkpoint Kinases – Chk1

**Claudia Tapia-Alveal and Matthew J. O'Connell**


### Abstract

Attempts to passage through mitosis with unrepaired DNA damage or incompletely replicated DNA leads to genome instability and/or cell death. To prevent this from occurring, an ancient checkpoint (known as the G2 DNA damage checkpoint) that inhibits the activation of the mitotic cyclin-dependent kinase is activated to hold cells in the G2 phase of the cell cycle. The effector of this checkpoint is Chk1, a protein serine-threonine kinase. Chk1 contains an N-terminal catalytic domain, and C-terminal regulatory domain. Within the regulatory domain there are two residues, Serine-317 (S317) and Serine-345 (S345), which are phosphorylated in active Chk1 molecules, and subsequently dephosphorylated to inactivate Chk1 and allow mitotic entry. Phospho-specific antibodies can be used to detect these activating phosphorylations, and this provides a simple and sensitive marker of Chk1 activation.


**Key words:** Chk1, checkpoint, DNA damage, Phosphorylation, Western blotting.

## 1. Introduction

The G2 DNA damage checkpoint holds cells in interphase to allow DNA repair to be completed prior to mitosis. This p53-independent checkpoint is highly conserved in eukaryotes, being essentially identical from fission yeast to humans. While different forms of DNA damage activate lesion-specific repair pathways, all lesions that occur in S- and G2-phases activate the same checkpoint-signaling cascade, culminating in the activation of its effector kinase, Chk1. Chk1 then directly phosphorylates regulators of the mitotic cyclin-dependent kinase Cdc2, and this ensures that Cdc2 remains in an inactive state (1). Upon completion of repair, Chk1 is inactivated, and the cells move into mitosis (2).

W.X. Li (ed.), *Cell Cycle Checkpoints*, Methods in Molecular Biology 782,
DOI 10.1007/978-1-61779-273-1_12, © Springer Science+Business Media, LLC 2011

Common to the diverse array of lesions that activate Chk1 is single-stranded DNA (ssDNA) that is either a direct result of the lesion, or an intermediate produced by lesion processing. The ssDNA is rapidly coated by replication protein A (RPA), which then acts as a point of assembly for checkpoint protein complexes, including either of the two PI-3 K-related protein kinases (PIKKs), ATM (Ataxia Telangiectasia Mutated) or ATR (ATM and Rad3-related) (3). Inactive Chk1 is recruited into these complexes via interaction with mediator proteins and is then activated by ATM/R-catalyzed phosphorylation in two C-terminal serines in the regulatory domain, S317 (4) and S345 (5). This results in a rapid and reversible activation of Chk1 kinase activity (6), which can be assayed in vitro by transfer of $^{32}P$ to substrate (7–10). However, a far simpler, highly reproducible, and non-radioactive assay is the detection of these phosphorylated residues by Western blotting using commercially available phospho-specific antibodies.

## 2. Materials

### 2.1. Cell Culture and Checkpoint Activation

1. Media (Hepes-buffered Dulbecco's Modified Eagle's Medium), Phosphate-buffered saline (PBS), 1× Trypsin/EDTA, fetal calf serum (Each from Invitrogen, Carlsbad, CA), and tissue-culture grade 10 cm petri dishes (BD Falcon, Franklin Lanes, NJ). *See* **Note 1**.
2. $^{137}Cs$ source irradiator (J.L. Sheppard and Associates, San Fernando, CA). *See* **Note 2**.

### 2.2. Extract Preparation

1. Chk1 extraction buffer: 10 mM Tris–HCl (pH 7.4), 30 mM $Na_4P_2O_7$, 150 mM NaCl, 50 mM NaF, 1% NP-40, 0.5 mM $Na_3VO_4$, 60 mM β-glycerophosphate, 15 mM ρ-nitrophenol phosphate, 1 mM dithiothreitol (DTT), and 1 μg/mL protease inhibitor cocktail (Sigma, St. Louis, MO). *See* **Note 3**.
2. Dc Protein Quantification Kit (BioRad, Hercules, CA). *See* **Note 4**.
3. 5× Sample buffer: Stock is 60 mM Tris–HCl (pH 6.8), 25% glycerol, 2% sodium dodecyl sulfate (SDS), and 0.1% bromophenol blue. Immediately before use add β-mercaptoethanol to final concentration of 5%. All chemicals must be electrophoresis grade.
4. 50-mL Falcon tubes (BD Falcon, Franklin Lanes, NJ).
5. 1.5-mL microfuge tubes (Fisher).

6. Liquid nitrogen in a Dewer Flask.
7. 1-mL syringe fitted with a 26-gauge needle (BD Falcon, Franklin Lanes, NJ).

### 2.3. SDS-Polyacrylamide Gel Electrophoresis (SDS-PAGE)

1. SDS-PAGE apparatus; we use the Mini-Protean system (Bio-Rad, Hercules, CA) and 0.75-mm-thick gels, but any common SDS-PAGE system can substitute. The power supply must be capable of at least 300 mA of power, e.g., Power Pac 300 (Bio-Rad, Hercules, CA).
2. Separating Gel and Stacking gel Solutions: 40% Acrylamide/Bisacrylamide mix (29:1) (Bio-Rad, Hercules, CA), 1.5 M Tris–HCl (pH 8.8), 1.0 M Tris–HCl (pH 6.8), 10% SDS, isopropanol.
3. Gel polymerizing agents, *N,N,N,N'*-tetramethyl-ethylenediamine (TEMED, Bio-Rad, Hercules, CA) and 10% ammonium persulfate. The latter is made in water, but is effervescent – therefore, ensure vessel is large enough to accommodate bubbles. Store at 4°C for up to 1 month.
4. 1× SDS-PAGE running buffer (made from 10× stock): 25 mM Tris, 192 mM glycine, 0.1% (w/v) SDS. All chemicals must be electrophoresis grade.
5. Molecular weight marker – Precision Plus Dual Color (Bio-Rad, Hercules, CA). *See* **Note 4**.

### 2.4. Western Blotting for Activated Chk1

1. Western blotting cassette and sponges. We use the Mini-Protean system (Bio-Rad, Hercules, CA) and wet (submerged) transfer, but any common Western blotting unit (wet or semi-dry) can substitute.
2. 3 mm chromatography paper and nitrocellulose (0.2 μm pore size) (Whatman, Florham Park, NJ) cut slightly larger than the gel.
3. Transfer buffer: 10 mM *N*-cyclohexyl-3-aminopropanesulfonic acid (CAPS), pH 11 (made from 1 M stock stored in the dark), plus 10% (v/v) methanol.
4. 0.1% (w/v) Ponceau S dissolved in 5% (v/v) acetic acid.
5. 1× Tris-buffered saline plus Tween-20 (TBS-T made from 20× stock): 50 mM Tris, 150 mM NaCl, 0.1% Tween-20, pH 7.5.
6. Blocking Buffer; TBS-T plus 5% (w/v) dried skim milk powder – this can be purchased at any supermarket.
7. Primary: Rabbit phospho-Chk1 (Ser317, Cell Signaling Technologies, Danvers, MA) and secondary antibodies (HRP-conjugated Donkey anti-Rabbit IgG, GE Healthcare, Piscataway, NJ). *See* **Note 4**.

8. Enhanced chemiluminescence detection (GE Healthcare, Piscataway, NJ) and autoradiography film (Fuji Super RX from Fisher). *See* **Note 4**.

#### 2.5. Stripping and Reprobing Western Blots for Total Chk1 Levels

1. Stripping buffer: 62.5 mM Tris–HCl, pH 6.8 (made out of 1 M stock), 2% SDS (10% stock), 100 mM 2-mercaptoethanol (made out of 14.2 M stock electrophoresis grade, Sigma, St. Louis, MO).
2. Primary Antibody: Mouse Anti-Chk1 (G4, Santa Cruz Biotechnology, Santa Cruz, CA) and secondary antibody (HRP-conjugated Donkey anti-Mouse IgG, GE Healthcare, Piscataway, NJ). *See* **Note 4**.
3. Enhanced chemiluminescence detection (GE Healthcare, Piscataway, NJ) and autoradiography film (Fuji Super RX from Fisher). *See* **Note 4**.

## 3. Methods

#### 3.1. Cell Culture, Checkpoint Activation, and Extract Preparation

1. Grow cells to 70% density in DMEM + 10% FCS, at 37°C in the presence of 5% $CO_2$. Remove media, wash in PBS, and harvest cells with 1 mL trypsin/EDTA. Inactivate the trypsin with fresh media, and determine the cell density with a hemocytometer or automated cell counter (e.g., Coulter Counter).
2. For each desired sample, plate $1\times10^7$ cells per 10 cm plate in 10 mL DMEM + 10% FCS and grow for 24 h at 37°C. A separate plate is required for each sample; samples can vary for dose of irradiation and time of recovery after irradiation. One additional plate of cells is required for an unirradiated control.
3. Irradiate cells with desired dose of ionizing radiation using the manufacturer's instructions. Robust Chk1 activation is achieved by a dose of 10 Gy. Alternatively, *see* **Note 2**.
4. At the desired timepoints, cells are harvested with trypsin as above. Importantly, the media must be retained to save cells that are not attached, including mitotic and apoptotic cells. Therefore, remove media and save in a 50-mL Falcon tube, wash cells in PBS, but retain that PBS and add to the saved media. Add 1 mL trypsin, and incubate at 37°C until all cells are unattached. Add saved media/PBS to the plate, and recover all cells to a single 50-mL falcon tube.
5. Spin the cells at $2000\times g$ for 2 min. Resuspend in 10 mL PBS, and spin again at $2000\times g$ for 2 min. Remove PBS, and

resuspend cells in 1 mL of PBS, and transfer to a 1.5-mL microfuge tube. Spin at 2000×*g* for 2 min, remove PBS, and snap freeze the pellet in liquid nitrogen. Save at −80°C until extract preparation.

### 3.2. Extract Preparation

1. Resuspend frozen cell pellet (cells from one 10-cm plate) in 250 μL of ice-cold Chk1 extraction buffer. Disrupt the pellet by repeated passage in and out of a 1-mL syringe fitted with a 26-gauge needle. The extract may be very viscous initially, but this will shear the DNA in the extract.
2. Centrifuge the extracts at 16,100×*g* in a microfuge at 4°C for 10 min. Remove 200 μL of the supernatant to a fresh tube, and keep on ice. Discard pellet.
3. Determine the concentration of protein using the instructions provided with the protein assay kit and a standard curve generated with a BSA control.
4. Add 50 μL of 5× sample buffer to each tube, and boil (100°C) for 3 min.
5. Briefly spin the tubes to collect any condensation on the lids.

### 3.3. SDS-PAGE

1. Ensure the glass plates for pouring the gels are clean. If needed, polish with 100% ethanol and allow to air dry.
2. Assemble the glass plates for gel pouring. Details of assembly come with the apparatus. Make sure the bottom is tightly sealed.
3. Wearing gloves, mix the separating gel to a final concentration of 10% acrylamide. This contains (10 mL is sufficient for 2 gels) 2.5 mL 1.5 M Tris–HCl (pH 8.8), 2.5 mL 40% acrylamide/bisacrylamide (29:1), 100 μL 10% Ammonium Persulfate, 100 μL of 10% SDS, and 4.8 mL of water. Once mixed, add 5 μL of TEMED, gently mix and pour the gels avoiding bubbles. Fill the plates until ~2 cm from the top, and gently layer 200 μL of isopropanol on top. Leave undisturbed at room temperature until the excess (unpoured) separating gel polymerizes. *See* **Note 5**.
4. Mix the stacking gel. This contains (5 mL is sufficient for 2 gels) 630 μL of 1 M Tris–HCl (pH 6.8), 630 μL of 40% acrylamide/bisacrylamide (29:1), 50 μL 10% Ammonium Persulfate, 50 μL of 10% SDS, and 3.64 mL of water. Leave at room temperature.
5. Once the separating gel has polymerized, pour off isopropanol, and rinse the top of the gel several times in pure water. Drain out the excess liquid.
6. Add 5 μL of TEMED to the stacking gel mix, pour on top of the separating gel to the top of the glass plates, and insert a 10-well comb.

7. Once the stacking gel is polymerized, remove the comb and rinse the wells with pure water.
8. Assemble the gels in the electrophoresis tank and load 50 μg of each sample. Include a pre-stained molecular weight marker.
9. Run gel at 200 V until the bromophenol blue runs off the bottom of the gel.

### 3.4. Western Blotting for Activated Chk1

1. Prepare 1 L of transfer buffer, the 3 mm paper, and the nitrocellulose.
2. Disassemble glass plates, and transfer the gels into a container (such as a large weigh dish) containing ~50–100 mL of transfer buffer. Incubate at room temperature for 5 min.
3. Pour remaining transfer buffer into a container such that the buffer is approximately 5 cm high. Insert the plastic Western blotting cassette, clear side down. Assemble the Western sandwich in the following order: sponge, 3 mm paper, nitrocellulose, gel, 3 mm paper, and sponge. Close and clamp the cassette. *See* **Note 6**.
4. Ensure the clear side is positioned against the anode (red) side of the transfer apparatus. Transfer at 70 V for 40 min.
5. Disassemble the Western blot, and stain the nitrocellulose in Ponceau S; incubate in Ponceau S for 1 min, and then rinse the blot in pure water. The abundant proteins should be clearly visible, and can be documented for a record of loading and transfer.
6. Transfer the blot to a vessel suitable for blocking and probing the membrane. Options include a small box, sealing in a plastic bag, or placing in a 50-mL Falcon tube that can then roll on a tube roller.
7. Add sufficient blocking buffer (TBS-T plus 5% skim milk) to completely cover the membrane. Incubate with agitation at room temperature for 1 h.
8. Add primary antibody (Rabbit phospho-Chk1 (Ser317, Cell Signaling Technologies, Danvers, MA) at a 1/200 dilution. Incubate at room temperature overnight. *See* **Note 7**.
9. Rinse the blot thoroughly in TBS-T, and then wash five times with TBS-T for 5–10 min at room temperature with agitation.
10. Incubate with blocking buffer containing a 1/2000 dilution of secondary antibody (HRP-conjugated Donkey anti-Rabbit IgG, GE Healthcare, Piscataway, NJ) for 4 h at room temperature with agitation.
11. Wash as in Step 9.

12. Prepare ECL reagents (2 mL per blot is sufficient) and incubate with blot for 1 min at room temperature.
13. Drain off ECL and wrap blot in clear plastic wrap (e.g., Glad Cling Wrap).
14. Expose to X-ray film under a safe light. In most cases, an exposure of 5–10 min is sufficient. Ensure a signal is seen at ~55 kD in the irradiated sample, but not in the unirradiated control.

### *3.5. Stripping and Reprobing Western Blots for Total Chk1 Levels*

1. Disassemble the blot and rinse in TBS-T.
2. Incubate in stripping buffer at 50°C for 30 min. Rinse extensively in TBS-T. *See* **Note 8**.
3. Re-block the blot and process with the non-phospho-specific anti-Chk1 antibody (G4, Santa Cruz Biotechnology, Santa Cruz, CA) diluted 1/100. Incubate overnight at room temperature with agitation.
4. Rinse the blot thoroughly in TBS-T and then wash five times with TBS-T for 5–10 min at room temperature with agitation.
5. Incubate with blocking buffer containing a 1/2000 dilution of secondary antibody (HRP-conjugated Donkey anti-Mouse IgG, GE Healthcare, Piscataway, NJ) for 2 h at room temperature with agitation.
6. Wash as in Step 9.
7. Prepare ECL reagents (2 mL per blot is sufficient), and incubate with blot for 1 min at room temperature.
8. Drain off ECL and wrap blot in clear plastic wrap (e.g., Glad Cling Wrap).
9. Expose to X-ray film under a safe light. In most cases, an exposure of 2–5 min is sufficient. Ensure a signal is seen at ~55 kD in all samples.

## 4. Notes

1. The culture conditions described are standard for many cell lines derived from human tumors and for cultures of fibroblasts of murine origin. The actual culture conditions are not critical for the subsequent assays and other conditions optimized for a particular cell line can be substituted.
2. The source of DNA damage is not important. Other vendors for $^{137}$Cs sources can be used, as can other isotopes. Ionizing radiation can be substituted for any radiomimetic drugs.

We routinely use the Topoisomerase II inhibitor Etoposide (1–2 μM for 12–24 h) as an alternative. However, particularly for cells that express wild-type p53, it is important to harvest cells prior to widespread apoptosis.

3. 0.5 mM $Na_3VO_4$ (sodium orthovanadate) is diluted from a 100-mM stock and must be activated by the following method: dissolve 0.92 g $Na_3VO_4$ in 40 mL of pure water. Adjust pH to 10.0 using HCl. Boil solution for 5 min, cool on ice, and repeat until the solution is colorless. Adjust pH to 10.0 with NaOH. Repeat cycles of boiling and cooling until the solution is again colorless, and the pH stabilizes at 10.0. Bring to 50 mL, aliquot, and freeze at −20°C.
4. There are numerous commercial alternatives for these reagents.
5. Unpolymerized acrylamide is a cumulative neurotoxin, and so gloves must be worn at all times. The time of polymerization of both stacking and separating gels is temperature dependent.
6. By assembling the Western blot submerged in the transfer buffer, air bubbles cannot form between the gel and the nitrocellulose.
7. Although the supplier recommends lower dilution (1/1000), this gives less reproducible results. Choosing the smallest possible vessel for incubation will limit the amount of antibody needed.
8. The 2-mercapethanol in stripping buffer has a very strong and harmful odor. Stripping should be done in a sealed container. Ensure rinsing is extensive enough to leave no detectable odor.

## Acknowledgments

This work was supported by NIH grant GM087326.

### References

1. Kuntz, K., and O'Connell, M. J. (2009) The G(2) DNA damage checkpoint: Could this ancient regulator be the achilles heel of cancer? *Cancer Biol Ther 8*, 1433–1439.
2. Calonge, T. M., and O'Connell, M. J. (2008) Turning off the G2 DNA damage checkpoint, *DNA Repair (Amst) 7*, 136–140.
3. O'Connell, M. J., and Cimprich, K. A. (2005) G2 damage checkpoints: what is the turn-on? *J Cell Sci 118*, 1–6.
4. Gatei, M., Sloper, K., Sorensen, C., Syljuasen, R., Falck, J., Hobson, K., Savage, K., Lukas, J., Zhou, B. B., Bartek, J., and Khanna, K. K. (2003) Ataxia-telangiectasia-

mutated (ATM) and NBS1-dependent phosphorylation of Chk1 on Ser-317 in response to ionizing radiation, *J Biol Chem 278*, 14806–14811.

5. Liu, Q., Guntuku, S., Cui, X. S., Matsuoka, S., Cortez, D., Tamai, K., Luo, G., Carattini-Rivera, S., DeMayo, F., Bradley, A., Donehower, L. A., and Elledge, S. J. (2000) Chk1 is an essential kinase that is regulated by Atr and required for the G(2)/M DNA damage checkpoint, *Genes Dev 14*, 1448–1459.
6. Tapia-Alveal, C., Calonge, T. M., and O'Connell, M. J. (2009) Regulation of Chk1, *Cell Div 4*, 8.
7. Capasso, H., Palermo, C., Wan, S., Rao, H., John, U. P., O'Connell, M. J., and Walworth, N. C. (2002) Phosphorylation activates Chk1 and is required for checkpoint-mediated cell cycle arrest, *J Cell Sci 115*, 4555–4564.
8. Harvey, S. H., Sheedy, D. M., Cuddihy, A. R., and O'Connell, M. J. (2004) Coordination of DNA damage responses via the Smc5/Smc6 complex, *Mol Cell Biol 24*, 662–674.
9. Latif, C., Elzen, N. R., and O'Connell, M. J. (2004) DNA damage checkpoint maintenance through sustained Chk1 activity, *J Cell Sci 117*, 3489–3498.
10. Mochida, S., Esashi, F., Aono, N., Tamai, K., O'Connell, M. J., and Yanagida, M. (2004) Regulation of checkpoint kinases through dynamic interaction with Crb2, *Embo J 23*, 418–428.

# Chapter 13

# A Human Cell Extract-Based Assay for the Activation of ATM and ATR Checkpoint Kinases

**Bunsyo Shiotani and Lee Zou**

## Abstract

The *Ataxia telangiectasia*-mutated (ATM) and the ATM-Rad3-related (ATR) kinases are master regulators of the DNA damage-signaling pathways that respond to a wide variety of DNA damage. In this chapter, we describe an in vitro biochemical assay to study the activation of ATM and ATR by double-stranded DNA breaks (DSBs) (Shiotani and Zou, 2009, *Mol Cell 33*, 547–58). In this assay, DNA fragments with different structural features are used to activate ATM and ATR in human cell extracts, and the activation of ATM and ATR is monitored by the phosphorylation of specific ATM and ATR substrates. Importantly, in this assay both ATM and ATR are activated in a DNA structure-regulated manner, providing a useful tool to characterize the DNA structural determinants for their activation. The four primary steps of this assay are as follows: (1) preparation of nuclear extracts from cultured human cells; (2) generation of various DNA fragments using DNA oligonucleotides or plasmids; (3) incubation of DNA fragments in extracts; (4) analysis of the phosphorylation of specific ATM or ATR substrates.

**Key words:** Checkpoint, ATM, ATR, DNA breaks.

## 1. Introduction

Genomic integrity is constantly challenged by both extrinsic and intrinsic genotoxic stresses. To safeguard the genome, eukaryotes have evolved a sophisticated DNA damage-signaling pathway to regulate and coordinate the cellular processes that are important for the maintenance of genome stability (1, 2). These processes include, but are not limited to, DNA replication, DNA repair, cell cycle progression, apoptosis, and senescence. The DNA damage-signaling pathway, which is often referred to as the checkpoint, is comprised of sensors, transducers, and effectors (3). The initial

W.X. Li (ed.), *Cell Cycle Checkpoints*, Methods in Molecular Biology 782,
DOI 10.1007/978-1-61779-273-1_13, © Springer Science+Business Media, LLC 2011

step of checkpoint response is the recognition of DNA damage by sensors, which is followed by the activation of the signal-initiating kinases, ATM and ATR. ATM and ATR are two protein kinases related to the phosphoinositide-3 kinase. Although ATM and ATR share many of their substrates and some of their functions, they have clearly distinct DNA damage specificities. ATM is primarily activated by DSBs caused by genotoxic stresses such as ionizing radiation (IR) and radiomimetics, whereas ATR responds to a wide variety of DNA damage including DSBs and different types of DNA replication problems (4, 5).

The activation of ATM by DSBs is regulated by the Mre11-Rad50-Nbs1 (MRN) complex, a putative sensor of DSBs. In vitro, MRN directly associates with DSBs and activates ATM (6). During ATM activation, ATM is autophosphorylated at Ser1981, leading to the conversion of ATM oligomers to monomers (7). After the initial ATM activation, the function of ATM at DSBs is further regulated by chromatin-mediated mechanisms involving H2AX, Mdc1, and other factors (1). Like ATM, ATR is also activated by DSBs. Replication protein A (RPA)-coated single-stranded DNA (ssDNA), a common protein–DNA structure induced by DSBs and replication stress, plays an important role in ATR activation (8). RPA-coated ssDNA directly interacts with ATRIP, the regulatory partner of ATR, and recruits the ATR-ATRIP complex to sites of DNA damage (9). In the presence of RPA, the RFC-like Rad17 complex recognizes the junctions between single- and double-stranded DNA and recruits the PCNA-like Rad9-Rad1-Hus1 (9-1-1) complex onto damaged DNA (10, 11). The co-localization of 9-1-1 complexes and ATR-ATRIP on damaged DNA may enable TopBP1, a protein that interacts with Rad9, to activate the ATR kinase (12, 13).

Despite the many details of ATM and ATR activation that have been revealed by previous studies, these two kinases have rarely been studied together. Exactly how the DNA damage specificities of ATM and ATR differ from each other was unclear. To address this important question, we sought to develop an in vitro assay in which both ATM and ATR can be activated in a DNA structure-regulated manner. We decided to focus on DSBs in this assay because it is a unique context in which both ATM and ATR are activated (**Fig. 1**; and *see* **Fig. 3**). Using the assay that we developed, we found that the activation of ATM and ATR by DSBs is orchestrated by the single-stranded overhangs (SSOs) of DNA breaks (14). ATM is activated by DSBs with no SSOs or short SSOs, and the lengthening of SSOs promotes an ATM-to-ATR switch at DSBs. Here, we will describe the protocols that we used to analyze the activation of ATM and ATR by various defined DNA structures. This protocol can be adopted to study how the activation of ATM and ATR is regulated by various DNA damage sensors, DNA repair proteins, and chromatin structures.

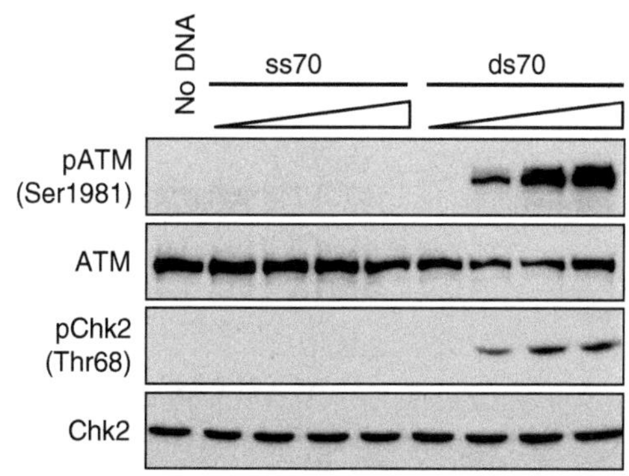


Fig. 1. Double-stranded DNA-induced ATM activation in human cell extracts. Increasing concentrations (1.25, 12.5, 125, and 1250 nM) of ssDNA (ss70) or dsDNA (ds70) was incubated in HeLa cell nuclear extracts. The levels of phospho-ATM (Ser1981), ATM, phospho-Chk2 (Thr68), and Chk2 were analyzed by Western blotting.

## 2. Materials

### *2.1. Preparation of Nuclear Extracts from Human Cells*

1. Dulbecco's Modified Eagle's Medium (DMEM, Fisher) supplemented with 10% fetal bovine serum (FBS, Valley Biomedical).
2. Trypsin-ethylenediamine tetra-acetic acid (EDTA, 0.05%) from Invitrogen.
3. Phosphate-buffered saline (PBS, Invitrogen).
4. Exponentially growing human cells (*see* **Note 1**).
5. Buffer A: 10 mM HEPES–NaOH (pH 7.9), 10 mM KCl, and 1.5 mM $MgCl_2$. Store at room temperature. Add 0.5 mM DTT, 0.5 mM PMSF, and 1× Protease inhibitor cocktail (Sigma, *see* **Note 2**) immediately before use.
6. Buffer C: 20 mM HEPES–NaOH (pH 7.9), 600 mM KCl, 1.5 mM $MgCl_2$, 0.2 mM EDTA, and 25% glycerol. Store at room temperature. Add 0.5 mM DTT, 0.5 mM PMSF, and 1× Protease inhibitor cocktail immediately before use.
7. Buffer D: 20 mM HEPES–NaOH (pH 7.9), 100 mM KCl, 0.2 mM EDTA, and 20% glycerol. Store at room temperature. Add 0.5 mM DTT and 0.5 mM PMSF immediately before use.
8. Bradford Dye reagent (Bio-Rad) to estimate protein concentration of nuclear extracts.
9. Slide-A-Layzer® Dialysis Cassette 3500 MWCO (Thermo scientific).
10. Needle (18 G, Becton Dickinson).
11. Syringe (10 mL, Becton Dickinson).

### 2.2. Generation of DNA Fragments

1. To generate blunt-ended double-stranded DNA (dsDNA) fragments, the following DNA oligonucleotides were synthesized by Invitrogen.
   ds70-1:
   5′-TGCAGCTGGCACGACAGGTTTTAATGAATCGGCC AACGCGCGGGGAGAGGCGGTTTGCGTATTGGGCG CT-3′
   ds70-2:
   3′-ACGTCGACCGTGCTGTCCAAAATTACTTAGCCGG TTGCGCGCCCCTCTCCGCCAAACGCATAACCCGCG A-5′.
2. Annealing buffer (10×): 100 mM Tris–HCl (pH 7.5) and 1 M NaCl.
3. Plasmids that can be linearized with a single restriction enzyme (i.e., Hpa I) to generate blunt-ended dsDNA fragments.
4. T7 exonuclease (5′ to 3′ resection) and Exonuclease III (3′ to 5′ resection) from New England Biolabs.
5. Agarose gel electrophoresis apparatus.

### 2.3. Activation of ATM and ATR in Extracts

1. Nuclear extracts from **Section 2.1**.
2. HEPES–KOH (1 M, pH 7.6).
3. KCl (2.5 M).
4. PMSF (100 mM) dissolved in ethanol. Store at –20°C.
5. DTT (1 M). Store at –20°C.
6. Protease inhibitor cocktail from Sigma. Store at –20°C.
7. ATP (100 mM). Store at –20°C.
8. $MgCl_2$ (10 mM).
9. Creatine kinase (1 mg/mL, Sigma). Store at –20°C.
10. Phosphocreatine (250 mM, Sigma). Store at –20°C.
11. Reaction buffer (4×): 40 mM HEPES–KOH (pH 7.6), 200 mM KCl, 4 mM PMSF, 2 mM DTT, and 1× protease inhibitor cocktail. Prepare this buffer immediately before use.
12. ATP buffer (5×): 5 mM ATP, 0.5 mM $MgCl_2$, 50 μg/mL Creatine kinase, and 25 mM phosphocreatine. Prepare this buffer immediately before use.
13. Laemmli buffer (6×): 0.35 M Tris–HCl (pH 6.8), 36% glycerol, 11% SDS, 0.6 M DTT, and 0.012% bromophenol blue. Store at –20°C.
14. Oligonucleotide- or plasmid-derived DNA fragments from **Section 2.2**.

15. KU55933 (10 mM, ATM inhibitor from KuDOS) dissolved in DMSO. Store at −20°C.
16. NU7026 (10 mM, DNA-PK inhibitor from KuDOS) dissolved in DMSO. Store at −20°C.

### 2.4. Detection of Substrate Phosphorylation by Western Blotting

1. SDS-PAGE apparatus
2. Precision Plus Protein™ Standards (Bio-rad)
3. Transfer apparatus (Owel)
4. Transfer buffer: 48 mM Tris and 390 mM glycine, and 10% methanol
5. Immobilon-P transfer membrane (PVDF, Millipore)
6. Tris-buffered saline with Tween-20 (TBST): 25 mM Tris–HCl (pH 7.6), 137 mM NaCl, 2.7 mM KCl, and 0.1% Tween-20
7. Blocking buffer: 5% Skim milk in TBST
8. Primary antibodies: The antibodies to phospho-Chk2 (Thr68) and phospho-ATM (Ser1981) are from Cell Signaling Technology. The antibodies to ATM and phospho-RPA32 (Ser33) are from the Bethyl Laboratory. The antibody to Chk2 is from Upstate Biotechnology. The antibody to RPA32 is from Thermo Scientific
9. Secondary antibodies: Anti-mouse and anti-rabbit IgG conjugated with horseradish peroxidase (Jackson ImmunoResearch)
10. Western Lightning plus-ECL reagent from PerkinElmer
11. X-OMAT™ Blue XB film from Kodak

## 3. Methods

### 3.1. Preparation of Nuclear Extracts from Human Cells

1. Culture HeLa cells in 150-mm dishes (30 dishes) to 80–90% confluency (*See* **Note 3**).
2. Collect cells by trypsinization and centrifugation (1000 rpm, 3 min at room temperature).
3. Wash the cell pellets with PBS and sediment cells by centrifugation (1000 rpm, 3 min at room temperature).
4. Estimate the cell pellet volume (CPV; approx. 4–5 mL). After this step, the rest of the procedure should be carried out at 4°C.
5. Resuspend cells in 5× CPV of Buffer A and leave the cell suspension on ice for 5 min.

6. Sediment cells by centrifugation (2000 rpm, 3 min at 4°C).
7. Estimate the cell pellet volume again (CPV-A). CPV-A will be approximately twice the volume of the original CPV.
8. Resuspend cells in 2× CPV-A of buffer A.
9. Homogenize cells by Dounce homogenizer (type A pistol, 50 strokes, *see* **Note 4**).
10. Sediment nuclear pellets by centrifugation (4750 rpm, 5 min at 4°C).
11. Estimate the nuclear pellet volume (NPV; approx. 3–4 mL).
12. Resuspend nuclei in 1× NPV of Buffer C (*See* **Note 5**).
13. Leave samples on a shaker at 4°C for 30 min, and mix the samples with occasional tapping.
14. Centrifuge the suspension to obtain nuclear extracts (15,000 rpm 10 min at 4°C).
15. Collect nuclear extract (supernatant) by using a 10-mL syringe with 18-G needle, and transfer the extract into a Slide-A-Layzer® Dialysis Cassette, which is prepared according to manufacture's instructions (*See* **Note 6**)
16. Dialyze the nuclear extracts against Buffer D for 4 h (>100× volume).
17. Collect the dialyzed nuclear extracts using a 10-mL syringe with 18-G needle, centrifuge the samples to remove precipitates, and collect clear supernatant (clarified nuclear extracts).
18. Estimate protein concentration using the Bradford reagent.
19. Add protease inhibitor cocktail to nuclear extracts to the 1× concentration.
20. Aliquot the nuclear extracts.
21. Freeze and keep the samples at −80°C until use (*See* **Note 7**).

### 3.2. Generation of DNA Fragments

Oligonucleotide-derived DNA structures

1. Dissolve ds70-1 and ds70-2 in sterilized water at 250 μM (*See* **Note 8**).
2. Take 1 vol of ds70-1, 1 vol of ds70-2, 1 vol of 10× annealing buffer and 7 vol of sterilized water and mix them in a tube.
3. Incubate the mixture at 95°C for 3 min, 60°C for 3 min, and 37°C for 30 min. This procedure generates dsDNA of 70 bp (ds70) at 25 μM.
4. Dilute ds70 with 1× annealing buffer if required.

5. Use the resulting dsDNA in the in vitro ATM/ATR activation reactions immediately (**Section 3.3**) or store it at –20°C until use (*See* **Note 9** and **Fig. 1**).

   *Plasmid-derived DNA structures:* In this example, a 6-kb plasmid is used to generate dsDNA fragments with SSOs (*See* **Note 10**).

6. Linearize the plasmid with a restriction enzyme that generates blunt ends (i.e., Hpa I). Make sure the restriction enzyme used is a single cutter. The recommended DNA concentration after this digestion step is 400 ng/μL.
7. Before adding DNA to exonucleases, mix exonucleases and the appropriate buffer. T7 exonuclease (10 Units) or Exonuclease III (10 Units) is mixed with 1 μL of the 10× NEB buffer, and water is added to bring the total volume up to 8 μL.
8. Add 2 μL of linearized plasmid (400 ng/μL) to the exonuclease in buffer from Step 7 to start reaction. Final DNA concentration is 80 ng/μL.
9. Incubate the samples for 1–5 min at room temperature (T7) or 37°C (Exo III). The individual reactions are stopped at different time points by dipping the tubes into liquid nitrogen (*See* **Note 11**).
10. To inactivate exonucleases, incubate the samples at 70°C for 20 min.
11. Analyze 2 μL of the resulting DNA structures in 0.8% agarose gels (**Fig. 2**).

### 3.3. In Vitro ATM/ATR Activation Reaction

1. Mix 2.5 μL of 4× reaction buffer and 2 μL of 5× ATP buffer (*see* **Section 2.3**).
2. Add 0.5 μL of DNA sample (from **Section 3.2**) to the reaction mix.
3. Start reactions by adding 5 μL of nuclear extracts (from **Section 3.1**). To detect ATR-specific substrate phosphorylation, pre-treat nuclear extracts with 10 μM of KU-55933 and NU7026 for 15 min on ice to inhibit ATM and DNA-PK (*See* **Note 12** and **Fig. 3**).
4. Incubate samples at 37°C for 30 min.
5. Stop the reactions by adding 2 μL of 6× Laemmli buffer and heat the samples at 95°C for 5 min.

### 3.4. SDS-PAGE and Western Blotting

1. Proteins are separated on an 8% SDS–PAGE gel.
2. Transfer the proteins to PVDF membrane.
3. Incubate the PVDF membrane in blocking solution for 30 min at room temperature.

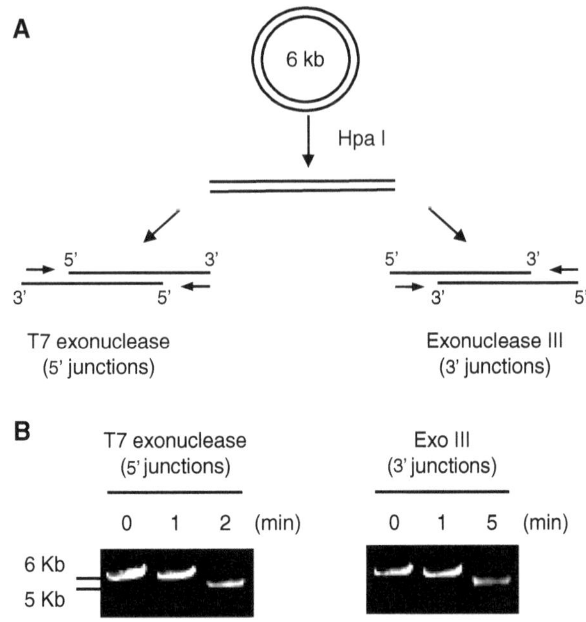


Fig. 2. Generation of SSOs by exonuclease resection. (**a**) Experimental scheme for generation of SSOs by exonulease resection. (**b**) A 6-kb plasmid was linearized with HpaI and then resected with T7 exonuclease or Exonuclease III for the indicated length of time to generate SSOs. The resulting DNA structures were analyzed in 0.8% agarose gels.

4. Incubate the membrane with primary antibodies for 3 h at room temperature or overnight at 4°C. Dilute primary antibodies in TBST with 5% milk as follows:
   1:1000 for phospho-ATM (Ser1981)
   1:1000 for phospho-Chk2 (Thr68)
   1:5000 for phospho-RPA32 (Ser33)
   1:1000 for ATM
   1:2000 for Chk2
   1:1000 for RPA32.
5. Wash the membrane with TBST three times (5 min each time).
6. Incubate the membrane with secondary antibody (1:5000 in TBST with 5% milk) for 30 min at room temperature.
7. Wash the membrane with TBST four times (5 min each time).
8. Mix equal volumes of ECL reagents A and B.
9. Incubate the membrane with the ECL mixture for 1 min.
10. Remove the ECL reagents, and place the membrane between two pieces of Saran wrap.
11. Expose the membrane to X-ray film in a film cassette (**Figs. 1** and **3**).

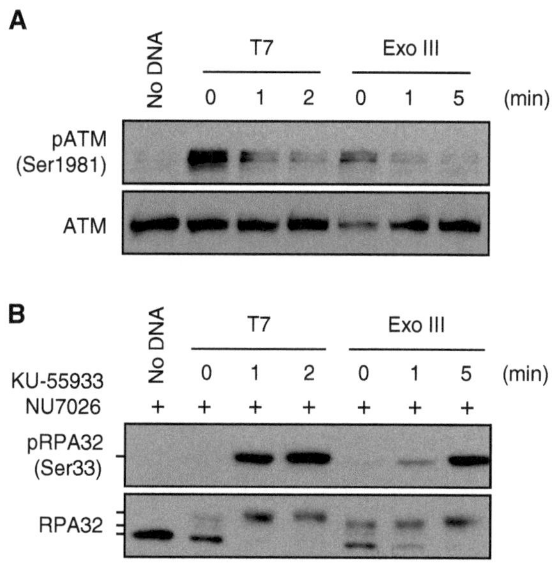


Fig. 3. Resection of DNA ends promotes an ATM-to-ATR switch in vitro. (**a**) Inhibition of ATM activation by resection of DNA ends. A plasmid linearized with HpaI was treated with T7 exonuclease (T7) or Exonuclease III (III) to generate SSOs. The resulting DNA (12 ng/μL) was incubated in extracts. The levels of phospho-ATM (Ser1981) and ATM were analyzed by Western blotting. (**b**) Induction of RPA32 phosphorylation by resection of DNA ends. Linear plasmid was resected as in (**a**) and incubated in extracts in the presence of KU-55933 and NU7026 (10 μM). The levels of phospho-RPA32 (Ser33), which reflect the ATR-dependent RPA32 phosphorylation, were analyzed by Western blotting.

## 4. Notes

1. We routinely use HeLa cells (purchased from ATCC) to prepare nuclear extracts. However, nuclear extracts derived from 293E and U2OS cells also work in this assay. To investigate the functions of various proteins in ATM or ATR activation, it is possible to use cells lacking particular proteins (such as the ATM-deficient AT cells). Alternatively, siRNA knockdown or overexpression of genes of interest in HeLa cells can also be used to analyze the functions of specific proteins.
2. The protease inhibitor cocktail from Sigma is 100×.
3. Preparations at this scale typically produce 4–5 mL of nuclear extracts at concentrations around 4–5 mg/mL. If CPV is too small, the quality and quantity of nuclear extract become variable. We therefore recommend that each preparation uses at least five dishes (150 mm) of HeLa cells.

4. The number of strokes needs to be optimized for each cell line. To assess the efficiency of cell breakage and if nuclei are intact, a drop of the resulting homogenate can be sampled with Trypan blue (Invitrogen) to stain the nuclei.
5. Make sure that all nuclei are well suspended because the lysate in buffer C is very viscous.
6. Pre-hydrate dialysis membrane with Buffer D. Remove any buffer inside the dialysis cassette before nuclear extracts are added.
7. If properly stored, nuclear extracts are active for the ATM/ATR activation assay for up to 1 year. Avoid repeated cycles of freezing and thawing.
8. Oligonucleotides are dissolved at 250 μM and stored at –20°C.
9. The DNA structures generated by annealing oligonucleotides are relatively unstable. We recommend that DNA structures are freshly prepared every 2 weeks.
10. We recommend using >4 kb dsDNA for ATR activation because long SSOs promote ATR activation more efficiently.
11. The conditions for DNA end resection need to be carefully optimized. The efficiency of resection is influenced by the activities of the enzymes and the concentration of DNA in the reactions. T7 exonuclease is relatively unstable and tends to lose its activity.
12. The use of KU-55933 and NU-7026 eliminates the phosphorylation of RPA32 by ATM and DNA-PK. This is necessary for the detection of ATR-dependent Rpa32 phosphorylation.

## Acknowledgments

This work is supported by a NIH grant (GM076288) to Lee Zou. Bunsyo Shiotani was partly supported by a fellowship form the Tosteson Foundation.

### References

1. Harper, J. W., and Elledge, S. J. (2007) The DNA damage response: ten years after. *Mol Cell 28*, 7309–45.
2. Jackson, S. P., and Bartek, J. (2009) The DNA-damage response in human biology and disease. *Nature* ***461***, 1071–8.

3. Zhou, B. B., and Elledge, S. J. (2000) The DNA damage response: putting checkpoints in perspective. *Nature* ***408***, 433–9.
4. Zou, L. (2007) Building of a trigger of ATR-mediated DNA damage response. *Genes Dev* *21*, 879–85.
5. Cimprich, K. A., and Cortez, D. (2008) ATR: an essential regulator of genome integrity. *Nat Rev Mol Cell Biol 9*, 616–27.
6. Lee, J. H., and Paull, T. T. (2005) ATM activation by DNA double-strand breaks through the Mre11-Rad50-Nbs1 complex. *Science* ***308***, 551–4.
7. Bakkenist, C. J., and Kastan, M. B. (2003) DNA damage activates ATM through intermolecular autophosphorylation and dimer dissociation. *Nature* ***421***, 499–506.
8. Shiotani, B., and Zou, L. (2009) ATR signaling at a glance. *J Cell Sci 122*, 301–4.
9. Zou, L., and Elledge, S. J. (2003) Sensing DNA damage through ATRIP recognition of RPA-ssDNA complexes. *Science* ***300***, 1542–8.
10. Ellison, V., and Stillman, B. (2003) Biochemical characterizations of DNA damage checkpoint complexes: clamp loader and clamp complexes with specificity for 5′ recessed DNA. *PLoS Biol. 1*, e33.
11. Zou, L., Liu, D., and Elledge, S. J. (2003) Replication protein A-mediated recruitment and activation of Rad17 complexes. *Proc Natl Acad Sci USA* ***100***, 13827–32.
12. Kumagai, A., Lee, J., Yoo, H. Y., and Dunphy, W. G. (2006) TopBP1 activates the ATR-ATRIP complex. *Cell 124*, 943–55.
13. Delacroix, S., Wagner, J. M., Kobayashi, M., Yamamoto, K., and Karnitz, L. M. (2007) The Rad9-Hus1-Rad1 (9-1-1) clamp activates checkpoint signaling via TopBP1. *Genes Dev 21*, 1472–7.
14. Shiotani, B., and Zou, L. (2009) Single-stranded DNA orchestrates an ATM-to-ATR switch at DNA breaks. *Mol Cell 33*, 547–58.

# Chapter 14

# Analyzing p53 Regulated DNA Damage Checkpoints by Flow Cytometry

Lois Resnick-Silverman and James J. Manfredi

## Abstract

The most critical feature of the cellular response to DNA damage is the ability of the cell to pause and repair the damage so that detrimental mutations will not be passed along to future generations of cells. The cell cycle of mammalian cells is equipped with checkpoints that can prevent cell cycle progression. Cells can either be arrested before replication of the DNA when the cells have a 2 N amount of DNA or after replication and prior to cell mitosis when the cells have a 4 N amount of DNA. Flow cytometry is a standard technique that is used to 'sort' cells based on their DNA content. It uses the principles of light scattering, light excitation, and emission of fluorochrome molecules to generate data about individual cells. The cells are fixed and permeabilized so that the DNA can be stained with a fluorescent dye. Cells that have a 2 N amount of DNA can be separated from cells with a 4 N amount of DNA. Using this technique, changes in the profile of the G1, S, and G2/M phases of the cell cycle are readily seen after DNA damage.

**Key words:** p53, DNA damage, DNA damage checkpoints, ATM/Chk2/p53 axis, G1 arrest, G2 arrest, p21, cyclin-dependent kinase complexes, apoptosis, flow analysis, western analysis, immunofluorescence.

## 1. Introduction

In the cell cycle, transitions (G1/S, intra S, G2/M) that lead from DNA replication to mitosis are monitored for successful completion. In the event of DNA damage, genotoxic stress, or ribonucleotide depletion, cell cycle checkpoints prevent progression to the next phase of the cell cycle. The cells will arrest at a checkpoint until the damage is repaired, the stress is removed, or nutrients are replenished. If the damage is irreparable, other pathways may be activated that ultimately result in programmed cell death (1).

W.X. Li (ed.), *Cell Cycle Checkpoints*, Methods in Molecular Biology 782,
DOI 10.1007/978-1-61779-273-1_14, © Springer Science+Business Media, LLC 2011

The maintenance of genomic stability is critical for cell survival and the prevention of tumorigenesis. To prevent the passage of DNA damage to daughter generations, DNA damage must be repaired. The cellular response pathway is a network that involves sensors of damage that ultimately transmit signals to mediator proteins that regulate the transcription of effector proteins that play an important role in arresting the cell cycle. When there are defects in the cell cycle checkpoints, gene mutations, chromosome damage and aneuploidy can result. Ultimately, cell transformation can be a consequence of such defects (2).

p53, a tumor suppressor protein (3) and transcription factor (4, 5), can regulate the expression of proteins that play important roles in growth arrest and apoptosis (programmed cell death) (6). p53 plays a critical role in both the G1/S checkpoint, in which cells arrest prior to DNA replication and have a 2 N content of DNA and in the G2/M checkpoint, in which arrest occurs before mitosis and cells have a 4 N content of DNA. The activation of p53 following DNA damage upregulates the expression of many proteins which are important in cell cycle arrest, repair, and apoptosis (7).

The cyclin-dependent kinase inhibitor, p21, prevents passage to S phase by inhibiting cyclin-cdk complexes that mediate the transition from G1 to S by phosphorylating cell cycle progression proteins (8–10). As a result of inhibition, the retinoblastoma protein (pRB) remains hypophosphorylated, E2F remains bound to pRB and arrest occurs at the G1/S boundary. Additionally, p21 can induce an S checkpoint as well. Proliferating cell nuclear antigen (PCNA), a protein that plays a role in both DNA replication and repair, is a component of the cyclin-cdk complex. p21 binds to and inhibits PCNA from mediating elongation during replication thus prevents any replication in cells that have already entered S phase (11).

Although the G1/S checkpoint is considered to be entirely p53 dependent, the G2/M checkpoint can be accomplished as a result of multiple pathways. In the presence of DNA damage, p53 dependent and independent pathways converge to inhibit the mitosis promoting activities of cyclin B and Cdc2 (12, 13). Phosphatases in the cdc25 family that normally activate Cdc2 become inactivated by downregulation or cytoplasmic sequestration as a result of the activation of the ATM/CHK2/cdc25 or ATR/CHK1/cdc25 pathways (14). p53 is also phosphorylated by the kinases in these pathways and in turn becomes stabilized and active. Maintenance of the G2/M checkpoint by p53 is accomplished by transcriptional repression of both cdc25C and cyclin B (15), upregulation of p21 that can inhibit cyclin B-cdk1 complexes (16), 14-3-3 sigma proteins that target cdc25C proteins to the cytoplasm (17), and GADD45, a protein that can inhibit cyclin B/Cdc2 complexes (18).

Cells in which p53 is mutated or where p53 is deleted lose the G1 checkpoint and can no longer arrest at the G1/S transition. Although they maintain a G2 arrest, this arrest can decay thus allowing cells to enter mitosis with unrepaired DNA damage and increase the risk of progression to malignancy. Li-Fraumeni, an inherited cancer prone syndrome, in which one allele of the p53 gene is mutated, is susceptible to sarcomas, leukemias, brain, and adrenal tumors. Often the remaining allele of p53 is deleted (loss of heterozygosity) in these tumors, highlighting the importance of the role of p53 in genomic stability (19).

The following method, flow cytometry, is a technique that can be used for the analysis cell cycle checkpoints following DNA damage. In this method, propidium iodide, a fluorescent dye that intercalates into the DNA of all cells is used (*see* **Fig. 1**). This property can be used to evaluate the DNA content of cells because prior to DNA replication the cells possess a 2 N content of DNA whereas after DNA replication, they will possess a 4 N amount of DNA. This will be reflected in the intensity of the fluorochrome molecules. The cells are drawn up into a chamber where they pass single file in front of a laser. This feature allows you to measure the fluorescence of single cells, rather than a population. The laser that is used in flow cytometry emits a light that is 488 nM. The excitation of propidium iodide can be detected with 562–588 nM band pass filter. An advantage of this technique is that more than one fluorochrome can be used if they are also excited at 488 nM but emit at a different wavelengths. This is demonstrated in

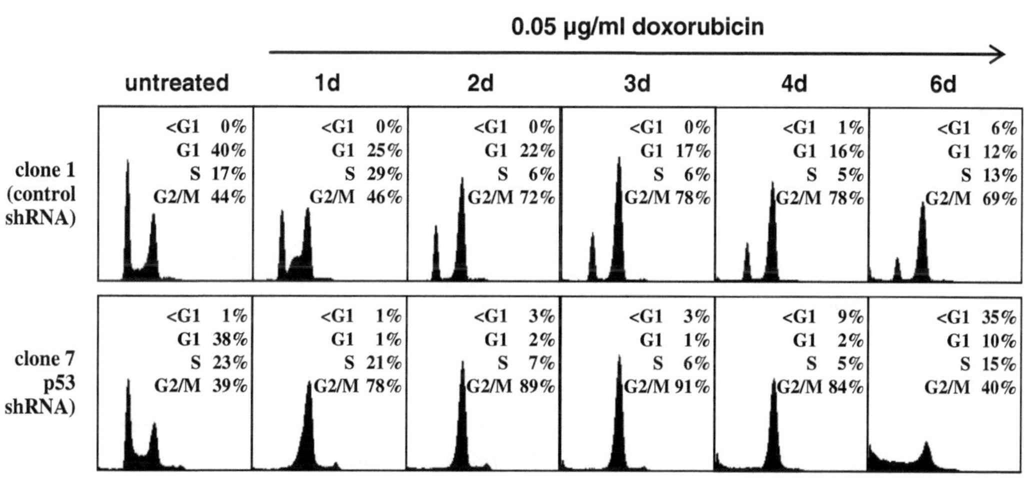


Fig. 1. After DNA damage with doxorubicin, wild-type p53 osteosarcoma cells maintain a G2 arrest whereas p53 knock-down cells ultimately lose the G2 arrest resulting in a hypodiploid population of cells. U2OS clone 1 (control-shRNA) and clone 7 (p53-shRNA) cells were treated with doxorubicin (0.05 μg/mL) for times ranging from 1 day to 6 days and cell profiles were analyzed by flow cytometry. The percentage of cells in G1, S, and G2/M or with a hypo-diploid DNA content (<G1) are indicated in the *upper right hand* corner of each histogram.

the second method, dual parameter analysis that uses a FITC-conjugated antibody to bromodeoxyuridine (BrdU) in addition to PI. BrdU, a synthetic analog of thymidine, is incorporated into replicating DNA. The antibody that detects BrdU is conjugated to a fluorescent dye that can be detected at a peak emission of 520 nM. Histograms, which are dot plots of each event, can be constructed to illustrate both parameters. The actively replicating cells (S phase) can be distinguished because of the BrdU uptake. Cells that have arrested in either G1 or G2/M should have a low percentage of cells in S phase. The histogram will also reflect the number of cells in G1, S, and G2/M because all cells incorporate PI (*see* **Fig. 2**). The third method is also dual parameter. A FITC conjugated antibody is also used; however, the antibody detects phosphorylated histone H3, a marker for mitotic cells. Once cells exit mitosis, H3 becomes dephosphorylated. When using only PI stain, one cannot distinguish between G2 and mitotic cells since both have a 4 N amount of DNA. This technique broadens the cell cycle information by specifically detecting mitotic cells (*see* **Fig. 3**).

Although FACS analysis is the only method described here it should be noted that other tools that include Western analysis and

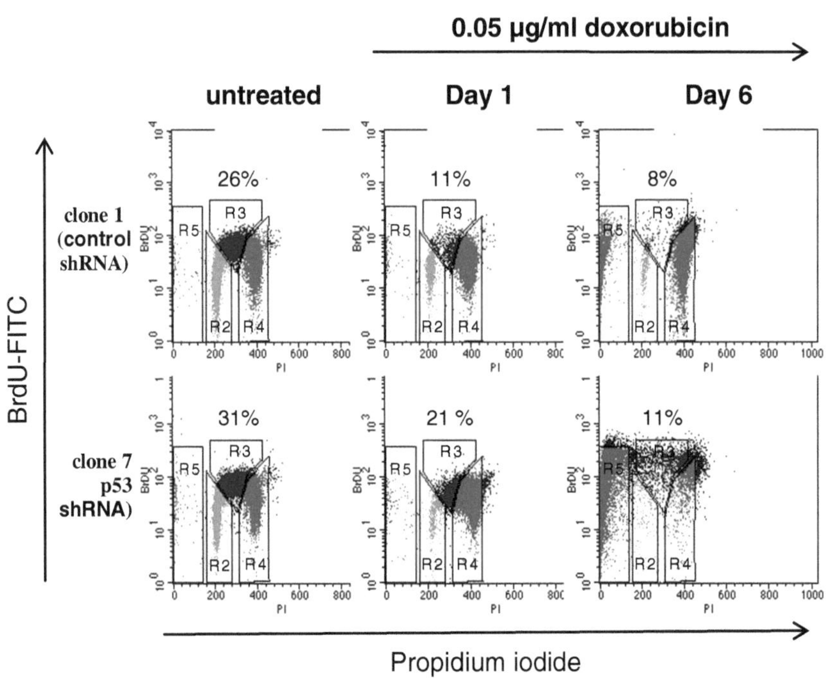


Fig. 2. DNA damage with doxorubicin in wild-type p53 osteosarcoma cells results in a loss of S-phase cells with a concomitant arrest in the G2/M phase of the cell cycle. Clone 1 (control-shRNA) and clone 7 (p53 shRNA) U2OS cells were treated with 0.05 μg/mL doxorubicin for 1 day or continuously for 6 days. The cells were pulsed with 10 μM BrdU for 45 min prior to harvesting. Following BrdU and propidium iodide staining the cells were analyzed by flow cytometry. The percentage of cells in S phase is indicated above the R3 region of the histogram.

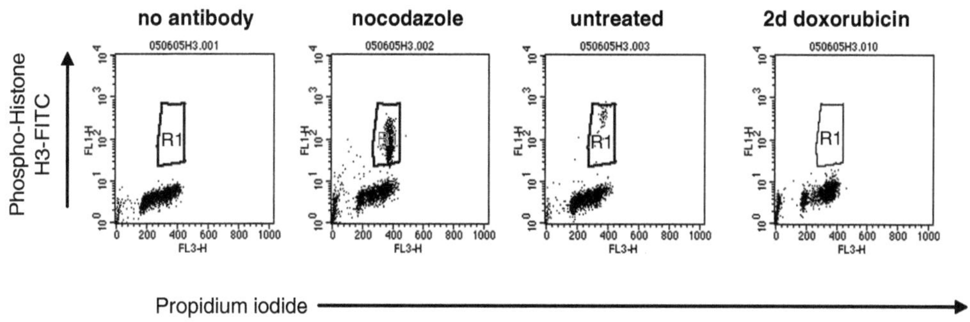


Fig. 3. No mitosis is observed after wild-type p53 osteosarcoma cells are DNA damaged with doxorubicin. Clone 1 (control-shRNA) U2OS cells were treated with 0.05 μg/mL doxorubicin for 48 h and stained for Ser10-phosphorylated histone H3 and DNA content (propidium iodide). No antibody samples denote cells stained only with secondary antibody and no primary antibody and nocodazole treatment (0.4 μg/mL) is a positive control for cells in mitosis. Mitotic cells are boxed and denoted R1.

immunofluorescence, used to examine the expression of specific proteins, can also be employed to complement flow cytometry in the analysis of cell cycle checkpoints after DNA damage.

## 2. Materials

### *2.1. Cell Culture, Lysis, and DNA Damage*

1. Dulbecco's Modified Eagle's Medium (DMEM with High Glucose) supplemented with 10% fetal bovine serum (FBS).
2. 0.05% Trypsin-ethylenediamine tetraacetic acid (EDTA).
3. Doxorubicin (Sigma-Aldrich, St. Louis, MO) is dissolved in dimethyl sulfoxide (DMSO) (Sigma-Aldrich, St. Louis, MO) and passed through a 0.2-μm filter to give a concentration of 1 mg/mL. Sterile aliquots are stored at –20°C. A 1:10 dilution of this stock (0.1 mg/mL) can be prepared in sterile $H_2O$ and is also stored at –20°C. This can be added directly to tissue culture cells at the appropriate concentration.

### *2.2. Propidium Iodide (PI) Staining for FACS Analysis*

1. Dulbecco's Modified Eagle's Medium (DMEM with High Glucose) supplemented with 10% fetal bovine serum (FBS)
2. 0.05% Trypsin-ethylenediamine tetraacetic acid (EDTA)
3. 70% ethanol (ETOH) prepared from reagent grade, 190 proof (95%) ethanol diluted with distilled water
4. 1× phosphate buffered saline without calcium and magnesium (PBS)
5. Propidium iodide (PI) (Sigma-Aldrich, St. Louis, MO). A 100× stock (2 mg/mL in PBS) is prepared and stored at 4° in the dark.

6. Bovine pancreatic ribonuclease A (RNase A) (Sigma, St. Louis, MO), stored at –20°C
7. Polystyrene round bottom 12×75 mm Falcon tubes (BD Falcon, Franklin Lakes, NJ)

#### 2.3. (+)-5-Bromo-2-deoxyuridine (BrdU) Incorporation

1. BrdU (Sigma-Aldrich, Fairlawn, NJ)
2. 1 × PBS
3. 1% Bovine Serum Albumin (BSA)/PBS
4. 70% ETOH
5. FITC-conjugated Anti-BrdU antibody (Becton Dickenson)
6. PI
7. 2 N HCl with 0.5% Triton X-100 (v/v)
8. 0.1 M $Na_2B_4O_7$ (sodium tetraborate), pH 8.5
9. 0.5% Tween-20/1% BSA/PBS
10. 12×75 polystyrene Falcon tubes

#### 2.4. Phosphorylated Histone H3 Assay

1. 0.5% formaldehyde in PBS
2. 90% methanol
3. Anti-phosphorylated ser 10 histone 3 antibody (Cell Signaling)
4. RNase A, stored at –20°C
5. FITC-conjugated secondary antibody (Jackson Laboratories)
6. BSA

#### 2.5. Equipment

1. $CO_2$ incubators, 37°C
2. Low speed centrifuge with swinging bucket rotor.
3. Freezer at –20°C
4. Refrigerator at 4°C
5. FACS brand flow cytometer (BD BioScience)

## 3. Methods

#### 3.1. Treatment with DNA Damaging Agent

Day 0. Cell lines containing wild-type, mutant, or no p53 are trypsinized and plated at an appropriate number into 60- or 100-mm tissue culture dishes. It is best to start by seeding enough dishes to do a dose–response curve of the drug.

Day 1. Doxorubicin (0.1 mg/mL) is thawed and added to cell cultures at a concentration of 0.05–0.5 μg/mL to establish a suitable curve. The cell cultures are maintained for an additional 24 h at 37°C before harvesting (*see* **Note 1**).

### 3.2. Staining with Propidium Iodide and Preparation for FACS Analysis

Day 0 and Day 1. Follow the above given protocol.

*Day 2.*

1. The cells are harvested by trypsinization. Do not discard the medium that might contain floating cells that have died as a result of the drug treatment. You can neutralize the trypsin with this medium.
2. The cell suspensions are centrifuged at 300×*g* for 5 min in order to pellet the cells.
3. The cell pellets are resuspended in 1× PBS and spun again at 300×*g* for 5 min.
4. The supernatant is removed and the pellet is gently dislodged. Add 1 mL of 70% ETOH to the cells while vortexing to disrupt any clumping of the cells (*see* **Notes 2** and **3**).
5. The fixed cells can be stored at –20°C for a minimum of 12 h or up to 2 weeks prior to PI staining.

*Day 3.*

1. The fixed cells are pelleted by centrifugation at 300×*g* for 5 min at 4°C.
2. The supernatant is discarded and the cells are resuspended in 20 μg/mL PI in PBS containing 1 mg/mL of RNase A. The PI is prepared from the 100× stock (2 mg/mL) but the RNase is added fresh each time. Generally, 1–2 mL of the PBS containing PI and RNase A is added per $10^6$ cells. The stained cells are transferred to polystrene 12×75 mm Falcon tubes and stored in the dark at room temperature for 2 h prior to FACS analysis. They may also be stored overnight at 4°C in the dark.
3. Samples are analyzed for cell cycle position using a BD BioSciences flow cytometer. The FL2 laser is used to detect the propidium iodide staining of the DNA. CellQuest software is used to generate the acquisition and analysis plots of 10,000 cells.

### 3.3. Dual Parameter, BrdU Incorporation Assay, and Propidium Iodide Staining

Day 0. Cells are plated at an appropriate number for optimal growth in 100 mm tissue culture dishes.

Day 1. The doxorubicin stock (0.1 mg/mL) is thawed and added to cell cultures. If this is the first time the cells are being treated, a curve must be established for obtaining a suitable cell cycle arrest.

Day 2.

1. The cells are pulsed with 10 μM BrdU. Do not wash the cells prior to addition of BrdU as this will slow the growth of the cells and the incorporation of the analogue. Add BrdU directly to culture medium. Include a dish of cells that will not be pulsed with BrdU (*see* **Note 4**).
2. Continue to incubate cells at 37°C for 45–60 min.
3. The cells are harvested by trypsinization and spun for 5 min at 300×*g*.
4. The cell pellet is resuspended in 1× PBS, adjusting the cell concentration to $10^6$ cells/sample, and spun again for 5 min at 300×*g*.
5. The supernatant is removed and the pellet is resuspended in 1 mL 70% ETOH by gently vortexing to disrupt any clumping of cells.
6. The fixed cells may be stored at −20°C for a minimum of 12 h before proceeding.
7. The cells are pelleted at 300×*g* for 5 min and resuspended in 1 mL of 2 N HCL/0.5%Triton X-1000 (v/v) by slowly adding a few drops at a time while maintaining a vortex. This treatment will denature the DNA and create single-stranded molecules.
8. Incubate at room temperature for 30 min.
9. The cells are centrifuged at 300×*g* for 5 min. After aspiration of the supernatant, the cells are resuspended in 1 mL of 0.1 M sodium tetraborate, pH 8.5, in order to neutralize the acid.
10. (You can proceed to the next step or store the BrdU-labeled cells at −20°C after centrifuging at 300×*g* for 5 min and resuspending the pellet in 70% ETOH.)
11. The cells are centrifuged at 300×*g* for 5 min. After removal of the supernatant the cells are resuspended in 500 μL 0.5% Tween-20/1% BSA/PBS containing and appropriate dilution of anti-BrdU-FITC secondary antibody.
12. The cells are incubated in the dark for 30 min.
13. After centrifugation at 300×*g* for 5 min, the cells are resuspended in 1 mL of PBS containing 10 μg/mL propidium iodide and transferred into 12×75 polystyrene Falcon tubes. They are stored at room temperature in the dark for 2 h.
14. The samples are ready for analysis on a FACS brand flow cytometer. The laser excitation is at 488 nm. The FL1

laser is used to detect the green fluorescent FITC stain and the FL3 laser is used to detect the red propidium iodide staining of the DNA. CellQuest software is used to generate the acquisition and analysis plots of 10,000 cells.

### 3.4. Phosphorylated Histone H3 Assay

Day 0. Cells are plated at an appropriate number for optimal growth in 100-mM tissue culture dishes.

Day 1. Doxorubicin (0.1 mg/mL) is thawed and added to cell cultures at a concentration of 0.05–0.5 μg/mL to establish a suitable curve. The cell cultures are maintained for an additional 24 h at 37°C.

Day 2.

1. The cells are harvested by trypsinization and spun for 5 min at 300×*g*.
2. The cells are resuspended in PBS and centrifuged for 5 min at 300×*g*.
3. The supernatant is removed and the cells are fixed in 1 mL of 0.5% formaldehyde for 10 min at 37°C.
4. The cells are centrifuged at 300×*g* for 5 min and resuspended in PBS to remove the fixative.
5. The cells are centrifuged again at 300×*g* for 5 min and resuspended in 90% methanol for permeabilization.
6. The permeabilized cells are kept on ice for 30 min or can be stored at –20°C for up to 10 days.
7. The cells are washed with PBS and centrifuged at 300×*g* for 5 min.
8. After removal of the supernatant, the cells are resuspended in the primary antibody, anti-phosphorylated ser 10 histone 3. The antibody is diluted 1:100 in PBS that contains 100 μg/mL RNase A.
9. The cells are incubated for 1 h in the dark.
10. The cells are centrifuged and washed twice with 1%BSA/PBS.
11. After removal of the BSA/PBS the cells are resuspended in the secondary antibody, anti-FITC-conjugated antibody that is diluted 1:100 in PBS.
12. The cells are incubated for 1 h in the dark.
13. Once again the cells are washed in 1% BSA/PBS and centrifuged for 5 min at 300×*g*.
14. The pellets are resuspended in PBS containing 20 μg/mL propidium iodide and transferred into polystyrene 12×75 mm Falcon tubes.

15. The stained cells are stored in the dark at room temperature for 30 min.
16. Samples are analyzed for cell cycle position using a BD BioSciences flow cytometer. The laser excitation is at 488 nm. FL1 and FL 3 detectors are used to distinguish between the excitation of PI and FITC. CellQuest software is used to generate the acquisition and analysis plots of 10,000 cells.

## 4. Notes

1. Each cell line might respond differently to a particular DNA damaging drug. Therefore it is important to empirically derive which concentration will produce the biologic effect and response that you desire. It is important to establish a concentration curve of each drug for each cell line.
2. Cells are impermeable to PI. Therefore for this assay, the cells must be fixed so that the PI stain can penetrate the cells and intercalate into the DNA. Although ETOH is used, paraformaldehyde can also be used and might be preferable, when you want to prevent transfected proteins, like green fluorescent protein (GFP), from leaking out of cell.
3. Some cell lines tend to be very sticky and are prone to clumping. Use of 12×75 mm Falcon polystyrene tubes with strainer caps will filter out clumps of cells from the final cell suspension.
4. An unlabeled sample will be helpful when it comes to the analysis portion of the BrdU assay. It will allow you to set the proper gating for BrdU incorporation.

## Acknowledgments

The authors would like to thank Dana Lukin for contributing figures from his PhD dissertation. The authors are supported by grants from the National Cancer Institute (P01 CA080058, R01 CA125741, and R01 CA086001).

### References

1. Hartwell, L. H., and Weinert, T. A. (1989) Checkpoints: controls that ensure the order of cell cycle events, *Science 246*, 629–634.
2. Paulovich, A. G., Toczyski, D. P., and Hartwell, L. H. (1997) When checkpoints fail, *Cell 88*, 315–321.

3. Finlay, C. A., Hinds, P. W., and Levine, A. J. (1989) The p53 proto-oncogene can act as a suppressor of transformation, *Cell 57*, 1083–1093.
4. Kern, S. E., Kinzler, K. W., Bruskin, A., Jarosz, D., Friedman, P., Prives, C., and Vogelstein, B. (1991) Identification of p53 as a sequence-specific DNA-binding protein, *Science 252*, 1708–1711.
5. Pietenpol, J. A., Tokino, T., Thiagalingam, S., el-Deiry, W. S., Kinzler, K. W., and Vogelstein, B. (1994) Sequence-specific transcriptional activation is essential for growth suppression by p53, *Proc Natl Acad Sci USA 91*, 1998–2002.
6. Burns, T. F., and El-Deiry, W. S. (1999) The p53 pathway and apoptosis, *J Cell Physiol 181*, 231–239.
7. Bargonetti, J., and Manfredi, J. J. (2002) Multiple roles of the tumor suppressor p53, *Curr Opin Oncol 14*, 86–91.
8. el-Deiry, W. S., Tokino, T., Velculescu, V. E., Levy, D. B., Parsons, R., Trent, J. M., Lin, D., Mercer, W. E., Kinzler, K. W., and Vogelstein, B. (1993) WAF1, a potential mediator of p53 tumor suppression, *Cell 75*, 817–825.
9. Harper, J. W., Adami, G. R., Wei, N., Keyomarsi, K., and Elledge, S. J. (1993) The p21 Cdk-interacting protein Cip1 is a potent inhibitor of G1 cyclin-dependent kinases, *Cell 75*, 805–816.
10. Waldman, T., Kinzler, K. W., and Vogelstein, B. (1995) p21 is necessary for the p53-mediated G1 arrest in human cancer cells, *Cancer Res 55*, 5187–5190.
11. Waga, S., Hannon, G. J., Beach, D., and Stillman, B. (1994) The p21 inhibitor of cyclin-dependent kinases controls DNA replication by interaction with PCNA, *Nature 369*, 574–578.
12. Agarwal, M. L., Agarwal, A., Taylor, W. R., and Stark, G. R. (1995) p53 controls both the G2/M and the G1 cell cycle checkpoints and mediates reversible growth arrest in human fibroblasts, *Proc Natl Acad Sci USA 92*, 8493–8497.
13. Flatt, P. M., Tang, L. J., Scatena, C. D., Szak, S. T., and Pietenpol, J. A. (2000) p53 regulation of G(2) checkpoint is retinoblastoma protein dependent, *Mol Cell Biol 20*, 4210–4223.
14. Matsuoka, S., Huang, M., and Elledge, S. J. (1998) Linkage of ATM to cell cycle regulation by the Chk2 protein kinase, *Science 282*, 1893–1897.
15. Taylor, W. R., DePrimo, S. E., Agarwal, A., Agarwal, M. L., Schonthal, A. H., Katula, K. S., and Stark, G. R. (1999) Mechanisms of G2 arrest in response to overexpression of p53, *Mol Biol Cell 10*, 3607–3622.
16. Bunz, F., Dutriaux, A., Lengauer, C., Waldman, T., Zhou, S., Brown, J. P., Sedivy, J. M., Kinzler, K. W., and Vogelstein, B. (1998) Requirement for p53 and p21 to sustain G2 arrest after DNA damage, *Science 282*, 1497–1501.
17. Chan, T. A., Hermeking, H., Lengauer, C., Kinzler, K. W., and Vogelstein, B. (1999) 14-3-3Sigma is required to prevent mitotic catastrophe after DNA damage, *Nature 401*, 616–620.
18. Kastan, M. B., Zhan, Q., el-Deiry, W. S., Carrier, F., Jacks, T., Walsh, W. V., Plunkett, B. S., Vogelstein, B., and Fornace, A. J., Jr. (1992) A mammalian cell cycle checkpoint pathway utilizing p53 and GADD45 is defective in ataxia-telangiectasia, *Cell 71*, 587–597.
19. Malkin, D. (1993) p53 and the Li-Fraumeni syndrome, *Cancer Genet Cytogenet 66*, 83–92.

# Chapter 15

## Using *Drosophila* S2 Cells to Measure S phase-Coupled Protein Destruction via Flow Cytometry

**Jean M. Davidson and Robert J. Duronio**

### Abstract

Cell proliferation depends on the timely synthesis and destruction of proteins at specific phases of the cell cycle. Recently it was discovered that the destruction of several key cell cycle regulatory proteins during S phase is coupled directly to DNA replication. These proteins harbor a motif called a PIP degron that mediates binding to chromatin bound PCNA at replication forks and recruits the $CRL4^{Cdt2}$ E3 ubiquitin ligase. These interactions comprise an elegant mechanism for coupling DNA replication with ubiquitylation and subsequent proteolysis by the 26S proteasome. Here we describe a flow cytometry-based method using *Drosophila* S2 cells that recapitulates S phase-specific protein proteolysis. Because of the high degree of evolutionary conservation of the PIP degron and $CRL4^{Cdt2}$ and the ease of culturing and inhibiting gene function by RNAi in S2 cells, our flow cytometric method should serve as a general tool for determining whether any eukaryotic protein is subject to replication-coupled protein destruction.

**Key words:** *Drosophila* S2 cells, flow cytometry, S phase, E2F, cell cycle, CRL4, ubiquitin-mediated proteolysis, RNAi.

### 1. Introduction

Accurate progression through the cell cycle depends on the timely synthesis and destruction of a myriad of proteins. A classic example is the B-type cyclins, which are targeted for destruction via the APC/C E3 ubiquitin ligase complex during the metaphase to anaphase transition to ensure completion of mitosis. Cell cycle regulators are also targeted for destruction in other phases of the cell cycle. Recently it was discovered that the proteolysis of a small set of proteins is coupled directly to DNA synthesis occurring during S phase or DNA repair. These proteins

W.X. Li (ed.), *Cell Cycle Checkpoints*, Methods in Molecular Biology 782,
DOI 10.1007/978-1-61779-273-1_15, 

include the pre-replicative complex component Cdt1 (1–7), the cyclin-dependent kinase inhibitor p21 (8–10), the *Caenorhabditis elegans* translesion DNA polymerase Pol-H (11), the histone methyltransferase Set8 (12–15), and the *Drosophila melanogaster* transcription factor E2f1 (15). There is increasing evidence that these proteins must be destroyed during S phase to ensure normal cell cycle progression.

The mechanism of destruction used by this group of proteins couples DNA replication with ubiquitin-mediated proteolysis. Each of these five proteins contains a short (approximately 14 amino acid) motif called a PIP degron that is required for replication-coupled destruction. The PIP degron has two critical features: a PIP box that confers binding to a hydrophobic pocket on PCNA (proliferating cell nuclear antigen) and key basic residues flanking the PIP box that recruit the multi-subunit $CRL4^{Cdt2}$ E3 ubiquitin ligase, presumably through direct interaction with the substrate receptor, Cdt2 (16). PCNA loads onto DNA as a homotrimer and serves both as a processivity clamp for DNA polymerase during DNA synthesis and as a platform to recruit various PIP box-containing proteins to replicating DNA (e.g., topoisomerase and histone deacetyltransferases) (17). Current data support a model whereby proteins with a PIP degron interact with PCNA and become ubiquitylated by $CRL4^{Cdt2}$ only when PCNA is assembled on chromatin at replication forks or at sites of DNA repair synthesis. This results in an elegant mechanism for coupling ongoing DNA replication with ubiquitylation and subsequent proteolysis by the 26S proteasome (**Fig. 1**).

*Drosophila* E2f1 is a member of the E2F family of transcriptional activators, which play a pivotal role in the progression of the G1-S transition (18). Like mammalian E2Fs, E2f1 is inhibited prior to S phase via interaction with the retinoblastoma protein (pRb) homolog, Rbf1. Activation of cyclin-dependent kinases during G1 results in hyper-phosphorylation of Rbf1, thereby relieving E2f1 repression and activating a transcription program

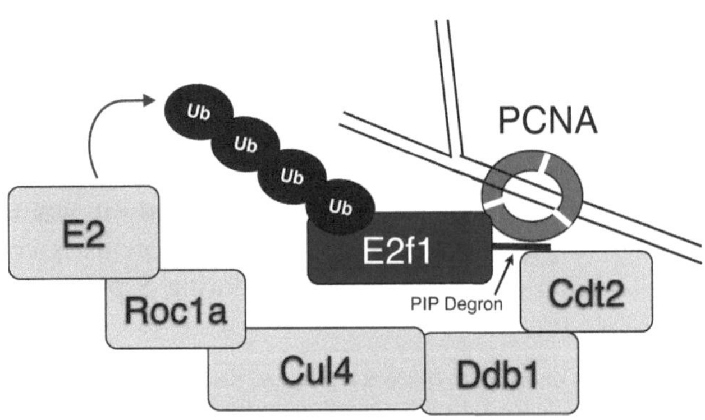


Fig. 1. A model of $CRL4^{Cdt2}$-mediated destruction of E2f1.

that promotes entry into S phase. E2f1 is then rapidly destroyed during S phase (19–22). We hypothesized that S phase-specific destruction provides an important Rbf1-independent negative regulation of E2f1. To test this, we developed a facile method for using exogenous GFP-E2f1 fusion proteins and flow cytometry of cultured S2 cells that recapitulates the S phase-specific destruction of E2f1 observed in vivo. This method has the capacity to rapidly test whether specific domains within E2f1 or trans-acting factors (e.g., components of $CRL4^{Cdt2}$) are required for destruction of E2f1 during S phase. Our assay allowed us to discover a PIP degron within E2f1 and to demonstrate that PCNA and $CRL4^{Cdt2}$ are necessary for S phase-specific destruction of E2f1 (15).

In this chapter we describe our method for determining whether a protein is destroyed specifically during S phase and whether this destruction requires components of the $CRL4^{Cdt2}$ E3 ligase. We discuss the method of generating stably transfected S2 cell lines, of treating those cell lines with dsRNA to knockdown components of the $CRL4^{Cdt2}$ E3 ligase, of inducing expression of GFP-fusion proteins, and of submitting fixed and stained cells to flow cytometric analysis to ascertain their cell cycle profile and quantitatively measure S phase-specific protein destruction. Because of a high degree of evolutionary conservation and the ease of culturing and inhibiting gene function by simply treating S2 cells with exogenous dsRNA, our flow cytometric method should serve as a general tool for determining whether any eukaryotic protein is subject to the replication-coupled, PIP degron/$CRL4^{Cdt2}$-mediated destruction mechanism.

## 2. Materials

### 2.1. Drosophila S2 Cell Culture

1. *Drosophila* S2 Cells (Invitrogen)
2. SF-900 III SFM culture medium (Gibco, Invitrogen)
3. Penicillin-Streptomycin (Sigma, St. Louis, MO). Solution stabilized, sterile-filtered, with 10,000 units penicillin and 10 mg streptomycin/mL
4. 25 $cm^2$ Corning Cell Culture Flasks (Sigma). Canted neck, vented caps
5. Incubators with proper humidity (28°C/100% air incubator for normal growth, 37°C/100% for heat shock induction)

### 2.2. Creating Double-Strand RNA for RNAi

1. T7 RNA Polymerase (New England Biolabs)
2. Rnasein® Ribonuclease Inhibitor (Promega)

3. 10× Transcription Buffer: 400 mM Tris, pH 8.0, 150 mM $MgCl_2$, 50 mM DTT, 0.5 mg/mL BSA
4. DNase 1U/μL (Promega)
5. NTPs set 100 mM Solutions (Fermentas)
6. Phenol:Chloroform:Isoamyl Alcohol 25:24:1 saturated with 10 mM Tris, pH 8.0, 1 mM EDTA (Sigma Aldrich)
7. dsRNA ladder (New England Biolabs)
8. AlphaImager™ 2200 (Alpha Innotech) and ImageQuant 5.2 software (GE Life Science)

### 2.3. Generating Stable Lines

1. pHGW: Gateway-compatible destination vector developed by Dr. Terence Murphy (http://www.ciwemb.edu/labs/murphy/Gateway%20vectors.html). Available upon request. This vector allows $NH_2$-terminal GFP fusion driven by the *Hsp70* promoter, with Ampicillin resistance. Other vectors allow COOH-terminal fusion, which may be necessary for certain proteins depending on the stability of the fusion protein and the location of the PIP degron (e.g., $NH_2$-terminal GFP fusion may mask an $NH_2$-terminal PIP degron).
2. Gateway® LR Clonase™ II enzyme (Invitrogen).
3. pCoHygro plasmid (Invitrogen) 4526 bp with Ampicillin resistance.
4. Hygromyosin B (Invitrogen).
5. Amaxa Nucleofectin® V for *Drosophila* S2 Cell transfection (Lonza).
6. Nucleofector® apparatus (Lonza).
7. 6-well culture plate (Corning).
8. 2-mL round bottom tubes (Eppendorf).
9. 25-$cm^2$ cell culture flasks (Corning). Canted neck, vented caps.

### 2.4. Fixing S2 Cells

1. 32% Paraformaldehyde (Electon Microscope Sciences)
2. 1× PBS-T: 1× phosphate-buffered saline with 1%/vol Tween-20 (Promega)
3. DNase-free RNase (Roche)
4. Propidium iodide: 1.0 mg/mL solution in water (Invitrogen), store at 4°C protected from light

### 2.5. Flow Cytometric Analysis

1. Dako CyAn Flow Cytometry System (Dako, Beckman-Coulter).
2. Summit 4.3 Flow Cytometer Data Collection software (Dako, Beckman-Coulter).

3. Falcon 35-2002 round bottom 6-mL polystyrene tube 12 × 75 $mm^2$. If using a different flow cytometer system, use appropriate tube required for that cytometer.
4. ModFit LT™ (Verity Software House).
5. Required parameters: height and area of fluorescence emission signal for GFP or propidium iodide signal. Height of forward and side scatter.
6. Nomenclature: Forward Scatter (FS); parameter to measure light scattered less than 10° FS measures relative cell size. Side Scatter (SS); light scattered at a 90° angle. Measures complexity of cellular membranes combined to measuring clumping of cells. Height (H) is the maximum amplitude of the emission signal. Area (A) is the area of the signal. Event Count measures the number of cells within to a certain parameter.

## 3. Methods

### 3.1. Maintaining Drosophila S2 Cells in Culture

1. Cells were routinely grown at room temperature in SF-900 *Drosophila* medium (Gibco) without serum and 1:200 Penicillin-Streptomycin (5 units/mL Penicillin, 5 μg/mL Streptomycin, Sigma). Schneider's *Drosophila* medium (Gibco) supplemented with 10% fetal bovine serum (Sigma) can also be used, but may interfere with RNAi and transfection efficiency (**Note 1**).
2. Passage cells when density is between 6 and 20 × $10^6$ cells/mL and split at 1:2 to 1:5 dilution. S2 cells do not grow well when seeded at a density below 5 × $10^5$ cells/mL. Cells will being to disassociate from the flask when overcrowded; however, these cells are still alive. It is a good idea to avoid allowing the cultures to become overcrowded (**Note 2**).
3. When removing cells from the flask, tap the flask several times to dislodge adherent cells. Use a pipette to wash the surface where cells adhere before transferring cells to a new flask.

### 3.2. Creating Double-Strand RNA and RNAi on S2 Cells

1. Generate primers to amplify target genes using primers that contain a T7 promoter sequence (labeled in Italics). Below are the sequences used to generate dsRNA against components of the $CLR4^{Cdt2}$-mediated destruction machinery. Note that an extended knockdown of PCNA is predicted

to eventually result in an S phase arrest, thereby precluding detection of S phase-coupled destruction because of a depletion of S phase cells from the population. However, we were able to find conditions of PCNA knockdown that stabilized E2f1 during S phase prior to extensive cell cycle arrest (15). A similar situation occurs with Cul4 knockdown, which eventually will result in G1 arrest.

a. Cul4 dsRNA Forward: *TAATACGACTCACTATAGGG*TTGGCCAAACGATTACTTGTGGG

b. Cul4 dsRNA Reverse: *TAATACGACTCACTATAGGG*GAGAAGATTATGGCTCAGCG

c. Cdt2 dsRNA Forward: *TAATACGACTCACTATAGGG*GCGGGCTCCGGCATACGCGGC

d. Cdt2 dsRNA Reverse: *TAATACGACTCACTATAGG*CGTGGCTGGAGCCCCAGGCCACG

e. PCNA dsRNA Forward: *TAATACGACTCACTATAGGG*CAGGCCATGGACAACTCCCATG

f. PCNA dsRNA Reverse: *TAATACGACTCACTATAGGG*TGTCTCGTTGTCCTCGATCTTGGG

2. Double-stranded RNAs (dsRNAs) are transcribed from PCR products generated by gene-specific primers containing T7 promoter sequences. PCR products are amplified from genomic DNA or cDNA clones. cDNA clones are available from the Drosophila Genomics Resource Center (https://dgrc.cgb.indiana.edu/) (**Note 3**).

3. Transcription reaction is performed by combining the following: 2 mM (final concentration) each NTP, 10 μL 10× Transcription Buffer, 0.5 μL RNasein, 1 μg purified DNA template (e.g., via phenol:chloroform extraction), 1 μL T7 RNA polymerase with sterile water to make 100 μL total reaction volume. Incubate at 37°C for 6 h (reaction may become cloudy). Add 1 μL DNase and incubate at 37°C for 30 min.

4. Annealed RNA strands are generated by boiling samples in a beaker of water to remove secondary structure and slowly returning to room temperature. Purify dsRNA reaction using phenol:chloroform:isoamyl alcohol 25:24:1.

5. Approximate concentrations of the dsRNAs are determined by comparing the intensity of band of 1 μL of RNA sample

in an agarose gel using a dsRNA ladder as a standard. The intensity of the RNA bands is analyzed using by imaging the AlphaImager™ and quantified with the ImageQuant 5.2 software.

### 3.3. Generating Stable Lines

1. The open reading frame of wild-type Drosophila *E2f1* was amplified from pUAST-*E2f1* and cloned into pENTR™/D-TOPO©. The forward primers used to amplify E2f1 are designed with a CACC at the 5′ end for directional cloning into pENTR.
2. To create expression vector plasmids, pENTR *E2f1* was recombined with the Gateway expression vector pHGW, which contains an $NH_2$-terminal GFP tag and a *Hsp70* promoter, using the Gateway® LR Clonase™ II enzyme mix (**Note 4**). The use of the heat shock-inducible *Hsp70* promoter allows one to analyze proteins (e.g., those containing a mutation that inactivate the PIP degron) that might otherwise be toxic when expressed from a constitutive promoter.
3. Split cells 2–3 days prior to transfection to ensure they are growing well and not overcrowded.
4. Ensure that the entire Nucleofector® Supplement is added to the Nucleofector® Transfection solution and allowed to reach room temperature prior to beginning transfection protocol. The transfection solution should be kept at 4°C otherwise and will expire within 3 months of the supplement being added.
5. Prepare 6-well plates by filling appropriate number of wells with 1 mL of SF-900 SFM culture media and allow it to come to room temperature.
6. Count an aliquot of cells and determine density. Use $1 \times 10^6$ cells per sample, and spin down at $500 \times g$ for 5 min at room temperature. Completely remove supernatant and flick tube to break up pellet.
7. Resuspend cells in 100 μL room-temperature Nucleofector® Solution per sample. Avoid leaving cells in this solution for longer than 15 min.
8. Combine cell solution with 2 μg pENTR *E2f1* and 0.5 μg pCoHygro and transfer solution to cuvette supplied in Nucleofection® Kit V, ensuring cell/DNA solution covers bottom of the cuvette with no air bubbles. Close cuvette with cap.
9. Select Nucleofector® Program G-030. Insert cuvette into apparatus and apply the program by pressing the X-button. Take cuvette out once transfection is successfully completed.

10. Immediately add 500 μL of room temperature SF-900 SFM media to cuvette and gently transfer to prepared 6-well plate, using supplied pipettes. Final volume is 1.6 mL/well.
11. Incubate cells in a humidified 25°C/100% air incubator without $C0_2$ for 2–3 days.
12. Collect cells and centrifuge for 5 min at 500 × *g*. Replace with 2 mL of fresh SF-900 SFM media and transfer to a new 6-well plate. Let grow 2–3 days in a humidified 25°C/100% air incubator.
13. Collect cells into two 2 mL round bottom Eppendorf tubes and centrifuge for 5 min at 500 × *g*. To one sample, add 2 mL of SF-900 SFM with 50 μg/mL Hygromyosin B. To the other, add only SF-900 SFM, in case selection kills the culture. Grow in 6-well plate for 2–3 days in humidified 25°C/100% air incubator.
14. Once cells reach confluency within several days, transfer to 25-$cm^2$ cell culture flask and split cells 1:5 every week. Maintain 50 μg/mL Hygromyosin B in SF-900 SFM media. Initial transfection efficiency should be around 50–60%; this can be checked by removing an aliquot, subjecting the cells to a 37°C heat shock for 30 min, waiting an hour, and visualizing GFP signal under a microscope. Once population is nearly homogeneous for containing the plasmid, experiments may begin.

### 3.4. RNAi and Heat Shock Induction

1. 3–4 days prior to the experiment, count and plate stably transfected pHGW-*E2f1* and pHGW-only cells at $1\times 10^6$ cells/mL. Add 1 mL to each well of a 6-well plate. Let cells grow overnight in a humidified 28°C/100% air incubator.
2. The following day, add the appropriate amount of dsRNA. The normal range of dsRNA is 2–30 μg dsRNA per treatment (e.g., Cul4 knockdown required 2 μg whereas Cdt2 and PCNA required 10 μg). Treat each sample with dsRNA two consecutive days while incubating at 28°C/100% air. A non-specific control (e.g., LacZ) should also be used to measure effect of dsRNA treatment. A subset of the cells may be collected and lysed for western blot analysis to measure efficiency of dsRNA knockdown. Optimization of the amount of dsRNA, and length of treatment, required to sufficiently knockdown a protein of interest may be required (**Note 5**).
3. Prior to heat shock induction, a control sample of cells may be taken to measure baseline cell cycle profile.

4. Transfer the rest of the cells to a 37°C/100% air incubator for 30 min to induce *Hsp70* expression.
5. After 30 min, return cells to 28°C/100% air incubator and remove cells for appropriate time course.

### 3.5. Fixing S2 Cells for Flow Cytometric Analysis

1. Collect all the cells from each well by pipetting gently up and down and transfer each to a labeled 2-mL round bottom Eppendorf tube.
2. Centrifuge tubes for 5 min at 500 × *g*. Pour out supernatant, flick tube gently to break up pellet. Wash cells with 500 μL 1× phosphate-buffered saline solution (PBS), centrifuge 5 min at 2300 rpm and gently tap tube to break up pellet.
3. Add 500 μL 1% paraformaldehyde in ice-cold 1× PBS and keep on ice for 30 min. Make the paraformaldehyde solution immediately prior to use by combining 700 μL 10× PBS, 218 μL 32% paraformaldehyde stock solution and bringing total volume to 7 mL.
4. Centrifuge tubes at 4° for 5 min at 500 × *g*. Remove paraformaldehyde solution into proper waste container. Gently tap tube to break up pellet.
5. Wash cells with ice-cold 1× PBS. Centrifuge tubes at 4° for 5 min at 500 × *g*, pour out supernatant, and flick to break up pellet.
6. Add 500 μL 1× PBT (PBS with 0.1% Tween-20), let sit 15 min at room temperature. This increases the permeability of the cell membranes.
7. Centrifuge for 5 min at 500 × *g* at room temperature. Remove supernatant and tap tube to break up pellet.
8. Add 3 μL DNase-free RNase. Incubate for 30 min at 37°C. It is critical to degrade RNA as propidium iodide binds equally to DNA and RNA.
9. Add 5 μL 15 mM Propidium iodide (**Note 6**) and 500 μL ice-cold 1× PBS. Leave at 4°C overnight. Fixed and stained cells can be kept up to a week until flow cytometric analysis, kept at 4°C, and protected from light.

### 3.6. Flow Cytometric Analysis

1. We recommend proper training from be received prior to beginning a flow cytometry experiment. The following method outlines a protocol for measuring cell cycle profiles in S2 cells on a CyAn (Dako, Beckman Coulter) (**Note 7**), but does not address all the complexities of flow cytometric analysis or the various other operating systems that can be used.
2. Transfer all cells to Falcon tubes appropriate for the flow cytometer being used. Break up any clumps of cells prior to analysis by gently vortexing tubes.

3. Establish the correct protocol using Summit 4.3 software. GFP detection uses a 488-nm laser and detection through a 530/30 bandpass filter. Propidium iodide uses a 488-nm laser and detection through a 630/30 bandpass filter.
4. Create an acquisition template with plots by selecting "Create Histograms" from the menu options and create the following plots measuring the following parameters (*x* and *y* axis, respectively): SS Area vs. FS Area, PI Lin vs. PI Area, PI Area vs. GFP Log, Event Count vs. GFP Log, Event Counts vs. PI Area, Even Count vs. PI Area (of only GFP positive gate). Save this template as a new protocol (**Fig. 2**).
5. Using the Gate Logic Builder and the Gate Scheme applications, create the following gates on your histograms, as follows:
   a. SS Area vs. FS Area: Avoid bottom left corner where debris will be measured (R1).

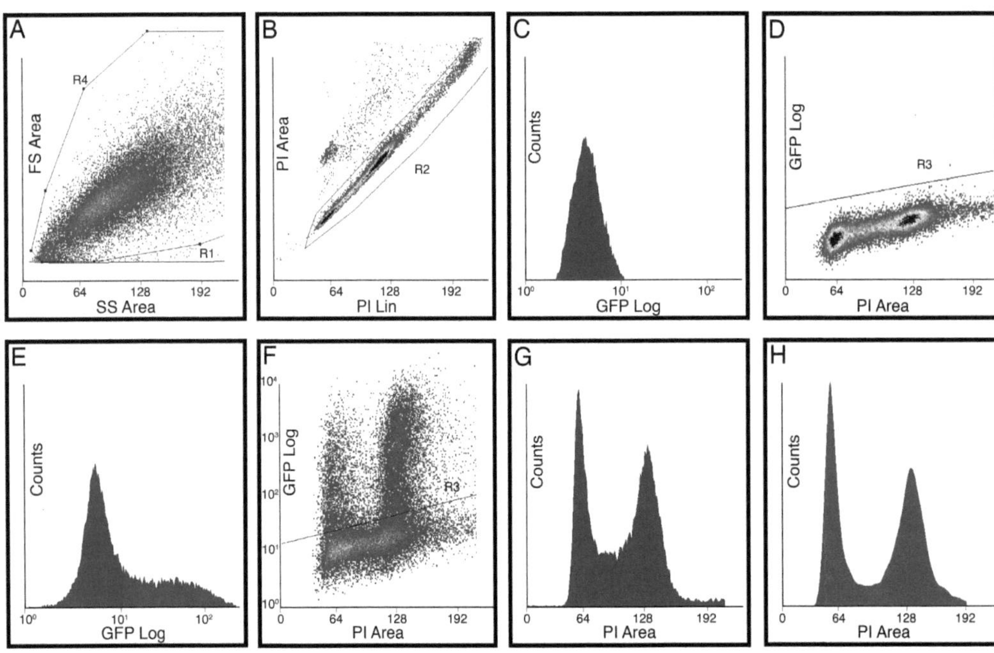


Fig. 2. Gating protocol for generating a cell cycle profile of *Drosophila* S2 cells, and measuring GFP fluorescence intensity using Summit software (*x* and *y* axis, respectively). **a**: SS Area vs. FS Area. Gate R1 and R4 to collect only cells, not debris. **b**: PI Linear vs. PI area. Gate R2 to only collect single cells, not doublets or triplets, which may skew laser intensity readings. **c**: GFP log vs. Counts on untransfected control cells. GFP laser intensity should be set so the peak falls between $10^0$ and $10^1$. **d**: PI Area vs. GFP log on untransfected control cells. Set R3 gate so that no more than 0.05% of untransfected cells falls in the GFP positive range. **e**: GFP log vs. Counts for GFP-E2f1 transfected cells. Note shift of GFP intensity to $10^2$ range. **f**: PI area vs. GFP log measurement of GFP-E2f1 transfected cells. Note shift of cell population now within the GFP positive gate (27% of all cells). **g**: PI area vs. Counts of all cells measured. First peak (64) represents G1; second peak (128) represents G2/M. **h**: PI area vs. Counts of GFP positive cells. Note that fewer cells appear in the area between the G1 and G2/M peaks, reflecting the targeted destruction in S phase of GFP-E2f1.

b. PI Lin vs. PI Area: Avoid doublets and clumped cells (R2).

c. PI Area vs. GFP Log: Create a gate that will select only GFP positive cells (the threshold will be determined by running the untransfected control).

6. Under the Sample menu, assign a root file name, a starting file number and a data storage location. Limit total events collected (<1,000,000). Under the Preferences menu, set parameters to "Auto-Start" and "Auto-Boost." However, occasionally manual Boost will be required to activate the collection.
7. Run the untransfected control cells to set the GFP negative parameter. Aim for less than 0.05% of cells falling above the GFP cut off in the untransfected samples. Adjust laser voltage to generate single peak in the GFP Log vs. Counts histogram that falls between $10^0$ and $10^1$. Adjust gates to ensure that no debris and only single cells are being collected. Once parameters are set for negative control, do not adjust for the remainder of the experiment.
8. When acquiring cells, ensure that cells following "Boost" are not collected and cells run through the analysis between 200 and 400 cells per second. Running the cells too quickly will decrease the accuracy of the data due to decreasing the time for excitation and detection.
9. Run pHGW-only (GFP only) control to measure efficiency of transfection and heat shock induction. Aim to collect at least 10,000 experimental events (GFP positive cells after all the gating). The minimum number needed for statistical analysis is 4000. 20,000 cells is optimal.
10. Ensure that data are properly stored after each collection before starting the next. Take care that the gates ensure that only the experimental data are saved, not debris, which could quickly overfill your saving capacity.
11. In between samples, activate "Backflush" under the Acquisition menu to clear out any residual cells. Rinse the CyAn uptake nozzle with distilled, deionized water.

### 3.7. Data Analysis

1. There are two options for analyzing flow cytometry data to visualize cell cycle coupled destruction, such as is seen with *Drosophila* E2f1. Visually, two histograms can be overlaid, using the Summit 4.3 software, and color labeled to distinguish the total cell population as compared to the GFP-positive population alone. This provides an easy way to quickly visualize S phase-specific destruction (**Fig. 3**).
2. To observe more subtle changes in the cell cycle, or for a more quantifiable approach, statistical software can calculate

the percentages of cells in each phase of the cycle, in addition to apoptotic (i.e., sub-$G_1$) cells. This will be discussed further.

3. Open ModFit LT software and open the file to analyze. Select Area PI as the Parameter for Analysis. Define three gates as follows (*x* and *y* axis, respectively). Gate 1: SS Area vs. FS Area. Gate 2: Linear PI vs. Area PI. Gate 3: Area PI vs. Log GFP. Adjust the gates to closely reflect how the data were collected (**Fig. 3**).
4. Initially, include all cells in Area PI vs. Log GFP gate. This will give you the whole cell cycle profile. Open Mod window to set the parameters. For S2 cells, use the following constraints: Linearity: 1.93, Standards: 0, Number of cycles: 1, Model Template: Diploid, Range Positions: Compute Range Positions. This will allow the computer to appropriately predict G1 and G2 peaks. Utilizing Auto-Aggregation, Auto-Debris and Apoptosis modeling functions is depended on

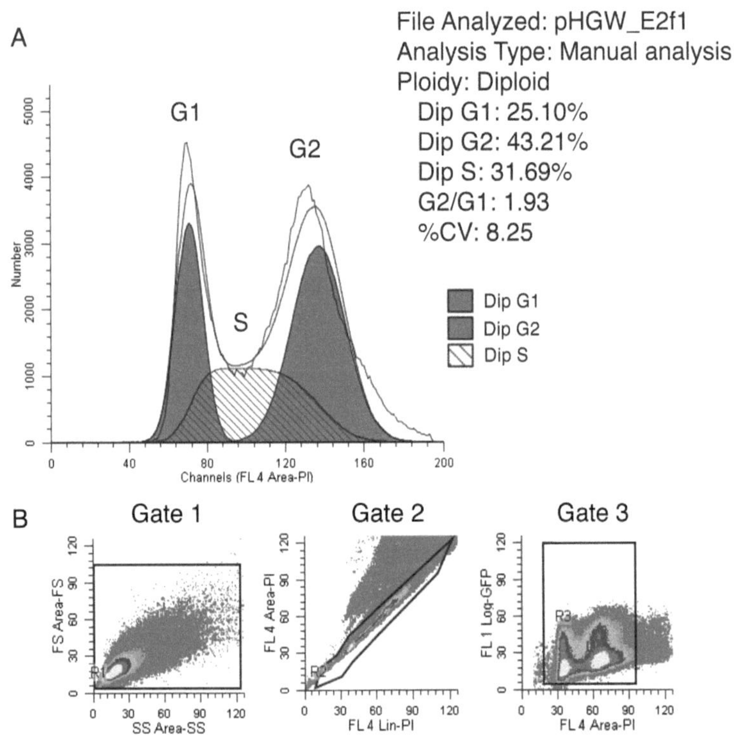


Fig. 3. ModFit LT analysis. **a**: Histogram representing the calculated percentage of cells in each phase of the cell cycle and the confidence in these values. Shaded peaks represent the calculated profile of G1 and G2 peaks; *dashed lines* represent the calculated percentage of cells in S phase. **b**: The gating required to analysis flow cytometry samples (*x* and *y* axis, respectively). Gate 1: SS area vs. FS area. Gate 2: PI Linear vs. PI Area. Gate 3: Area PI vs. GFP Log. Include all cells in this gate to get a total cell population profile, and gate only the GFP positive cells to measure the cell cycle profile of GFP-E2f1 transfected cells.

the samples and the data required. Select Range, and adjust the automatically applied G1 and G2 ranges to most accurately fit the data. Then select Fit, and the software will calculate the percentage of cells in $G_1$, S, and $G_2/M$. The strength of confidence is an %CV value, which should be under 10 for high confidence in the data. Low experimental cell numbers can reduce this confidence value.

5. Repeat this with the GFP positive only cell population to compare the values. A S phase-coupled destruction will result in a significant decrease in GFP positive cells in S phase. RNAi that depletes a component of the destruction mechanism will result in a flow cytometry profile that resembles a GFP-only control. Basic spreadsheet software (e.g., Excel) can create graphical representations of the percentages of cells in each phase (**Fig. 4**).

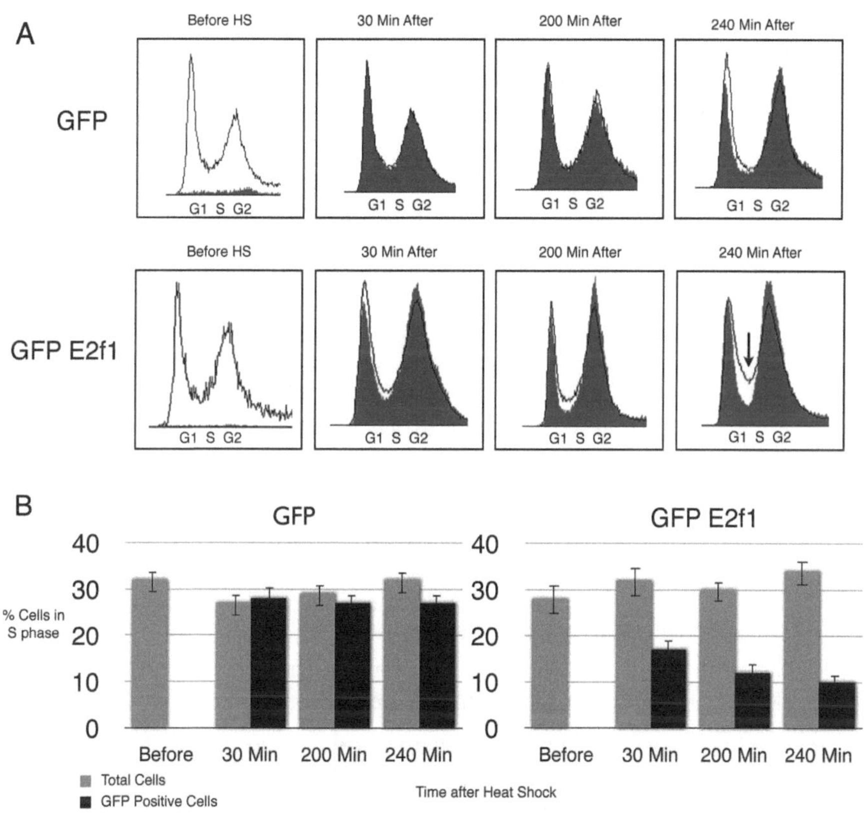


Fig. 4. Representations of flow cytometric data. **a**: Time course of heat shock induction of GFP and GFP-E2f1 expression in stable S2 cell lines. Overlay of total cell population (*unfilled histogram*) with GFP positive cells (*filled histogram*). Samples taken before heat shock induction and for the indicated times after heat shock. Note that by 240 min after heat shock expression of GFP-E2f1, there is a reduction in the number of S phase cells that are GFP positive (*arrow*), which represents S phase-coupled destruction of GFP-E2f1. **b**: Graphical representation of data collected in ModFit analysis. Percentages of S phase cells in the total cell population compared to the percentage of S phase cells in the GFP positive cell population at different times after induction of GFP-E2f1. Note the sharp decline in GFP-E2f-expressing S phase cells. Error bars represent standard error.

## 4. Notes

1. There are many multiparameter flow cytometer systems available that are functional to collect cell cycle data. The main requirement is that the system be able to excite and detect GFP signal vs. Propidium iodide (PE) staining. Similarly, many operating systems are publicly available for cytometric analysis, including FlowJo (TreeStar) and TestDNA (Verity Software).
2. Propidium iodide is a DNA intercalating agent and must therefore be treated appropriately and discarded in a suitable manner.
3. Fetal calf serum (FCS) must be heat-inactivated prior to use by heating in a water bath for 30 min at 56°C. It can then be placed in 60-mL aliquots and stored at –20°C.
4. Simple laminar flow hoods are sufficient when working with *Drosophila* S2 cells, as they are not thought to carry vectors of human disease. However, good sterile technique should be utilized to prevent contamination of the cells. Once contaminate, there is little to no possibility of recovery. Therefore, it is a good idea to freeze an aliquot of all cell lines generated. We recommend freezing $2 \times 10^7$ cells from a single plate in 2.5 mL of freezing medium in $5 \times 0.5$ mL aliquots and storing in liquid nitrogen.
5. Since RNase is highly ubiquitous and ssRNA is unstable, we recommend wearing gloves and maintaining a clean environment when working with RNA. DEPC-treated water and RNAse-free plastics should be used when generating dsRNA. dsRNA itself is fairly stable; however, multiple freeze–thaw cycles should be avoided.
6. RNAi causes a reduction in gene expression and is not equivalent to a complete loss of function. Western blot analysis after dsRNA treatment is recommended to determine the efficiency of knockdown. Various responses to RNA interference occur depending on the threshold of activity necessary for each particular gene product.
7. Other selection methods are available. Methotrexate resistance plasmids: pHGCO, pHCO, p8hCHO, actDHFR, (available from Drosophila Genomics Resource Center (DGRC)). Methotrexate (Sigma) (L-amethopterin) used at a final concentration of $2 \times 10^{-7}$ M (store stock solution protected from light at –20°). α-Amantinin resistance plasmids: pPC4 is available at DGRC. α-Amantinin is used at a final concentration of 5–10 μg/mL. Blastocidin S resistance: pCoBlast (available from Invitrogen). Blastocidin S used at approximately 5 μg/mL (per Invitrogen).

## References


1. Jin, J., Arias, E. E., Chen, J., Harper, J. W., and Walter, J. C. (2006) A family of diverse Cul4-Ddb1-interacting proteins includes Cdt2, which is required for S phase destruction of the replication factor Cdt1. *Mol Cell 23*, 709–21.
2. Higa, L. A., Banks, D., Wu, M., Kobayashi, R., Sun, H., and Zhang, H. (2006) L2DTL/CDT2 interacts with the CUL4/DDB1 complex and PCNA and regulates CDT1 proteolysis in response to DNA damage. *Cell Cycle 5*, 1675–80.
3. Nishitani, H., Sugimoto, N., Roukos, V., Nakanishi, Y., Saijo, M., Obuse, C., Tsurimoto, T., Nakayama, K. I., Nakayama, K., Fujita, M., Lygerou, Z., and Nishimoto, T. (2006) Two E3 ubiquitin ligases, SCF-Skp2 and DDB1-Cul4, target human Cdt1 for proteolysis. *Embo J 25*, 1126–36.
4. Senga, T., Sivaprasad, U., Zhu, W., Park, J. H., Arias, E. E., Walter, J. C., and Dutta, A. (2006) PCNA is a cofactor for Cdt1 degradation by CUL4/DDB1-mediated N-terminal ubiquitination. *J Biol Chem 281*, 6246–52.
5. Hu, J., and Xiong, Y. (2006) An evolutionarily conserved function of proliferating cell nuclear antigen for Cdt1 degradation by the Cul4-Ddb1 ubiquitin ligase in response to DNA damage. *J Biol Chem 281*, 3753–6.
6. Arias, E. E., and Walter, J. C. (2006) PCNA functions as a molecular platform to trigger Cdt1 destruction and prevent re-replication. *Nat Cell Biol 8*, 84–90.
7. Hall, J. R., Lee, H. O., Bunker, B. D., Dorn, E. S., Rogers, G. C., Duronio, R. J., and Cook, J. G. (2008) Cdt1 and Cdc6 are destabilized by rereplication-induced DNA damage. *J Biol Chem 283*, 25356–63.
8. Kim, Y., Starostina, N. G., and Kipreos, E. T. (2008) The CRL4Cdt2 ubiquitin ligase targets the degradation of p21Cip1 to control replication licensing. *Genes Dev 22*, 2507–19.
9. Abbas, T., Sivaprasad, U., Terai, K., Amador, V., Pagano, M., and Dutta, A. (2008) PCNA-dependent regulation of p21 ubiquitylation and degradation via the CRL4Cdt2 ubiquitin ligase complex. *Genes Dev 22*, 2496–506.
10. Nishitani, H., Shiomi, Y., Iida, H., Michishita, M., Takami, T., and Tsurimoto, T. (2008) CDK inhibitor p21 is degraded by a proliferating cell nuclear antigen-coupled Cul4-DDB1Cdt2 pathway during S phase and after UV irradiation. *J Biol Chem 283*, 29045–52.
11. Kim, S. H., and Michael, W. M. (2008) Regulated proteolysis of DNA polymerase eta during the DNA-damage response in C. elegans. *Mol Cell 32*, 757–66.
12. Centore, R.C., Havens, C.G., Manning, A.L., Li, J.M., Flynn, R.L., Tse, A., Jin, J., Dyson, N.J., Walter, J.C., and Zou, L. (2010) CRL4$^{Cdtw}$-mediated destruction of the histone methyltransferase Set8 prevents premature chromatin compaction in S phase. *Mol Cell 40*, 22–33.
13. Abbas, T., Shibata, E., Park, J., Jha, S., Karnani, N., and Dutta, A. (2010) CRL4(Cdt2) regulates cell proliferation and histone gene exptression by targeting PR-Set7/Set8 for degradation. *Mol Cell 40*, 9–21.
14. Oda, H., Hubner, M.R., Beck, D.B., Vermeulen, M., Hurwitz, J., Spector, D.L., and Reinberg, D. (2010) Regulation of the histone H4 monomethylase PR-Set7 by CRL4(Cdt2)-mediated PCNA-dependent degradation during DNA damage. *Mol Cell 40*, 364–76.
15. Shibutani, S. T., de la Cruz, A. F., Tran, V., Turbyfill, W. J., 3rd, Reis, T., Edgar, B. A., and Duronio, R. J. (2008) Intrinsic negative cell cycle regulation provided by PIP box- and Cul4Cdt2-mediated destruction of E2f1 during S phase. *Dev Cell 15*, 890–900.
16. Havens, C. G., and Walter, J. C. (2009) Docking of a specialized PIP Box onto chromatin-bound PCNA creates a degron for the ubiquitin ligase CRL4Cdt2. *Mol Cell 35*, 93–104.
17. Moldovan, G. L., Pfander, B., and Jentsch, S. (2007) PCNA, the maestro of the replication fork. *Cell 129*, 665–79.
18. van den Heuvel, S., and Dyson, N. J. (2008) Conserved functions of the pRB and E2F families. *Nat Rev Mol Cell Biol 9*, 713–24.
19. Asano, M., Nevins, J. R., and Wharton, R. P. (1996) Ectopic E2F expression induces S phase and apoptosis in Drosophila imaginal discs. *Genes Dev 10*, 1422–32.
20. Reis, T., and Edgar, B. A. (2004) Negative regulation of dE2F1 by cyclin-dependent kinases controls cell cycle timing. *Cell 117*, 253–64.
21. Heriche, J. K., Ang, D., Bier, E., and O'Farrell, P. H. (2003) Involvement of an SCFSlmb complex in timely elimination of E2F upon initiation of DNA replication in Drosophila. *BMC Genet 4*, 9.
22. Shibutani, S., Swanhart, L. M., and Duronio, R. J. (2007) Rbf1-independent termination of E2f1-target gene expression during early Drosophila embryogenesis. *Development 134*, 467–78.

# Chapter 16

# Assessing $G_1$-to-S-Phase Progression After Genotoxic Stress


## Michael DeRan, Mary Pulvino, and Jiyong Zhao



## Abstract

Maintenance of genomic integrity is critical for the survival of organisms. Thus, mammalian cells employ a complex DNA damage response that can sense and repair DNA damage. One important aspect of the cellular DNA damage response is the activation of checkpoints that result in cell cycle arrest. In this chapter we present methods for the induction of genotoxic stress. Additionally, we describe methods for studying the progression of cells from $G_1$ to S phase after genotoxic stress.



**Key words:** Cell cycle, checkpoint, DNA double-strand breaks, $G_1$, S phase, propidium iodide, bromodeoxyuridine, γ-H2AX, immunofluorescence.


## 1. Introduction

The success of an organism requires the ability to transmit an intact copy of its genome to its progeny. This task is complicated by the sheer number of genotoxic insults that assail organisms from exogenous as well as endogenous sources including ultraviolet and ionizing radiation, oxidative metabolic byproducts, chemotherapeutics, and errors in replication (1–6). It is thus necessary for organisms to employ complex mechanisms of DNA surveillance and repair. These processes are coupled to a number of cell cycle checkpoints which allow for proliferative arrest, DNA repair, and, in cases of extreme genetic insults, apoptosis. Breakdown of these checkpoints can result in a number of human disorders often characterized by immunodeficiency and a predisposition to certain types of cancer (3–4).


W.X. Li (ed.), *Cell Cycle Checkpoints*, Methods in Molecular Biology 782,
DOI 10.1007/978-1-61779-273-1_16, © Springer Science+Business Media, LLC 2011

The generation of DNA double-strand breaks (DSBs), such as those caused by ionizing radiation or radiomimetic drugs, potently activates cell cycle checkpoints. In $G_1$ phase, DSBs are typically recognized by the multiprotein MRN (MRE11, Rad50, and NBS1) complex, which is responsible for activating the signaling kinase ATM (Ataxia Telangiectasia Mutated) (1). After activation, ATM is able to phosphorylate a number of targets including the histone H2A variant H2AX, producing γ-H2AX, and the signal transduction kinase CHK2 (1, 7). ATM and CHK2 cooperate to activate and stabilize the tumor suppressor p53 by directly targeting p53 as well as targeting MDM2, the ubiquitin ligase that is normally responsible for the rapid turnover of p53 protein (8–12). Among the transcriptional targets of p53 is the cyclin dependent kinase (CDK) inhibitor $p21^{CIP1/WAF1}$, which inhibits the activity of the S-phase promoting cyclin E/Cdk2 complex (1, 3–4). The inhibition of cyclin E/Cdk2 results in cell cycle arrest at the $G_1$/S phase boundary, as well as the down-regulation histone gene expression (13).

Experimentally, activation of the $G_1$ checkpoint can be studied by monitoring the inhibition of cell cycle progression from $G_1$ to S phase after genotoxic insults. These insults are commonly introduced using radiation (ionizing or ultraviolet) or radiomimetic drugs. After genotoxic insults, progression from $G_1$ to S can be monitored by flow cytometry to assay for DNA content or the incorporation of 5-bromo-2-deoxyuridine (BrdU), a synthetic nucleotide that is incorporated into cellular DNA during replication.

Here we provide a number of protocols for studying the progression of cells from $G_1$ to S phase after genotoxic stress. In these protocols, genotoxic insults are introduced by γ-irradiation or the use of the radiomimetic drug, bleomycin. These sources of genotoxic insults produce DNA double-strand breaks in addition to other genotoxic lesions. The induction of DNA double-strand breaks by these treatments can be assessed by the immunofluorescent staining of γ-H2AX, a marker for DSBs (**Fig. 1**) (14–15).

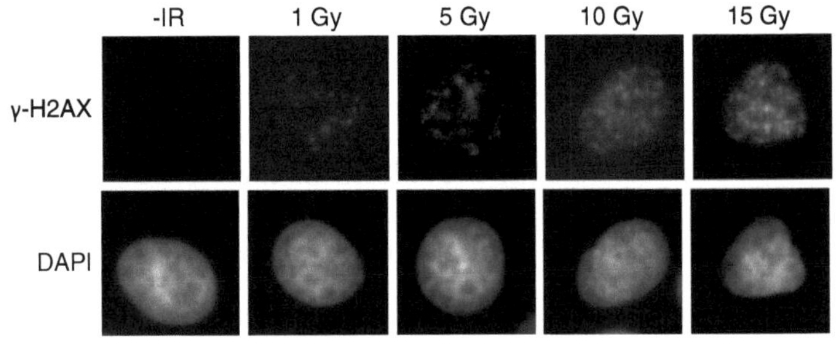


Fig. 1. U2OS cells were γ-irradiated with the indicated dose. 30 min after irradiation the cells were fixed and accumulation of γ-H2AX foci was examined by immunofluorescent staining. DAPI was employed to label the nucleus.

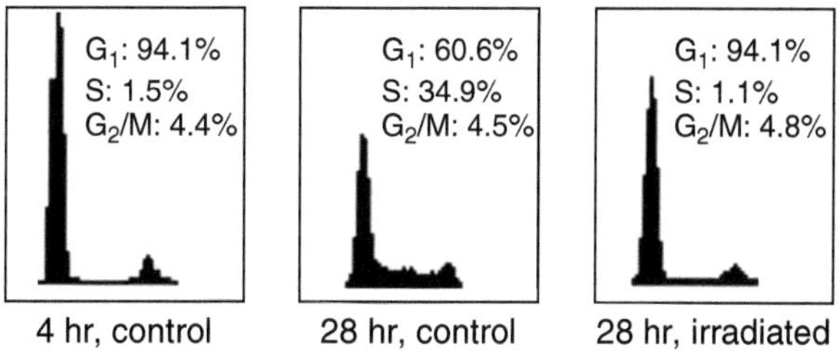


Fig. 2. WI38 cells were first arrested at $G_0$ by serum starvation and then stimulated to enter the cell cycle by serum addition. At 4 h after the addition of serum the cells were harvested (4 h, control), treated with IR (28 h, irradiated), or left untreated (28 h, control). The cell cycle profile of the cells is shown. (from Su et al. [13])

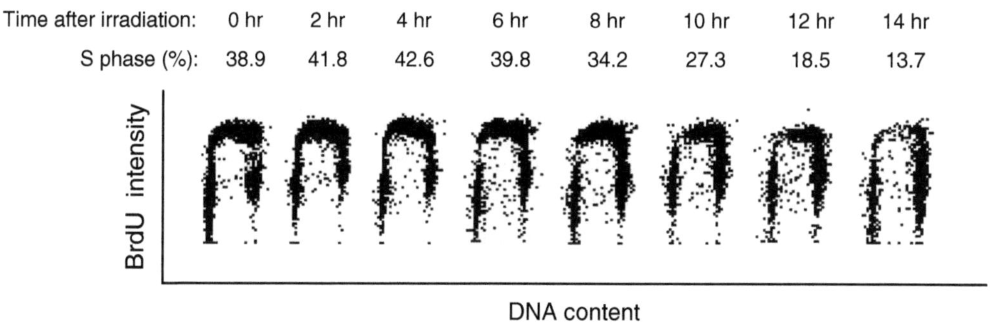


Fig. 3 Asynchronously growing HCT116 cells were γ-irradiated (12 Gy) and assayed at the indicated time points for DNA content and BrdU incorporation. (from Su et al. [13])

Genotoxic insults can be introduced into asynchronously growing cells. Alternatively, cells can be synchronized in $G_0$ first and then treated with DNA damaging agents. This synchronization allows for monitoring progression from $G_1$ to S phase in a majority of cells after genotoxic insults. Cell cycle progression is assessed by monitoring DNA content with the fluorescent DNA-binding dye propidium iodide (**Fig. 2**) or BrdU incorporation (**Fig. 3**).

## 2. Materials

### *2.1. Cell Cycle Synchronization by Serum Starvation*

1. Normal growth medium: Cells should be cultured in the appropriate medium for the cells of interest. Growth medium is usually supplemented with 10% fetal bovine serum and antibiotics.
2. Starvation medium: The appropriate medium for the cells of interest supplemented with 0.1% FBS.

### 2.2. Induction of DNA breaks

1. Cesium-137 irradiator in a shielded room. Use extreme caution when operating the irradiator. Be sure to read and follow all safety precautions for your instrument.
2. Bleomycin: A stock solution is prepared by dissolving bleomycin in phosphate-buffered saline (*see* **Section 2.3**, Step 4) at 10 mg/mL. This solution is filter sterilized and stored at –20°C for several months.

### 2.3. Assessment of DNA Breaks by Immunofluorescent Staining of γ-H2AX

1. 12-mm round glass coverslips
2. 24-well cell culture dish
3. Dumont No. 5 Jewelers Forceps
4. Phosphate-buffered saline (PBS): 137 mM NaCl, 2.7 mM KCl, 10 mM $Na_2HPO_4$, 1.5 mM $KH_2PO_4$, pH 7.4
5. Paraformaldehyde diluted to 4% (v/v) in PBS
6. Phosphate-buffered saline with Tween-20 (PBS-T): PBS containing 0.2% (v/v) Tween-20
7. Permeabilization solution: PBS-T containing 0.5% (v/v) Triton X-100
8. Blocking solution: PBS-T with 5% (v/v) horse serum, 5% (v/v) goat serum, 0.1% (v/v) fish gelatin
9. Primary antibody: anti-phospho-histone H2A.X (Ser139), clone JBW301 (Millipore, Billerica, MA)
10. Secondary antibody: Alexa Fluor 488 goat anti-mouse IgG (Invitrogen, Carlsbad, CA)
11. 4′,6-Diamidino-2-phenylindole dihydrochloride (DAPI): A stock solution of 2 mg/mL DAPI in PBS is stored at –20°C and diluted 1:40,000 when needed.
12. Vectashield mounting medium (Vector Laboratories, Burlingame, CA)
13. Glass microscope slides
14. Clear nail polish

### 2.4. Propidium Iodide Staining and Flow Cytometric Analysis of DNA Content

1. EDTA/PBS: 0.1% EDTA in PBS.
2. 80% Ethanol.
3. PBS with 0.1% FBS.
4. Propidium iodide (PI): A 50× stock solution is prepared by dissolving PI at 0.5 mg/mL in 38 mM sodium citrate, pH 7.0. The stock solution is then stored at 4°C in the dark. A working solution is obtained by diluting the stock solution 1:50 in PBS.
5. RNase A: A 40× stock solution is prepared by dissolving RNase A at 10 mg/mL in 10 mM Tris–HCl, pH 7.5, 15 mM NaCl. Boil for 15 min. After cooling to room temperature,

the solution is aliquoted and stored at –20°C. The working solution is prepared by diluting the stock solution 1:40 in 1× PI solution.

6. 40-μm cell strainer.
7. 5-mL polystyrene round-bottom tubes.
8. Flow cytometer.

***2.5. BrdU Labeling and Propidium Iodide Staining***

1. 5-Bromo-2′-deoxyuridine (BrdU): A stock solution is prepared by dissolving BrdU in PBS at a concentration of 10 mM. This solution is filtered through a 0.2-μm filter, aliquoted, and stored at –20°C.
2. PBS containing 0.5% bovine serum albumin (BSA).
3. 2 N HCl with 0.5% (v/v) Triton X-100.
4. Neutralization Buffer: 0.1 M $Na_2B_4O_7$.
5. PBS containing 0.5% BSA and 0.5% (v/v) Tween-20.
6. Anti-BrdU (Becton Dickson, Franklin Lakes, NJ).
7. Fluorescein-conjugated anti-mouse IgG (Vector Laboratories, Burlingame, CA).
8. Propidium iodide (PI): A 50× stock solution is prepared by dissolving PI at 0.5 mg/mL in 38 mM sodium citrate, pH 7.0. The stock solution is then stored at 4°C in the dark. A working solution is obtained by diluting the stock solution 1:50 in PBS.
9. RNase A: A 40× stock solution is prepared by dissolving RNase A at 10 mg/mL in 10 mM Tris–HCl, pH 7.5, 15 mM NaCl. Boil for 15 min. After cooling to room temperature the solution is aliquoted and stored at –20°C. The working solution is prepared by diluting the stock solution 1:40 in 1× PI solution.
10. 40-μm cell strainer.
11. 5-mL polystyrene round-bottom tubes.

## 3. Methods

***3.1. Cell Cycle Synchronization by Serum Starvation***

1. Grow cells to 40–50% confluence in normal growth medium.
2. Wash cells twice with PBS to completely remove serum.
3. Add starvation medium containing 0.1% serum.
4. Incubate cells in starvation medium for 48–72 h depending on the cells of interest.

5. After starvation, cells can be stimulated to reenter the cell cycle by replacing the starvation medium with normal growth medium.
6. Harvest or treat the cells at the desired time.

### 3.2. Induction of DNA Breaks by Ionizing Irradiation

1. Grow cells to 70–80% confluence in normal growth medium. Alternately, synchronized cells prepared as described in **Section 3.1** can be used.
2. Irradiate the cells with the desired dose (*see* **Note 1**), using a Cesium-137 irradiator. An unirradiated control should be included in all experiments.
3. Incubate the cells until the desired time point for analysis.

### 3.3. Induction of DNA Breaks by Bleomycin

1. Grow cells to 70–80% confluence in normal growth medium. Alternately, cells prepared as described in **Section 3.1** can be used.
2. Aspirate medium and replace with medium containing the desired concentration of bleomycin (*see* **Note 2**).
3. Incubate the cells with bleomycin for 30 min then wash the cells with PBS and add normal growth medium.
4. Harvest the cells at the desired time point for analysis.

### 3.4. Assessment of DNA Double-Strand Breaks by Immunofluorescent Staining of γ-H2AX

1. Grow cells to 70–80% confluence or as described in **Section 3.1** on round glass coverslips in plates.
2. DNA breaks are induced by the methods described in **Section 3.2** or **3.3**.
3. Transfer each coverslip to individual wells of a 24-well tissue culture plate using jewelers forceps (*see* **Note 3**). When transferring coverslips it is important to keep the side on which the cells are growing facing up. This allows for proper exposure of the cells to solutions added to the wells.
4. Wash once with PBS.
5. Fix the cells with 4% paraformaldehyde for 10 min at room temperature. Remove the fixing solution.
6. Wash the coverslips twice with PBS.
7. Permeabilize the cells by incubating with permeabilization solution for 10 min at room temperature. Remove the permeabilization solution.
8. Wash the coverslips twice with PBS-T.
9. To prevent non-specific binding of antibodies, incubate the cells with the blocking solution for 20 min at room temperature. Remove the blocking solution.

10. Add primary antibody diluted in blocking solution to each well (*see* **Notes 4** and **5**). Incubate for 1 h at room temperature.
11. Wash coverslips three times for 5 min each with PBS-T.
12. Add secondary antibody diluted in blocking solution to each well (*see* **Notes 5, 6**, and **7**). Incubate for 1 h in the dark at room temperature.
13. Wash coverslips three times for 5 min each with PBS-T.
14. Dilute the DAPI stock solution 1:40,000 in PBS and add the this solution to each well. Incubate 10 min in the dark at room temperature.
15. Wash coverslips two times for 5 min each with PBS.
16. Place a small drop of Vectashield mounting medium on a glass microscope slide.
17. Remove a coverslip from the plate using jewelers forceps.
18. Carefully remove excess solution from the coverslip by tapping the edge of the coverslip on a paper towel.
19. Mount the coverslip cell-side down in the drop of mounting medium.
20. Remove excess mounting solution by gently pressing down on the coverslip with a paper towel.
21. To prevent drying, seal the coverslips by applying clear nail polish to the edges.
22. View the slides on an epifluorescence microscope.

### 3.5. Propidium Iodide Staining and Flow Cytometric Analysis of DNA Content

1. Harvest cells by incubation with 0.1% EDTA/PBS for 5 min at 37°C.
2. Resuspend the cells by pipetting up and down several times (*see* **Note 8**).
3. Centrifuge the suspended cells at 250 × *g* for 5 min. Discard the supernatant.
4. Wash once with PBS.
5. Resuspend the cell pellet in 0.5 mL PBS.
6. Fix the cells by adding 6 mL of 80% ethanol dropwise while gently vortexing to prevent cell clumping.
7. Incubate the cells in ethanol for at least 15 min at 4°C. Cells can be stored this way for several weeks.
8. Centrifuge the fixed cells at 250 × *g* for 5 min. Discard the supernatant.
9. Wash the cells once with PBS containing 0.1% serum.

10. Resuspend the cells in propidium iodide/RNase A solution at a final concentration of approximately $10^6$ cells/mL.
11. Pass the cell suspension through a 40-μm cell strainer and collect the strained suspension in a 5-mL round-bottom tube.
12. Incubate the cells in the dark for 30 min at 37°C.
13. Analyze the cell cycle distribution of the cells by flow cytometry.

### 3.6. BrdU Labeling and Propidium Iodide Staining

1. Add 30 μM BrdU to the medium 30 min before harvesting cells.
2. Remove the medium and wash the cells twice with PBS.
3. Harvest the cells by incubating with 0.1% EDTA/PBS and incubate for 5 min at 37°C.
4. Resuspend the cells by pipetting thoroughly (*see* **Note 8**).
5. Centrifuge the cells at 250×*g* for 5 min and discard the supernatant.
6. Wash twice with PBS.
7. Resuspend the cells thoroughly in 0.5 mL PBS.
8. Fix the cells by the dropwise addition of 6 mL of 80% ethanol while gently vortexing.
9. Incubate the cells on ice for at least 15 min.
10. Wash once with PBS containing 0.5% BSA.
11. Denature cellular DNA by incubating the cells in 0.5 mL of 2 N HCl/0.5% Triton X-100 for 20 min at room temperature.
12. Centrifuge the cells at 250×*g* for 5 min and discard the supernatant.
13. Resuspend the cells in 1 mL of neutralization buffer (0.1 M $Na_2B_4O_7$) and incubate for 2 min at room temperature.
14. Centrifuge the cells at 250×*g* for 5 min and discard the supernatant.
15. Wash the cells once with PBS containing 0.5% BSA and 0.5% Tween-20.
16. Centrifuge the cells at 250×*g* for 5 min and discard most of the supernatant. Leaving 30–50 μL of wash solution can help in resuspending the pellet in the next step.
17. Tap the tube to loosen the cell pellet and add 10 μL of anti-BrdU antibody per $10^6$ cells. Tap again to mix. Incubate for 30 min at room temperature.
18. Wash the cells twice with PBS containing 0.5% BSA and 0.5% Tween-20.

19. Centrifuge the cells at 250×*g* for 5 min and discard most of the supernatant. Leaving 30–50 μL of wash solution can help in resuspending the pellet in the next step.
20. Tap the tube to loosen the cell pellet and add 1.5 μL fluorescein-conjugated anti-mouse IgG per $10^6$ cells. Tap the tube again to mix. Incubate for 30 min at room temperature.
21. Wash twice with PBS containing 0.5% BSA and 0.5% Tween-20.
22. Resuspend the cells at a concentration of approximately $10^6$ cells/mL in propidium iodide/RNase A solution.
23. Pass the cell suspension through a cell strainer and collect the strained suspension in a 5 mL round bottom tube.
24. Incubate the cells in the dark for 30 min at 37°C.
25. Analyze the cell cycle distribution of the cells by flow cytometry.

## 4. Notes

1. It may be helpful to test a range of doses in initial experiments. We generally use doses in the range of 3–20 Gy.
2. We suggest testing a range of doses initially. In general, we use doses between 5 and 50 μg/mL.
3. Twenty-four-well tissue culture plates are convenient because they allow for removal of solutions from wells by decanting without removing the coverslips. This is especially important for cells that are poorly adherent and may be lost during aspiration.
4. We typically us a dilution of 1:5000 for the primary antibody, but you may wish to test a range of concentrations.
5. Be sure to add enough antibody to completely cover the coverslips.
6. We typically us a dilution of 1:1000 for the secondary antibody.
7. After addition of the secondary antibody it is important to minimize exposure to light. This prevents fading of the fluorescent signal. This can be accomplished by covering the plates with aluminum foil during washes or placing the plates in a drawer during incubations.
8. In each resupsension step of the flow cytometry protocols it is important to pipette cells thoroughly in order to obtain a single cell suspension.

## Acknowledgments

This work was supported by grant R01 GM 65814 from the National Institutes of Health.

### References


1. Bartek, J., and Lukas, J. (2007) DNA damage checkpoints: from initiation to recovery or adaptation, *Curr Opin Cell Biol 19*, 238–245.
2. Harper, J. W., and Elledge, S. J. (2007) The DNA damage response: ten years after, *Mol Cell 28*, 739–745.
3. Kastan, M. B., and Bartek, J. (2004) Cell-cycle checkpoints and cancer, *Nature 432*, 316–323.
4. Massague, J. (2004) G1 cell-cycle control and cancer, *Nature 432*, 298–306.
5. Petrini, J. H., and Stracker, T. H. (2003) The cellular response to DNA double-strand breaks: defining the sensors and mediators, *Trends Cell Biol 13*, 458–462.
6. Zhou, B. B., and Elledge, S. J. (2000) The DNA damage response: putting checkpoints in perspective, *Nature 408*, 433–439.
7. Shiloh, Y. (2003) ATM and related protein kinases: safeguarding genome integrity, *Nat Rev Cancer 3*, 155–168.
8. Chehab, N. H., Malikzay, A., Appel, M., and Halazonetis, T. D. (2000) Chk2/hCds1 functions as a DNA damage checkpoint in G(1) by stabilizing p53, *Genes Dev 14*, 278–288.
9. Craig, A., Scott, M., Burch, L., Smith, G., Ball, K., and Hupp, T. (2003) Allosteric effects mediate CHK2 phosphorylation of the p53 transactivation domain, *EMBO Rep 4*, 787–792.
10. Hirao, A., Kong, Y. Y., Matsuoka, S., Wakeham, A., Ruland, J., Yoshida, H., Liu, D., Elledge, S. J., and Mak, T. W. (2000) DNA damage-induced activation of p53 by the checkpoint kinase Chk2, *Science 287*, 1824–1827.
11. Khosravi, R., Maya, R., Gottlieb, T., Oren, M., Shiloh, Y., and Shkedy, D. (1999) Rapid ATM-dependent phosphorylation of MDM2 precedes p53 accumulation in response to DNA damage, *Proc Natl Acad Sci USA 96*, 14973–14977.
12. Maya, R., Balass, M., Kim, S. T., Shkedy, D., Leal, J. F., Shifman, O., Moas, M., Buschmann, T., Ronai, Z., Shiloh, Y., Kastan, M. B., Katzir, E., and Oren, M. (2001) ATM-dependent phosphorylation of Mdm2 on serine 395: role in p53 activation by DNA damage, *Genes Dev 15*, 1067–1077.
13. Su, C., Gao, G., Schneider, S., Helt, C., Weiss, C., O'Reilly, M. A., Bohmann, D., and Zhao, J. (2004) DNA damage induces downregulation of histone gene expression through the G1 checkpoint pathway, *EMBO J 23*, 1133–1143.
14. Bonner, W. M., Redon, C. E., Dickey, J. S., Nakamura, A. J., Sedelnikova, O. A., Solier, S., and Pommier, Y. (2008) GammaH2AX and cancer, *Nat Rev Cancer 8*, 957–967.
15. Thiriet, C., and Hayes, J. J. (2005) Chromatin in need of a fix: phosphorylation of H2AX connects chromatin to DNA repair, *Mol Cell 18*, 617–622.

# Chapter 17

# Indirect Immunofluorescence for Monitoring Spindle Assembly and Disassembly in Yeast

Jacob W. P. Keeling and Rita K. Miller

## Abstract

In yeast like all eukaryotes, microtubules are a crucial element of the mitotic spindle that separates the genetic material during cell division. The assembly status and position of the mitotic spindle, as well as cytoplasmic microtubules, can be monitored easily using indirect immunofluorescence with antibodies against tubulin. A detailed protocol is described for *Saccharomyces cerevisiae* that involves the fixation of actively growing cells, removal of the cell wall by enzymatic digestion, post-fixation, and the application of tubulin antibodies. The use of secondary antibodies conjugated to a fluorescent moiety permit visualization of the mitotic spindle by fluorescence microscopy. Methods for the reduction of background and pre-absorption of antibodies are discussed.

**Key words:** Fluorescence microscopy, microtubules, mitotic spindle, mitotic apparatus, *Saccharomyces cerevisiae*, tubulin.

## 1. Introduction

The budding yeast *Saccharyomyces cerevisiae* has become a laboratory favorite for the study of several microtubule-dependent processes, including assembly of the mitotic spindle and spindle positioning. This is owing not only to the pliancy of its genetic system, but also the simplicity of its microtubule structures, which can be viewed by indirect immunofluorescence. *S. cerevisiae* contains two sets of microtubules, the intra-nuclear microtubules, which comprise the major element of the mitotic spindle, and the cytoplasmic (or astral) microtubules, which play an important role in positioning the mitotic spindle through interactions with the cortex (1–5). In yeast, the spindle is assembled within the nucleus,

W.X. Li (ed.), *Cell Cycle Checkpoints*, Methods in Molecular Biology 782,
DOI 10.1007/978-1-61779-273-1_17, 

as the nuclear envelope remains intact at all stages of the cell cycle (6). Cytoplasmic microtubules are attached to the nucleus at the yeast's microtubule organizing center, or spindle pole body (SPB), which is embedded in the nuclear envelope (7, 8).

Microtubules display a series of characteristic patterns at various stages of the cell cycle. With a little practice and a good microscope, the two sets of microtubules can be distinguished morphologically by indirect immunofluorescence with anti-tubulin (**Fig. 1**) (9–11). In the unbudded G1 cell, one or more cytoplasmic microtubules emanate from the SPB, which is located at the center of the array (**Fig. 1**, **G1**). During early S-phase and the formation of a small bud, the SPBs have duplicated but have not separated (6, 12) and a spindle is not yet discernable by immunofluorescence. At this stage, cytoplasmic microtubules can sometimes be seen projecting into the bud. However, because microtubules are dynamic structures, undergoing growth and shrinkage at their plus ends, the cytoplasmic microtubules can also be found entirely within the mother cell, as shown in a deletion strain of *KAR9*, a gene important for

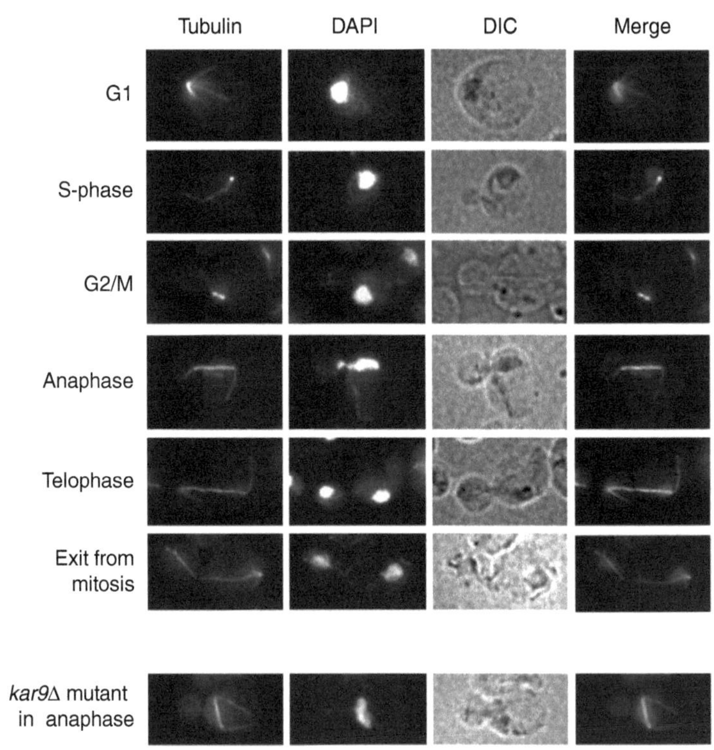


Fig. 1. Indirect immunofluorescence of microtubules. Wild-type cells were fixed and stained using rat anti-tubulin and goat anti-rat conjugated with Dy-Light as described in the protocol. The position of the nucleus was detected using DAPI. Cells representative of the various cell cycle stages were selected. A *kar9Δ* strain is shown the in the *bottom panel*.

orientation of cytoplasmic microtubules (*kar9*Δ panel of **Fig. 1**) (3). In yeast, the mitotic spindle assembles rapidly near the end of S-phase (12, 13) and is seen as a short bipolar spindle associated with the 4,6-diamidino 2-phenylindole (DAPI) stained nuclear region. The short spindle (**Fig. 1, G2/M**) is characterized by a 1.5- to 2.0-μM bar of immunofluorescence with a region of lower intensity at its center, corresponding to the two halves of the mitotic spindle. The break in fluorescence intensity is usually seen easily under the microscope, but is sometimes lost in published micrographs. Because the spindle contains multiple microtubules (14) and the cytoplasmic microtubule "bundle" is likely to contain fewer microtubules or perhaps even a single microtubule, cytoplasmic microtubules by immunofluorescence usually appear less intense than the spindle. At this stage, the cytoplasmic microtubules extend at varying angles from the end of the spindle. With the formation of the short spindle, the spindle begins to be positioned and oriented toward the bud neck through interactions of the plus-ends of the cytoplasmic microtubule with the bud neck (data not shown) (15, 16) and bud cortex (**Fig. 1, G2/M**) (2, 3–5). At anaphase, the nucleus is translocated and elongated through the mother–bud neck (17, 18). Using immunofluorescence, the elongated spindle spans the length of the DAPI mass. The cytoplasmic microtubules extend at an angle. At telophase, the spindle reaches the full length of the mother-bud axis and the two nuclear masses have separated. At this stage, the cytoplasmic microtubules are often seen as short extensions protruding at a distinct angle from the end of the spindle. At the end of mitosis, the spindle breaks down and the cells return to G1 (**Fig. 1, exit from mitosis**).

Microtubules are dynamically instable polymers and the images obtained of microtubules by indirect immunofluorescence represent an inherently different view of microtubule biology than live images of microtubules obtained using green fluorescent protein (GFP)-tubulin. Length measurements obtained by either method represent the average dynamicity of the microtubule, a measurement that is derived from a series of parameters that include the catastrophe frequencies, rescue frequencies, growth and shrinkage rates, and length of time spent in "pause" (2, 19). Using live microscopy, the parameter responsible for a difference in microtubule length can be deduced, whereas this information is not obtained from an indirect immunofluorescence image. In addition, live observations of GFP-tagged microtubules can reveal global patterns of microtubule movements such as "sweeping" and "sliding" of microtubules across the bud cortex that are not seen in static images (20). Another apparent difference between time-lapse observations of microtubules in live cells and the appearance of microtubules in fixed cells is the number of cytoplasmic microtubules observed. Live-cell images usually show multiple dynamic cytoplasmic microtubules (2). Fixed populations often show a single microtubule bundle. It may be possible that stable microtubules are uniquely retained during formaldehyde fixation.

Despite the several advantages of live microscopy, there are some important considerations in its use. Setting up a time lapse video microscopy system requires a substantial financial investment, followed by a significant time commitment in collecting and analyzing live images. Indirect immunofluorescence is a relatively simple technique, but nevertheless powerful. While a "snap shot in time," indirect immunofluorescence can provide a wealth of information about microtubules, especially when combined with a cell cycle analysis. As such, this technique has led to numerous advances in our understanding of both nuclear and cytoplasmic microtubules.

This chapter describes an indirect immunofluorescence method designed for the visualization of microtubules in the yeast, *S. cerevisiae*. Cells are fixed by treatment with formaldehyde. To allow antibodies access, the cell wall is removed by enzymatic digestion and post-fixed with methanol and acetone. Primary antibodies directed against tubulin are bound to microtubules. The signal is enhanced and visualized with secondary antibodies conjugated with a fluorescent moiety. This protocol is a modification of that described previously (21, 22), which was adapted from (9) and (10).

## 2. Materials

1. Anti-tubulin: YOL1/34 is a monoclonal tubulin antibody raised in rat and is available in the purified form (AbD Serotec, Raleigh, N.C., 0.5 mg, cat. no. MCA78G). Multiple freeze–thaw cycles have an adverse effect on most antibodies. On the initial thawing of the antibody, we recommend aliquoting it 100–150 μL per tube and storing it frozen at –80°C.
2. Beta mercaptoethanol (JT Baker, cat. no. JT4049-0).
3. Coplin jars: Two, tall enough to hold microscope slides (Wheatin #900570, VWR cat no. 25457-006).
4. Coverslips (Corning, No. 1 weight, 22 × 50 mM, VWR cat. no. 48396-068).
5. DAPI (4′, 6-diamidino-2-phenylindole) (Accurate Chemical and Scientific Corp., Westbury, NY, cat. no. 18860). To prepare a stock solution, dissolve 1 mg/mL in $H_2O$. This can be stored at –20°C for several years. We routinely re-freeze the remainder of the aliquot after use.

6. Formaldehyde: 37% solution, available from a number of manufacturers (JT Baker, cat. no. 2106-4). Because formaldehyde can oxidize at RT into formic acid, we prefer to use smaller 150-mL bottles. They are more likely to be used up in a shorter time period.
7. Gel loading pipette tips (Corning cat. no. 4853 or VWR cat. no. 29442-666).
8. Nail Polish: The use of clear or colored nail polish is a matter of personal preference. We prefer colored polish because it is easily distinguished from the mounting media that may ooze from beneath the coverslip.
9. Microscope slides, teflon coated (Polysciences, Inc., Warrington, PA cat. No.18357). We prefer the 10-well slides because we get less leakage of liquid between the wells. However, 14-well slides can also be obtained. If the amount of antibody available is limiting, then slides with more wells may be desirable because the volume per well is less.
10. Mounting media: *p*-phenylenediamine (Sigma, cat. no. P6001) functions as an anti-bleaching agent. In a 15-mL plastic conical tube, solubilize 10 mg of *p*-phenylenediamine in 1 mL of phosphate-buffered saline (PBS) by vortexing. Bring the volume to 10 mL with glycerol. Mix well. This should be stored wrapped in tin foil at –20°C. Use within 3 days, but should be discarded if it has turned light brown or honey colored or if it autofluoresces.
11. PBS: Dissolve 8 g of NaCl, 0.2 g of KCl, 1.44 g of $Na_2HPO_4$, and 0.24 g of $KH_2PO_4$ in 800 mL of $H_2O$, pH to 7.4 with HCl, and bring to a final volume of 1 L. This can be sterilized by autoclaving and stored at RT.
12. Poly-L-lysine (Molecular weight >300,000, Sigma cat. no. P-1524). To make a 0.5% stock solution, dissolve 50 mg in 10 mL of water. Aliquot and store frozen at –20°C. Thawed aliquots may be re-frozen.
13. Secondary antibody: Goat anti-rat conjugated with FITC (Jackson Immunoresearch West Grove, PA cat. no. 112-096-003). The new Dy-Light conjugated goat anti-rat secondary also works equally well, if not better (Jackson Immunoresearch West Grove, PA cat no. 112-486-003). Store as small aliquots at –80°C.
14. 1 M sorbitol/PBS: Dissolve 18.2 g of D-sorbitol (Sigma, cat. no. S-7547) in 100 mL PBS. This can be either filter sterilized or autoclaved. Store on a shelf at RT.
15. Zymolyase 100T (United States Biological, Swampscott, MA, cat. no. Z1004). The stock solution should be

made fresh. Dissolve 10 mg/mL Zymolyase in 1 M sorbitol/PBS. Mix well. The Zymolyase will not dissolve completely. To remove the particulate material, centrifuge at 13,500×*g* in micro-centrifuge for 3 min. Use the supernatant.

## 3. Methods

### 3.1. Growth and Fixation of Cells

1. Grow a 10-mL culture of the yeast strain of interest to early-exponential phase in the appropriate growth media ($4 \times 10^5 - 4 \times 10^6$ cells /mL) at 30°C. Yeast extract–peptone–dextrose (YPD) or synthetic-complete media work equally well in this protocol.
2. Transfer cells to a plastic conical tube with a screw cap. This decreases potential odor leaks from the formaldehyde fixation step and reduces the risk of broken culture tubes.
3. In a fume hood, fix the cells by adding formaldehyde directly to each culture to a final concentration of 3.7%. Formaldehyde is usually obtained as a 37% liquid solution. Adding one-tenth of the culture volume as formaldehyde is sufficient. Return the cultures to the shaking incubator at 30°C for 2 h (*see* **Notes 1** and **2**).
4. Wash the cells twice by collecting the cells at 2000×*g* in a tabletop centrifuge for 3 min. Discard the supernatant and resuspend the cells in PBS. Repeat. Cells can be stored overnight at 4°C at this step.
5. Collect the cells by centrifugation at 2000×*g* for 3 min and resuspend the cell pellet in 5 mL of 1 M sorbitol/PBS. Collect the cells at 2000×*g* for 3 min. Resuspend the cell pellet in 1 mL of 1 M sorbitol/PBS.

### 3.2. Digestion of Cells and Methanol–Acetone Fixation

1. Yeast cells adhere very poorly to untreated microscope slides. To circumvent this problem, microscope slides should be treated with poly-L-lysine. Add 10–15 μL of 0.5% poly-L-lysine to each well of a teflon-coated microscope slide. Let the poly-L-lysine drop evaporate. A hazy film will form on the slide. Wash the slide by soaking in a beaker of $dH_2O$ for 15 min. Swish the slide back and forth once or twice. Let the slide air dry. Slides can be prepared several hours in advance.
2. At least 2–3 h before they are needed, chill two Coplin jars at -20°C, one containing 100% methanol and the other with 100% acetone. Measure the level of each solvent so that it does not extend above the frosted end of the microscope slide reserved for writing. The ink of many common

laboratory "sharpies" is solubilized by methanol. The "VWR Lab Marker" is alcohol resistant, but smearing of the labeling can still occur.

3. Yeast cells have cell walls that must be digested in order for antibodies to gain access to the interior of the cell. This is accomplished by digestion with Zymolyase, an enzyme that hydrolyzes β-1, 3-glucan linkages of the glucose polymer within the yeast cell wall.
   Add 5 μL of β-mercaptoethanol to the resuspended cells. Add 30 μL of Zymolyase stock (10 mg/mL) at 30°C for 30–90 min. The optimal digestion can be determined by examining the cells under a phase contrast microscope. Properly digested cells will appear medium to dark gray. Bright glowing cells are incompletely digested. The pale, translucent cells are over digested and appear to have lost their internal organelles. These result in poor quality immunofluorescence. It is helpful to save out a 25-μL aliquot of undigested cells for comparison of digested versus undigested cells. Collect the resulting spheroplasts in a tabletop centrifuge at 2000×*g* for 3 min and discard the supernatant (*see* **Notes 3** and **4**).
4. Wash the cells by resuspending the pellet in 5 mL of 1 M sorbitol/PBS and centrifuging at 2000×*g* for 3 min. Resuspend the cells in 1 mL of 1 M sorbitol/PBS.
5. Attach the cells to the microscope slide by applying 15 μL of digested cells to each well of a Teflon-coated microscope slide. Let the cells adhere for 10–15 min.
6. Working quickly, pipette off the cells and immerse the slide in methanol (–20°C) for 6 min, prepared in Step 2 (**Fig. 2**). Remove the slide and very rapidly drain the excess methanol from the back of the slide by laying it on a Kimwipe for 1 s or less. Immediately immerse the slide in acetone (–20°C) for 30 s. Remove the slide and allow it to air dry (*see* **Notes 5** and **6**).

### 3.3. Antibodies

1. Centrifuge the YOL1/34 monoclonal tubulin antibody immediately prior to use at 13,500×*g* in a microcentrifuge in the cold for 2 min (*see* **Note 7**).
2. Apply 20–25 μL of the diluted tubulin antibody per well of the microscope slide. We usually use a dilution of 1:500 or 1:1000 in PBS. Incubate overnight at 4°C (*see* **Notes 8** and **9**). From this point on, it is very important to not let the slides dry out.
   To prevent the antibody from evaporating, incubations should be carried out in a humid chamber. This can be created by wetting several layers of Whatman filter paper or

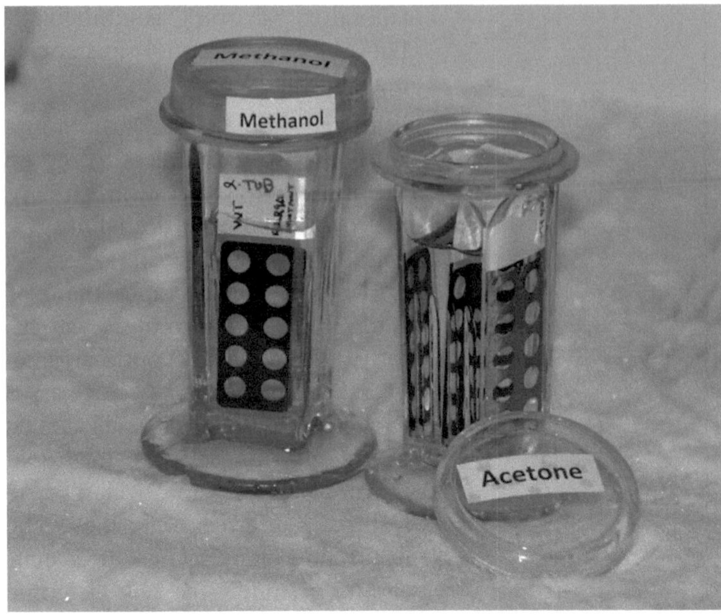


Fig. 2. Coplin jars used for post-fixation and permeabilization.

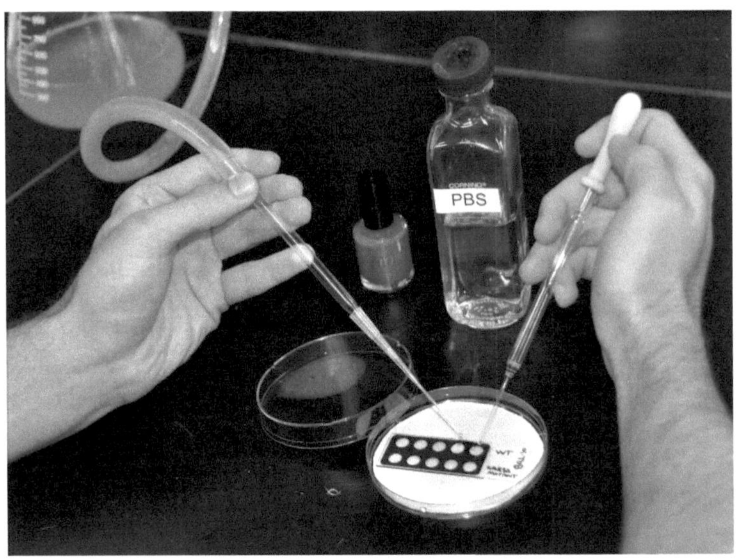


Fig. 3. Washing the slides with PBS. Work quickly so that the wells do not dry out.

paper towel that have been cut to fit the inverted lid of a petri dish (**Fig. 3**). Cover the microscope slide and wet blotting paper with the inverted petri dish.

3. Gently wash the cells four times with PBS by aspirating the antibody and quickly replacing the liquid with a drop of PBS. We use gel-loading pipette tips for the washing steps because

they are extra long and narrow. This reduces the force of the flow from the washes, resulting in fewer cells being washed off the slide.

4. Dilute the secondary antibody, in this case goat anti-rat conjugated with Dy-Light, 1:200 in PBS. Perform a clarifying spin at 13,500 g for 2 min. Remove the last PBS wash from the slide and apply 20–25 μL of diluted secondary antibody to each well. Incubate for 3–4 h at RT (*see* **Notes 9** and **10**).
5. Remove the secondary antibody and wash the cells four times with PBS, as described in Step 3.
6. In studying the position of the mitotic spindle, it is useful to identify the position of the nucleus. This is easily done using DAPI, a dye that intercalates into deoxyribonucleic acid (DNA) and fluoresces in the blue range. Thaw the stock solution of DAPI and dilute it 1:1000 in PBS. Perform a clarifying spin at 13,500×*g* for 2 min. Apply 20–25 μL to each well and incubate at RT for 15 min. Wash the cells twice with PBS (*see* **Note 11**).
7. The slides are now ready for mounting. Pipette off the excess PBS and apply one drop of mounting media to approximately every other well of the slide. A wooden applicator stick works well for this step. Carefully lay a coverslip over the wells. Using a dull object such as the reverse end of a yellow pipette tip, squeeze out the excess mounting media. Do not allow bubbles to form. Carefully wipe away the excess mounting media. Seal the edges of the coverslip/slide with nail polish and allow it to dry thoroughly. Rinse any extra mounting media from the surface of the coverslip with cool tap $H_20$ (*see* **Note 12**).

### *3.4. Microscopy*

The cells are now ready to be viewed under a microscope equipped with fluorescence optics. Most yeast work is done using a 100× oil objective with a 1.3 or 1.4 numerical aperture (NA). The cells shown in **Fig. 1** were photographed using an inverted IX81 Olympus microscope equipped with a UPlan Apo lens (1.4 NA), differential interference contrast optics, and an ORCA-ER2 CCD camera (Hamamatsu Corp., Hamamatsu City, Japan), also known as a charged-coupled device, which was controlled by Metamorph Premier version 7.6.3 software (Molecular Devices, Inc., Downingtown, PA). Image processing of contrast, brightness, and image overlays was performed using the Metamorph software, with minor adjustments in Adobe photoshop.

### *3.5. Controls*

Indirect immunofluorescence experiments are usually conducted with at least two controls, a no primary antibody control and a no antigen control.

1. For the no primary antibody control, simply omit the primary antibody from one well and add PBS in its place during that incubation period. This controls for the possibility that the observed signal could be due to an unexpected cross reactivity of the secondary antibody.
2. The no antigen control can take two forms. Yeasts strains are now available from Research Genetics, Inc., in which each open reading frame has been deleted. This is an important control that can be used when the protein of interest is non-essential for life. When epitope-tagged proteins (e.g., V5, HA, or myc) are used, the yeast strain containing non-epitope tagged version of the protein serves as the no antigen control.

### 3.6. Pre-absorption of Antibodies

For indirect immunofluorescence and other immunologically based applications, the use of monoclonal or affinity purified antibodies is usually preferred. However, there are instances in which this may not be feasible because the necessary antigen required for purification is not available. In which case, pre-absorption of the antibodies against a yeast strain lacking the antigen of interest can be a viable alternative, provided the necessary controls with deletion strains are carried out. Improved results for immunofluorescence have been seen when both the primary and secondary antibodies have been pre-absorbed against an "antigen-minus" strain (*see* **Note 13**).

This pre-absorption protocol has been used successfully on rabbit polyclonal antibodies. Rabbits can often have high background reactivity against yeast lysates because rabbits are susceptible to yeast infections.

1. Grow to saturation a 150- to 200-mL culture of a yeast strain that lacks the antigen of interest.
2. Fix the cells by the same protocol as you will use for indirect immunofluorescence. In this case, 2 h with formaldehyde and then digest with Zymolyzase, as described above in **Sections 3.1** and **3.2**. Wash the digested spheroplasts twice in PBS. Resuspend the pellet in approximately 3 mL PBS.
3. In microfuge tubes, collect three loose pellets of spheroplasted cells each with a volume of about 150 μL. Use a gentle centrifugation pulse at 1500×$g$ for 10–20 s. Remove the supernatant. If necessary, add additional spheroplast suspension and repeat the spin to generate a volume of approximately 150 μL. Remove the supernatant.
4. Dilute the antibody 1:5 in PBS. More or less dilution may be required depending on your conditions.
5. Apply 150 μL of diluted antibody to the first pellet. Gently mix by inverting the tube. Incubate 1 h at RT or overnight at

4°C. Invert the tube occasionally to remix the settled cells. Store the second and third tubes of cells on ice.

6. To compact the pellet to its smallest possible volume, spin the cells at 13,500×*g* for 10–15 min. The pellet of cells will be much smaller, approximately 60 μL. Transfer the supernatant to the second eppendorf tube containing a loose pellet (150 μL) of spheroplast-cells. Mix gently. Incubate at RT for 1 h. Invert the tube occasionally.
7. Compact the cells by centrifugation at 13,500×*g* for 10–15 min. Transfer the supernatant to the third eppendorf tube containing a loose pellet (150 μL) of spheroplast-cells. Mix gently. Incubate at RT for 1 h. Invert the tube occasionally.
8. Compact the cells by centrifugation at 13,500×*g* for 10–15 min. Transfer the supernatant to a fresh eppendorf tube. Perform a clarifying spin at 13,500×*g* for 2 min. Transfer supernatant to a new eppendorf tube. This volume will be significantly larger than the original 150 μL applied to the first pellet. Calculate the dilution of antibody, which is typically 2–4 times larger. The antibody is now ready to apply to fixed cells on a slide for indirect immunofluorescence.

## 4. Notes

1. Formaldehyde is a strong irritant and a carcinogen. Care should be taken to avoid inhalation or contact with the skin. Personal protective equipment (gloves, safety glasses, and lab coat) should be worn. Undiluted formaldehyde should be used in a fume hood. Some universities may require formaldehyde safety training prior to use. Dispose of used formaldehyde according to your institutions hazardous waste protocol.
2. For fixation of cold sensitive strains, we have found that formaldehyde fixation at 18°C for 3–4 h works well.
3. During the enzymatic digestion, 1 M sorbitol in PBS provides osmotic stability to cells that now lack their cells walls. The importance of this can be observed under a microscope. If the osmolarity is inadvertently too high during digestion, the cells will shrink in size. If distilled water is then added back under the microscope coverslip, the cells will quickly swell and burst. While this is can be an amusing aside in demonstrations for high school students, it is neither amusing nor beneficial if applied to one's entire experiment.

4. Once the cell wall has been digested, the resulting spheroplasts are especially fragile. Care should be taken to resuspend and handle them gently. Never vortex cells intended for immunofluorescence. Instead, resuspend them by gentle pipetting.
5. Do not forget to remove the Coplin jars of methanol and acetone from a non-explosion proof freezer. Otherwise, safety citations can be issued by the appropriate regulatory authorities. If necessary, it is also possible to chill the methanol and acetone in an ice bucket containing a mixture of dry ice and wet ice.
6. If there is a problem with too few cells sticking to the slide, one can determine whether it is occurring at this step by examining the slide under a microscope using a low power lens. Exercise care to make sure that the lens does not inadvertently contact any residual liquid and/or the cells.
7. For any indirect immunofluorescence application, a clarification spin of both the primary and secondary antibodies can greatly improve background that results from precipitated or tiny aggregates of antibodies.
8. The optimal times of incubation for immunofluorescence can vary depending on the antibody, ranging from 30 min at 37°C to overnight at 4°C. With the YOL1/34 antibody, we have had consistently good results with no background using an overnight incubation at 4°C. Immunofluorescence with other antibodies may require empirical determination for optimal incubation times.
9. For double label indirect immunofluorescence, application of the two primary antibodies serially is usually successful. After washing, the two secondary antibodies can then be mixed and applied simultaneously to the cells.
10. Secondary antibodies from several other manufacturers can be used successfully. The dilutions for these can be determined empirically.

    If the background is high, we have found that decreasing the incubation time for the secondary antibody often alleviates the problem. Too high of a concentration of secondary antibody also contributes to high background. To assure that the secondary antibody concentration is optimal, it is often beneficial to try a range of dilutions the first time that one uses a new secondary antibody.
11. DAPI staining of nuclear DNA appears as a very bright and large staining mass. In addition, DAPI also stains mitochondrial DNA. This staining is usually lighter and punctate and can occasionally be seen in rows (*see* **Fig. 1**, G2/M

and anaphase stages/DAPI panel). Distinguishing the two types of DNA staining is usually not a problem.

12. The cytostructure viewed by DIC optics and immunofluorescence is usually best if viewed soon after the nail polish dries. However, excellent results have also been obtained from slides that have been stored overnight at 4°C in the dark (to prevent bleaching of the fluorophore). For longer periods, slides can be stored in the dark at –20°C with only a minimal loss of signal. Extra care should be taken to be sure the nail polish has dried before viewing the slides; removing nail polish from a lens is a very difficult task and could potentially damage the lens.
13. It is not necessary to pre-absorb the rat anti-tubulin or the secondary from Jackson Immunoresearch used in this indirect immunofluorescence protocol. Both provide excellent results without pre-absorption.

## Acknowledgments

We thank Mark Rose for his continued support and helpful comments. We thank Nelda Driggs for administrative assistance. This work was supported in part by a Basil O'Connor Starter Scholar Research Award from the March of Dimes (#5-FY01-523), the Oklahoma Health Research Program of the Oklahoma Center for the Advancement of Science and Technology (OCAST #HR09-150S), and start-up funds to R.K.M from O.S.U. and the Oklahoma Agricultural Experiment Station. J.W.P.K. was supported by a Niblack Research Scholarship.

### References

1. Sullivan, D. S. and Huffaker, T. C. (1992) Astral Microtubules are not required for anaphase B in *Saccharomyces cerevisiae*. *J. Cell Biol.* **119**, 379–388.
2. Carminati, J. L. and Stearns, T. (1997) Microtubules orient the mitotic spindle in yeast through dynein-dependent interactions with the cell cortex. *J. Cell Biol.* **138**, 629–641.
3. Miller, R. K. and Rose, M. D. (1998) Kar9p is a novel cortical protein required for cytoplasmic microtubule orientation in yeast. *J. Cell Biol.* **140**, 377–390.
4. Miller, R. K., Cheng, S.-C., and Rose, M. D. (2000) Bim1p/Yeb1p mediates the Kar9p-dependent cortical attachment of cytoplasmic microtubules. *Mol. Biol. Cell* **11**, 2949–2959.
5. Beach, D. L., Thibodeaux, J., Maddox, P., Yeh, E., and Bloom, K. (2000) The role of the proteins Kar9 and Myo2 in orienting the mitotic spindle of budding yeast. *Curr. Biol.* **10**, 1497–1506.
6. Byers, B. (1981) Cytology of the yeast life cycle. *In* The Molecular Biology of the Yeast *Saccharomyces*: Life Cycle and Inheritance. J. Strathern, E. W. Jones, and J. R. Broach, editors. Cold Spring Harbor Laboratory, Cold Spring Harbor, NY. pp. 59–96.

7. Robinow, C. F. and Marak, J. (1966) A fiber apparatus in the nucleus of the yeast cell. *J. Cell Biol.* **29**, 129.
8. Byers, B. and Goetsch, L. (1974) Duplication of spindle plaques and integration of the yeast cell cycle. *Cold Spring Harbor Symp. Quant. Biol.* **38**, 123–131.
9. Adams, A. E. M. and Pringle, J. (1984) Relationship of actin and tubulin distribution to bud growth in wildtype and morphogenetic-mutant *Saccharomyces cerevisiae. J. Cell Biol.* **98**, 934–945.
10. Kilmartin, J. and Adams, A. E. M. (1984) Structural rearrangements of tubulin and actin during the cell cycle of the yeast *Saccharomyces. J. Cell Biol.* **98**, 922–933.
11. Botstein, D., Amberg, D., Mulholland, J., Huffaker, T., Adams, A., Drubin, D., and Stearns, T. (1997) The yeast cytoskeleton. *In* The Molecular and Cellular Biology of the Yeast *Saccharomyces.* J. R. Pringle, J. R. Broach, and E. W. Jones, editors. Cold Spring Harbor Laboratory, Cold spring Harbor, NY. pp. 1–90.
12. Byers, B. and Goetsch, L. (1975) Behavior of spindles and spindle plagues in the cell cycle and conjugation of *Saccharomyces cerevisiae. J. Bacteriol.* **124**, 511–523.
13. Lew, D. J., Weinert, T., and Pringle, J. R. (1997) Cell cycle control in *Saccharomyces cerevisiae. In* The Molecular and Cellular Biology of the Yeast *Saccharomyces.* J. R. Pringle, J. R. Broach, and E. W. Jones, editors. Cold Spring Harbor Laboratory, Cold Spring Harbor, NY. pp. 607–695.
14. Winey, M., Mamay, C. L., O'Toole, E. T., Mastronarde, D. N., Giddings, T. H., McDonald, K. L., and McIntosh, J. R. (1995) Three-dimensional ultrastructural analysis of the *Saccharomyces cerevisiae* mitotic spindle. *J. Cell Biol.* **129**, 1601–1615.
15. Segal, M. and Bloom, K. (2001) Control of spindle polarity and orientation in *Saccharomyces cerevisiae. Trends Cell Biol.* **11**, 160–166.
16. Kusch, J., Meyer, A., Snyder, M. P., and Barral, Y. 2002. Microtubule capture by the cleavage apparatus is required for proper spindle positioning in yeast. *Genes Dev.* **16**, 1627–1639.
17. Kahana, J. A., Schnapp, B. J., and Silver, P. A. (1995) Kinetics of spindle pole body separation in budding yeast. *Proc. Nat'l. Acad. Sci. USA* **92**, 9707–9711.
18. Yeh, E., Skibbens, R. V., Cheng, J. W., Salmon, E. D., and Bloom, K. (1995) Spindle dynamics and cell cycle regulation of dynein in the budding yeast, *Saccharomyces cerevisiae. J. Cell Biol.* **130**, 687–700.
19. Tirnauer, J. S., O'Toole, E., Berrueta, L., Bierer, B. E., and Pellman, D. (1999) Yeast Bim1p promotes the G1-specific dynamics of microtubules. *J. Cell Biol.* **145**, 993–1007.
20. Adames, N. R. and Cooper, J. A. (2000) Microtubule interactions with the cell cortex causing nuclear movements in *Saccharomyces cerevisiae. J. Cell Biol.* **149**, 863–874.
21. Rose, M. D., Winston, F., and Hieter, P. (1990) Methods of Yeast Genetics, a laboratory course manual. Cold Spring Harbor Laboratory, Cold Spring Harbor, NY. pp. 97–102.
22. Miller, R. K. (2004) Monitoring spindle assembly and disassembly in yeast by indirect immunofluorescence. *In* Cell Cycle Checkpoint Control Protocols. H. B. Lieberman, editors. Humana Press, Totowa, NJ. *Methods in Molecular Biology* series. Vol. 241. pp. 341–352.

# Chapter 18

# Detecting Recruitment of DNA Damage Response Factors Through the eChIP Approach

**Yucai Wang and Lei Li**


## Abstract

DNA interstrand crosslinks (ICLs) are lesions that covalently link the two strands of DNA. This type of DNA damage represents one of the most complex DNA lesions whose repair mechanisms remain largely unclear. Uncovering proteins involved in the processing of ICLs and understand how they interact with the damaged DNA in vivo is crucial for the understanding of DNA interstrand crosslink repair processes. Moreover, the presence of an ICL during S phase constitutes the most severe blockage to DNA synthesis and results in prolonged stall of replication forks. The mechanisms of resolving a stalled replication fork is poorly understood because proper experimental platforms are lacking. To enable detection of protein recruitment to site-specific ICLs and to ICL-stalled replication forks, we established a novel eChIP (abbreviation for *e*pisomal *ch*romatin *i*mmuno*p*recipitation) assay system to study the association of various DNA damage repair proteins with ICL lesions in vivo. This EBV episomal replication-based assay allows detection of protein enrichment at ICLs at the molecular level. Since ICLs cause replication fork blockage in an episomally replicating plasmid, the eChIP approach also allows the study of DNA damage response factor recruitment, such as checkpoint initiation factors, to stalled DNA replication forks. With proper adaptation, the eChIP approach may be employed to study other site-specific DNA lesions such as UV photoproducts and oxidative damage in vivo.


**Key words:** DNA interstrand crosslink, replication fork, Fanconi anemia, episomal replication.

## 1. Introduction

DNA interstrand crosslinks (ICLs) are complex DNA lesions that constitute absolute blockage to essential DNA functions such as replication and transcription (1, 2). A wide range of DNA damage response and DNA repair proteins have been implicated to function in the processing of ICLs. Typical examples of such proteins include structural-specific endonucleases,

W.X. Li (ed.), *Cell Cycle Checkpoints*, Methods in Molecular Biology 782,
DOI 10.1007/978-1-61779-273-1_18, 

nucleotide excision repair (NER) factors (3–5), mismatch repair factors (6–8), and Fanconi anemia (FA) proteins (2, 9–12). However, how these proteins directly interact with damaged DNA in vivo remains largely unclear. This major gap of knowledge arises partly from the lack of experimental approach to study protein recruitment to the sites of ICLs in vivo and at the molecular level. Contrasting to double strand breaks (DSBs), which could be readily introduced in a site-specific manner in vivo taking advantage of rare cutting endonuclease such as homing nucleases (13) or zinc-finger nucleases (14), site-specific DNA adducts and crosslinks are exceedingly difficult to introduce into genomic DNA with defined positions. To overcome this deficiency, we established an *e*pisomal *ch*romatin *i*mmuno*p*recipitation (eChIP) system (15), which allows examination of protein recruitment to a defined ICL site on an episomally replicating plasmid substrate.

The key to this system is a site-specific ICL lesion placed about 500 base pairs downstream of the Epstein-Barr virus (EBV) replication origin (OriP) in a plasmid substrate (pORIP). When introduced into cultured cells, pORIP was able to undergo Epstein-Barr nuclear antigen-1 (EBNA)-dependent replication (16). The EBV oriP-driven replication has two unique characteristics compared to most defined viral or lower eukaryotic replication origins. Once cleared from the starting site, the replication machinery constitutes exclusively of endogenous proteins, enabling the episomal replication to be reflective of mammalian replication (17). Secondary DNA structure derived from a stretch of repetitive sequences immediately on one side of the origin forms a physical barrier (18) to the newly formed elongation complex. Thus replication initiated from OriP is virtually unidirectional, allowing the recapitulation of the ICL lesion being encountered by one elongating replication fork on a circular plasmid substrate.

The presence of a site-specific ICL on pORIP enables the examination of protein recruitment to the defined ICL sites by chromatin immunoprecipitation (ChIP) followed by quantitative PCR (QPCR), in both non-replicative (in EBNA-negative cell line) and replicative (in EBNA-expressing cell lines) settings (**Fig. 1**). Using the eChIP approach, we have discovered the replication status governed recruitment of Fanconi anemia and breast cancer susceptibility proteins to defined ICL DNA damage sites (15). Notably, the eChIP approach also allows readily the study of recruitment of various DNA damage response factors to stalled replication forks, which may facilitate the understanding of early DNA damage response. For example, we are able to detect the recruitment of ATRIP and Rad17, two of the important factors in activating the ATR-mediated checkpoint, to stalled

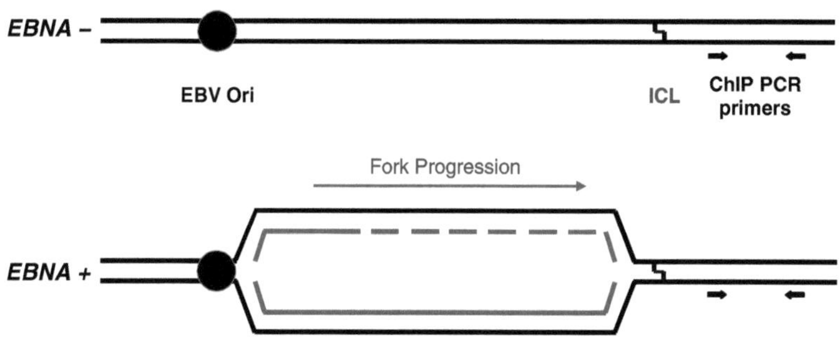


Fig. 1. Schematics of the pORIP DNA substrate under non-replicative and replicative states. Positions of primers for ChIP PCR amplification and the direction of replication are indicated.

replication fork (Wang and Li, unpublished data). In addition, the eChIP approach paves the way for the site-specific analysis of many different types of DNA lesions such as alkylation, bulky adducts, apurinic sites, and UV photoproducts that have gone largely unstudied in a site-specific manner in vivo.

Here we describe a detailed protocol for carrying out the eChIP assay, and provide examples showing recruitment of three FA proteins (FANCA, FAAP24, and FANCJ) to site-specific ICL lesions.

## 2. Materials

### 2.1. Cell Culture and Plasmid DNA Substrates

1. Dulbecco's Modified Eagle's Medium (DMEM, from Mediatech) supplemented with 10% fetal bovine serum (Sigma-Aldrich).
2. G418 sulfate solution (Mediatech).
3. 0.05% trypsin/0.53 mM EDTA solution (Mediatech).
4. TD buffer: 137 mM NaCl, 25 mM Tris–HCl, 0.7 mM $Na_2HPO_4$, 5 mM KCl, 0.1% dextrose, pH 7.4.
5. pORIP-ctrl unmodified control substrates (*see* **Note 1**).
6. pORIP-XL interstrand crosslink substrates (*see* **Note 1**).

### 2.2. Transfection and Cell Lysate Preparation

1. The Amaxa Nucleofector device and Amaxa Cell Line Nucleofector Kit V are from Lonza.
2. 1% formaldehyde solution. Prepare both the stock and working solution freshly. In a 50-mL Falcon tube, add 1.85 g paraformaldehyde (Sigma), 9 mL $H_2O$, and 70 μL 1 M KOH (Fisher Scientific). Microwave briefly to dissolve. The final volume will be 10 mL, hence the stock concentration

18.5%. Dilute with cold DMEM to reach the desired 1% working concentration just before use.

3. PBS buffer: 137 mM NaCl, 2.7 mM KCl, 10 mM $Na_2HPO_4$, 1.76 mM $KH_2PO_4$, pH 7.4.
4. Solution **1**: 10 mM HEPES (pH 7.5), 10 mM EDTA, 0.5 mM EGTA, 0.75% Triton X-100.
5. Solution **2**: 10 mM HEPES (pH 7.5), 200 mM NaCl, 1 mM EDTA, 0.5 mM EGTA.
6. Lysis buffer: 25 mM Tris–HCl (pH 7.5), 5 mM EDTA, 150 mM NaCl, 0.1% SDS, 0.5% deoxycholate, 1% Triton X-100. Add one Complete Mini protease inhibitor cocktail tablet (Roche) per 10 mL stock buffer before use (*see* **Note 2**).

### 2.3. Chromatin Immunoprecipitation

1. Protein A/G Sepharose beads (GE Healthcare).
2. Primary antibody against protein of interest (*see* **Note 3**).
3. RIPA buffer: 50 mM Tris–HCl (pH 8.0), 1 mM EDTA, 150 mM NaCl, 0.1% SDS, 0.5% deoxycholate, 1% NP-40.
4. High Salt buffer: 50 mM Tris–HCl (pH 8.0), 1 mM EDTA, 500 mM NaCl, 0.1% SDS, 0.5% deoxycholate, 1% NP-40.
5. LiCl buffer: 50 mM Tris–HCl (pH 8.0), 1 mM EDTA, 250 mM LiCl, 0.5% deoxycholate, 1% NP-40.
6. TE buffer: 10 mM Tris–HCl (pH 8.0), 1 mM EDTA.
7. Protease K buffer: 10 mM Tris–HCl (pH 8.0), 1 mM EDTA, 0.5% SDS, 0.25 mg/mL protease K.
8. Saturated phenol, chloroform, ethanol, and sodium acetate are from Fisher Scientific.

### 2.4. PCR and QPCR

1. GoTaq DNA polymerase (supplied with 5× green and colorless GoTaq buffer) is from Promega.
2. PCR primers are synthesized by Integrated DNA Technologies. The sequences of forward and reverse primers are 5′-GGGGCACGTGTACGTGAACCAT-3′ and 5′-CACCGATCGCCCTTCCCAACAGT-3′, respectively. The concentration of working solution is 10 μM for each primer.
3. Deoxynucleotide triphosphates (dNTPs) are from Invitrogen. The concentration of working solution is 2.5 mM for each type of dNTP.
4. UltraPure agarose (Invitrogen).
5. TAE Buffer: 40 mM Tris acetate, 1 mM EDTA.
6. SYBR Green JumpStart *Taq* ReadyMix (Sigma-Aldrich).

## 3. Methods

To allow examination of protein recruitment to defined stalled replication forks by eChIP in both non-replicative and replicative settings, ENBA negative and EBNA positive cell lines are needed. A pair of isogenic cell lines is preferentially used for this purpose. We commonly use the HEK293 and HEK293EBNA cell lines commercially available from ATCC. Since the experiments are carried out in parallel with identical unmodified plasmid DNA substrates along with crosslinked substrates in both ENBA-negative and EBNA-positive cell lines, the eChIP assay not only allows detection of crosslink-specific recruitment of a given protein of interest but also enables determination of the replication-independent or replication-dependent recruitment manner for the given protein.

### 3.1. Cell Culture and DNA Substrate Transfection

1. HEK293 and HEK293EBNA cells are maintained in DMEM medium supplemented with 10% FBS and grown in a humidified 5% $CO_2$-containing atmosphere at 37°C. To retain expression of EBNA1 in the HEK293EBNA cell line during passages, 0.25 mg/mL G418 was routinely supplemented in the culture medium. Cell passages are carried out at 1:4 to 1:10 split ratios.
2. One day prior to transfection, seed HEK293 and HEK293EBNA cells into 140-mm cell culture dishes at the density of $1.0 \times 10^7$ cells per dish. In every experiment, at least two dishes of cells are needed for each cell line.
3. 16–24 h after seeding, wash the monolayer cells once with TD buffer. Add 2 mL trypsin/EDTA, and incubate the cells at 37°C for 3 min. Add 5 mL culturing medium, resuspend the cells, and transfer to labeled 15-mL Falcon tubes. Spin down the cells at 1100 rpm for 5 min, and carefully aspirate all the supernatant.
4. Add 95 μL Solution V (supplied in the Amaxa Cell Line Nucleofector Kit V) and 2 μg of plasmid DNA substrate to the cell pellet, gently mix well, and transfer into Amaxa cuvettes (*see* **Note 4**). All together four sets of transfection mixtures need to be prepared: HEK293 cells with control substrates, HEK293 cells with crosslink substrates, HEK293EBNA cells with control substrates, and HEK293EBNA cells with crosslink substrates.
5. Insert the cuvette into the Amaxa Nucleofector, and use the HEK293 cell line specific program to electroporate the plasmid DNA into cells.

6. Quickly aspirate the electroporated cells out using a plastic pipette and transfer into 140-mm dishes containing pre-warmed culturing medium (*see* **Note 5**). Shake gently to homogenize, and return into cell culture incubator.

### 3.2. Preparation of Cell Lysates

1. 4–6 h after transfection, aspirate the culturing medium and wash the cells with TD buffer three times (*see* **Note 6**).
2. Add 10 mL freshly prepared 1% formaldehyde, and incubate for exactly 8 min at room temperature, then immediately remove the formaldehyde solution completely.
3. Add 5 mL cold PBS, scrape the cells off on ice, and transfer into 15-mL Falcon tubes. Spin down the cells at 1500 rpm for 10 min at 4°C (*See* **Note 7**).
4. Remove the supernatant, add 10 mL cold PBS to resuspend the cells, and spin down the cells at 1500 rpm for 10 min at 4°C. Repeat once.
5. Add 10 mL Solution 1 to resuspend the cells and rotate at 4°C for 10 min. Spin down the cells at 1500 rpm for 5 min at 4°C, and completely remove all the supernatant.
6. Repeat Step 5 using Solution 2.
7. Add 300 μL lysis buffer to dissolve the cell pellet, and transfer the cell lysates into 1.5 mL Eppendorf tubes. Bring the final volume close to but not more than 500 μL using lysis buffer; keep on ice for 10 min.

### 3.3. Chromatin Immunoprecipitation

1. Sonicate on ice to shear the genomic DNA into fragments of less than 0.8 kb in size (*see* **Note 8**).
2. Centrifuge at 14,000 rpm for 15 min at 4°C, and collect the supernatants. Repeat once.
3. Bring the final volume to 500 μL using lysis buffer. Save 50 μL cell lysate as input control (store at −20°C) and proceed to chromatin immunoprecipitation with the rest (450 μL) lysate.
4. Add 30 μL protein A or protein G beads (*see* **Note 9**) to the lysate, and slowly rotate at 4°C for 1 h for pre-clearing.
5. Pellet the beads at 10,000 rpm for 20 s at 4°C, and remove the supernatant.
6. Add 2 μg (*see* **Note 10**) of primary antibody against the protein of interest to the supernatant, and rotate at 4°C overnight.
7. Centrifuge at 14,000 rpm for 5 min at 4°C, and recover the supernatant.
8. Add 35 μL protein A or protein G beads (*see* **Note 9**) to the supernatant, and rotate at 4°C for 1 h.
9. Spin down the beads at 3000 rpm for 2 min at 4°C, and discard the supernatant.

### 3.4. DNA Recovery

1. Use 1 mL RIPA buffer to wash the beads by rotating at 4°C for 10 min. Spin down the beads at 3000 rpm for 2 min at 4°C and discard the supernatant.
2. Repeat Step 1 sequentially using High Salt buffer, LiCl buffer, and finally with TE buffer twice.
3. Add 300 μL protease K buffer to resuspend the beads. Take out previously saved input control lysates, and add 300 μL protease K buffer. Incubate the samples at 37°C for 6 h.
4. Transfer the samples to 65°C water bath and continue incubation overnight to reverse the DNA–protein crosslinking and to digest the protein.
5. Add 300 μL phenol/chloroform to each sample, vertex thoroughly, and centrifuge at 10,000 rpm for 5 min. Recover the aqueous layer and repeat the phenol/chloroform extraction one more time.
6. Add 0.1 volume of sodium acetate and 2.5 volumes of cold ethanol (*see* **Note 11**), mix thoroughly and centrifuge at 14,000 rpm for 15 min at 4°C. Discard the supernatant, and wash the DNA pellet with 70% cold ethanol several times.
7. Spin down briefly, and carefully aspirate all the supernatant. Air-dry the DNA pellet.
8. Dissolve the DNA pellet with 40 μL TE buffer. For DNA recovered from the input control lysates, dilute ten times with TE buffer before PCR.

### 3.5. PCR and QPCR

1. Perform a semi-quantitative PCR. The reaction mixture contains 5 μL 5 × buffer, 4 μL DNA template, 2 μL dNTP mixture, 1 μL primer mixture, 1 U of GoTaq polymerase, and the final volume is brought to 25 μL by ddH$_2$O. The cycling conditions include a denature at 95°C for 5 min, 25–30 cycles of 30 s denature at 95°C, 30 s anneal at 55°C and 30 s extension at 72°C, and a final extension at 72°C for 10 min.
2. Resolve the PCR products by 1% agarose gel electrophoresis. Examples of FANCA, FAAP24 and FANCJ eChIP-PCR are shown in **Fig. 2** (15).
3. Proceed to a quantitative real-time PCR in triplicates. Each reaction mixture contains 5 μL 2 × SYBR Green Jumpstart Taq Readymix, 2 μL DNA template, 1 μL primer mixture, and 2 μL ddH$_2$O. The cycling conditions are programmed as follows: a denature at 95°C for 2 min, followed by 40 cycles of 15 s denature at 95°C, 1 min anneal at 60°C, and 1 min extension at 72°C.
4. Perform quantitative analysis using the Rotor Gene software. Examples of FANCA, FAAP24, and FANCJ eChIP-Q-PCR are shown in **Fig. 3** (15).

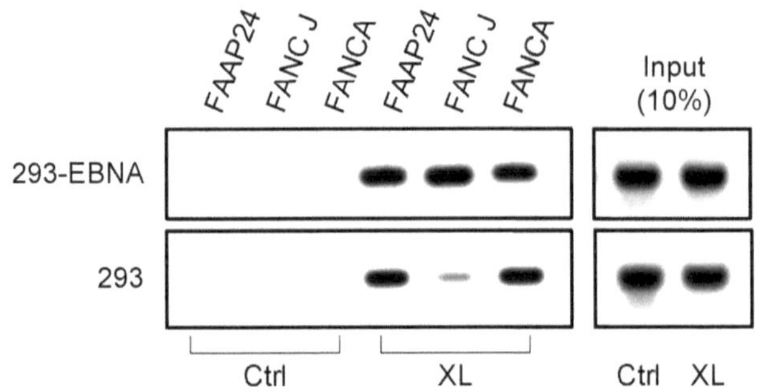


Fig. 2. Replication status governs differential recruitment of FANCA, FAAP24, and FANCJ to site-specific ICLs. Ctrl, unmodified pORIP substrate. XL, pORIP substrate with a single defined psoralen interstrand crosslink downstream of the EBV origin.

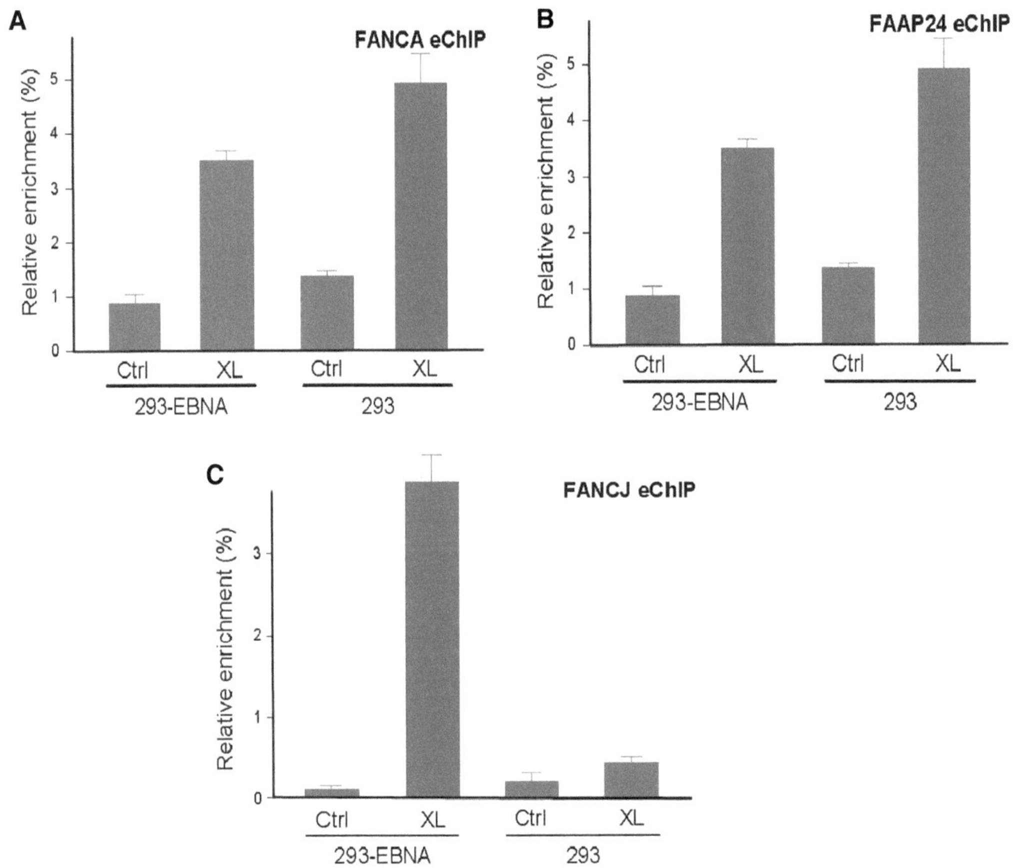


Fig. 3. Recruitment of FANCA, FAAP24 and FANCJ to site-specific ICLs as determined by eChIP and real-time PCR.

## 4. Notes

1. Both the control and crosslink substrates contain the EBV replication origin (OriP). In the crosslink substrate, a site-specific psoralen or mitomycin C interstrand crosslink was placed ~500 base pairs downstream of the EBV replication start site. For detailed information on substrate preparation, *see* Ref. (15).
2. ChIP grade antibody is preferentially used. An antibody proven to work well for Co-immunoprecipitation (Co-IP) or IP might also be suitable for eChIP. If applications for ChIP, Co-IP, and IP are not tested for candidate antibodies, affinity-purified polyclonal antibodies normally give better results. Otherwise, we pilot several commercial antibodies and it is not unusual that several antibodies have to be tested before a suitable one for eChIP is found.
3. Avoidance of bubbles is crucial for efficient electroporation. We usually gently swirl the pellets in the solution with pipette tip to homogenize, and then pipette up and down twice before transferring the mixture into cuvettes.
4. After electroporation, the cells need to be resuspended in pre-warmed culturing medium as quickly as possible. We usually keep four 140-mm cell culture dishes, each containing 30 mL culturing medium in the 37°C incubator before hand, and take them out just before electroporation. After starting the electroporation program, use the plastic pipette to aspirate out about 1 mL medium, and add it into the cuvette immediately after the electroporation discharge. Then quickly and gently aspirate all the cells and transfer into the pre-warmed/pre-filled cell culture plates and resuspend them. Using another 1 mL of media to recover any remaining cells from the cuvette is recommended.
5. The time point of cell collection varies depending on the nature of study. In our experience, the 4–6 h time window after transfection is good for the examination of FA proteins at the site of ICLs. When examining the recruitment of other proteins, the collection time point may need to be optimized. And keep in mind that the cells may not be attached firmly when collecting them shortly after electroporation, so you need to be gentle when washing the cell with TD buffer.
6. You may sometimes find the cell suspension sticky when scraping, and thus be concerned whether the cells are completely collected. We sometimes add another 5 mL cold PBS to wash the dish and collect any remaining cells, and

transfer them to the same Falcon tube. The cell pellets can be saved at −80°C before proceeding to the next step.

7. Using different devices, the sonication procedures may vary. We use a Branson Sonifier 150 to perform sonication at an output power of 7–10 W. The samples were sonicated on ice using 20 cycles of sonication (10 s) – incubation (10 s). This procedure is sufficient in shearing genomic DNA into less than 0.8-kb fragments. We recommend checking DNA fragment sizes by agarose gel to optimize the sonication procedures when using different devices.
8. The choice of protein A or G beads depend on the primary antibody used for immunoprecipitation. We refer to the "Choosing the correct beads-summary table" provided by Abcam. Wash protein A or G beads with lysis buffer for at least three times before use. The volumes specified in the protocol refer to the volumes of beads/buffer (50%/50% volume) slurry.
9. The amount of antibody needed for each immunoprecipitation varies with antibody specificity and strength. In our experiences, 2 μg is a good starting point in most cases. However, optimization of primary antibody quantity is likely to improve signal/noise ratio.
10. Ten to twenty micrograms of glycogen (20 mg/mL, from Roche) should be added to each sample to serve as a carrier for the precipitation of DNA before adding ethanol.
11. A semi-quantitative PCR is done first to make sure the efficient amplification of correct product, and a quantitative real-time PCR will follow if no background non-specific amplifications are observed.

## Acknowledgments

The authors are supported by grants (CA97175, Project 3, and CA127945) from the National Institute of Health. Permission of Figure reproduction from Elsevier Science is obtained under the license Number 2454350875135.

### References

1. McHugh PJ, Spanswick VJ, Hartley JA. Repair of DNA interstrand crosslinks: molecular mechanisms and clinical relevance. Lancet Oncol 2001;2:483–90.
2. Thompson LH, Hinz JM. Cellular and molecular consequences of defective Fanconi anemia proteins in replication-coupled DNA repair: mechanistic insights. Mutat Res 2009;668:54–72.

3. Sarkar S, Davies AA, Ulrich HD, McHugh PJ. DNA interstrand crosslink repair during G1 involves nucleotide excision repair and DNA polymerase zeta. EMBO J 2006;25:1285–94.
4. Zheng HY, Wang X, Warren AJ, et al. Nucleotide excision repair- and polymerase eta-mediated error-prone removal of mitomycin C interstrand cross-links. Mol Cell Biol 2003;23:754–61.
5. Wang X, Peterson CA, Zheng H, Nairn RS, Legerski RJ, Li L. Involvement of NER in a recombination-independent and error-prone pathway of DNA interstrand cross-link repair. Mol Cell Biol 2001;21:713–20.
6. Peng M, Litman R, Xie J, Sharma S, Brosh RM, Cantor SB. The FANCJ/MutL[alpha] interaction is required for correction of the cross-link response in FA-J cells. EMBO J 2007;26:3238–49.
7. Zhang N, Lu X, Zhang X, Peterson CA, Legerski RJ. hMutS{beta} is required for the recognition and uncoupling of Psoralen interstrand cross-links in vitro. Mol Cell Biol 2002;22:2388–97.
8. Wu Q, Christensen LA, Legerski RJ, Vasquez KM. Mismatch repair participates in error-free processing of DNA interstrand crosslinks in human cells. EMBO Rep 2005;6:551–7.
9. Wang W. Emergence of a DNA-damage response network consisting of Fanconi anaemia and BRCA proteins. Nat Rev Genet 2007;8:735–48.
10. Moldovan GL, D'Andrea AD. How the fanconi anemia pathway guards the genome. Annu Rev Genet 2009;43:223–49.
11. Muniandy PA, Liu J, Majumdar A, Liu ST, Seidman MM. DNA interstrand crosslink repair in mammalian cells: step by step. Crit Rev Biochem Mol Biol 2010;45:23–49.
12. Räschle M, Knipscheer P, Enoiu M, et al. Mechanism of replication-coupled DNA interstrand crosslink repair. Cell 2008;134:969–80.
13. Zeitlin SG, Baker NM, Chapados BR, et al. Double-strand DNA breaks recruit the centromeric histone CENP-A. Proc Natl Acad Sci USA 2009;106:15762–7.
14. Kandavelou K, Chandrasegaran S. Custom-designed molecular scissors for site-specific manipulation of the plant and mammalian genomes. Methods Mol Biol 2009;544:617–36.
15. Shen X, Do H, Li Y, et al. Recruitment of fanconi anemia and breast cancer proteins to DNA damage sites is differentially governed by replication. Mol Cell 2009;35:716–23.
16. Gahn TA, Schildkraut CL. The Epstein-Barr virus origin of plasmid replication, oriP, contains both the initiation and termination sites of DNA replication. Cell 1989;58:527–35.
17. Dhar SK, Yoshida K, Machida Y, et al. Replication from oriP of Epstein-Barr virus requires human ORC and is inhibited by geminin. Cell 2001;106:287–96.
18. Norio P, Schildkraut CL. Visualization of DNA replication on individual Epstein-Barr virus episomes. Science 2001;294:2361–4.

# Chapter 19

## Methods to Study Cancer Therapeutic Drugs That Target Cell Cycle Checkpoints

**Yun Dai and Steven Grant**

### Abstract

Cell cycle checkpoints operating through a network of multiple signaling pathways provide a key mechanism for self-defense of cells against DNA damage caused by various endogenous or environmental stresses. In cancer treatment, checkpoints are activated in response to diverse DNA-damaging agents and radiation, thus representing a critical barrier limiting therapeutic efficacy. To date, despite efforts to target other components of checkpoint signaling pathways (e.g., ATM, Chk2, Wee1), checkpoint kinase 1 (Chk1) remains the most important target for cancer treatment because of its functional association with essentially all cell cycle checkpoints. The primary goal in the development of therapeutic agents targeting cell cycle checkpoints continues to be improving the anti-cancer activity of chemo- and radiotherapy by abrogating checkpoints necessary for DNA repair, thereby killing cancer cells through engagement of the apoptotic machinery.

**Key words:** Cell cycle, checkpoint, DNA damage, Chk1 inhibitor.

## 1. Introduction

### *1.1. Signaling Cascades of Cell Cycle Checkpoints*

Proteins involved in the checkpoint signaling pathways are generally classified as sensors, mediators, transducers, and effectors (1). In response to DNA damage, "*sensor*" multiprotein complexes (e.g., Mre11-Rad50-Nbs1/MRN complex for ATM; Rad17 and Rad9-Rad1-Hus1/9-1-1 complex for ATR) recognize damage, recruit "*proximal transducers*" (ATM and ATR) to lesions where they are initially activated. In response to double-strand breaks (DSBs) (e.g., induced by ionizing radiation [IR] or radiomimetic agents), the MRN complex transiently interacts with DSB lesions to recruit ATM, which is initially activated through dissociation of

W.X. Li (ed.), *Cell Cycle Checkpoints*, Methods in Molecular Biology 782,
DOI 10.1007/978-1-61779-273-1_19, © Springer Science+Business Media, LLC 2011

dimers (inactive) and formation of monomers via intramolecular autophosphorylation on serine residues S1981, S367, and S1893. S1981 phosphorylation is a marker of ATM activation in vitro and in vivo. Activated ATM phosphorylates MDC1/Nfbd1 and H2A.X, leading to sustained recruitment of the MRN complex and BRCA1 at the site of lesions, resulting in further activation of ATM. ATR activation is initiated by single-strand DNA (ssDNA). Whereas ssDNA is present in single-strand breaks (SSBs) or gaps in DNA strands (2), it primarily occurs during DNA replication (3) or the DSB repair process (4). In S phase, endogenous and exogenous insults hinder replication fork progression, resulting in stalled forks, unstable structures that are prone to collapse or breaks. When a replication fork encounters a lesion, DNA polymerase stalls while helicase continues to unwind the DNA, generating a large stretch of ssDNA. In DSB repair, MRN and ATM mediate DSB resection, leading to formation of ssDNA as an intermediate structure for repair. ssDNA is coated with and stabilized by replication protein A (RPA), after which ATR-interacting protein (ATRIP) recognizes and binds to RPA-ssDNA, recruiting the ATR/ATRIP complex to the lesions, leading to ATR activation. Full activation of ATR/ATRIP requires the independent loading of another complex, Rad17/9-1-1, onto DNA lesions, which is essential for ATR signaling to its downstream targets (e.g., Chk1) but is not required for phosphorylation of other ATR substrates (e.g., ATRIP and H2A.X).

ATM and ATR transduce signals to "*distal transducer*" checkpoint kinases (i.e., Chk1 and Chk2). Although cross-talk between ATM and ATR occurs in Chk1/Chk2 activation, ATR and ATM primarily activate Chk1 and Chk2, respectively. In contrast to Chk2, which is largely inactive in the absence of DNA damage, Chk1 is active even in cells undergoing normal cell cycle progression, but further activated in response to DNA damage or replication stress. Chk1 and Chk2 are structurally unrelated serine/threonine kinases, which are differentially activated. Activation of ATM/ATR and recruitment of the multiple "sensor" proteins leads to relocalization of Chk1 and Chk2 to multiprotein complexes residing at the site of DNA lesions. ATM phosphorylates Chk2 at threonine 68, promoting its homodimerization and full activation via intramolecular trans-autophosphorylation at threonine 383 and 387, accompanied by spreading of Chk2 throughout the nucleoplasm. Chk1 is activated by both ATR and ATM (to a less extent, e.g., in response to IR) via phosphorylation at serine 317 and 345. Unlike Chk2, Chk1 activation does not require dimerization and transphosphorylation for full activation. Unphosphorylated Chk1 remains inactive via auto-inhibition by its C-terminal domain through intramolecular interactions with the N-terminal catalytic domain (5). Phosphorylations at the C-terminal serine residues (S317, S345, and most likely S366 as well) abrogates this inhibitory interaction, thereby opening the molecule to unmask the N-terminal kinase domain and allowing it to associate with trans-regulatory factors. S345 phosphorylation is essential for kinase activation and function of Chk1, which are promoted by phosphorylation of adjacent sites (e.g., S317 and S366). Moreover, release of Chk1 from chromatin (inactive state) leads to its accumulation at the centrosome (active state) where it prevents activation of Cdk1/cdc2 which drives cells entry into mitosis (6). Phosphorylation at S345, but not S317, enhances Chk1 accumulation in the nucleus via binding to 14-3-3 proteins. Notably, phosphorylation of Chk1 at S317 and S345 may account for "essential" cell survival and "non-essential" checkpoint functions of Chk1, respectively (7, 8).

ATM/ATR activation and resulting phosphorylation of "sensor" proteins lead to recruitment and phosphorylation of various "*mediator*" proteins (e.g., 53BP1, BRAC1, MDC1, SMC1, FANCD2, Claspin, TopBP1, and H2A.X, etc.), facilitating and optimizing activation of Chk1 and Chk2. Chk1 activation by ATR requires loading of the PCNA-like DNA sliding clamp 9-1-1 complex by the Rad17-RFC complex (together designated the Rad17/9-1-1 complex). Moreover, optimal activation of Chk1 requires interactions with multiple "mediator" proteins such as BRCA1, Claspin, and TopBP1, Timeless (Tim or Tim1) and Tipin. Claspin functions as a highly conserved, key "mediator" of ATR-mediated Chk1 activation. Claspin directly binds to Chk1, maintaining stability of both proteins and recruiting Chk1 to ATR. In response to DNA damage and replication stress, phosphorylation of Claspin also results in recruitment and phosphorylation of BRCA1, an event required for Chk1 activation. Whereas TopBP1 is directly involved in activation of the ATR/ATRIP kinase, it is also important for ATR-mediated Chk1 phosphorylation (9). Timeless (10) and Tipin (11) form a stable complex that associates with chromatin via binding of Tipin with RPA, which is essential for nuclear accumulation/chromatin association of Claspin and ATR-mediated Chk1 phosphorylation induced by replication stress.

Once activated, "mediator" proteins remain at the site of DNA lesions, whereas Chk1 and Chk2 are released to phosphorylate their soluble "*effector*" protein Cdc25s (i.e., Cdc25A, B, and C). Cdc25s are specialized phosphatases that activate cyclin-dependent kinases (Cdks, specifically Cdk1/cdc2 and Cdk2) via dephosphorylation on their inhibitory sites (i.e., tyrosine 15 and threonine 14). Phosphorylation of Cdc25s leads to nuclear exclusion or ubiquitin-proteasomal degradation, resulting in inhibition of their activity as Cdk phosphatases. Activated Chk1/Chk2 and ATM/ATR also phosphorylate another critical "effector," p53. Chk1 and Chk2 phosphorylate p53 at threonine 18, serine

20, and most likely other residues as well. ATM/ATR phosphorylate both p53 (e.g., serine 15 and 37) and its ligase Mdm2 (e.g., serine 395), a protein that binds to and promotes the rapid turnover of p53. Phosphorylation at these critical sites attenuates interactions between p53 and Mdm2, thereby promoting p53 stabilization. Accumulation/activation of p53, a transcription factor, results in transcriptional expression of endogenous Cdk inhibitors (e.g., $p21^{CIP1/WAF1}$, Gadd45 and 14-3-3σ), which bind to and inhibit Cdks. As a consequence, sequestration/down-regulation of Cdc25s and accumulation of p53 inactivate Cdks, kinases responsible for driving cell cycle progression, leading to cell cycle arrest at specific phases of the cell cycle. Whereas p53-dependent checkpoint pathways account for slower, irreversible cell cycle arrest, the ATM/ATR-Chkl/Chk2-Cdc25s signaling cascades leads to a fast, reversible response because it does not require de novo protein synthesis (12, 13).

### 1.2. Cancer Therapeutic Drugs That Target Cell Cycle Checkpoints

Cell cycle checkpoints are crucial for transformed cells to protect or rescue themselves from DNA damage induced by a majority of chemotherapeutic agents and radiation therapy (1). In response to these genotoxic insults, the DNA damage response (DDR) is initiated via activation of checkpoints, a mechanism which induces cell cycle arrest, allowing time for the DNA repair machinery to rectify the damage, or, if the damage is irreparable, trigger apoptosis to eliminate injured cells (2). When repair is successful, cells can survive and resume cell cycle progression through a checkpoint recovery mechanism. Therefore, the primary goal of the development of checkpoint-targeting agents is to improve the anti-cancer activity of chemotherapeutic agents and radiotherapy by abrogating checkpoints responsible for cell cycle arrest in response to DNA damage triggered by genotoxic insults. In so doing, transformed cells are not allowed to repair DNA damage, but instead undergo cell death via activation of apoptotic signaling pathways. Significantly, checkpoints are characteristically defective in transformed cells (14).

#### 1.2.1. Chk1 as a Primary Anti-cancer Target

Proximal (i.e., ATM/ATR) and distal (i.e., Chkl /Chk2) "transducer" proteins comprise the core of DDR signaling networks. Theoretically, inhibition of each protein could improve chemo- or radio-therapeutic efficacy. To date, no ATR-specific inhibitor has been developed. ATM is a rational candidate target, and several ATM inhibitors (e.g., KU-55933 and KU-60019 (15); Kudos) are currently at early preclinical stages of development. Whether targeting ATM, ATR, or both in combination with genotoxic agents will prove to be effective strategies remains to be determined. Despite similarities in substrate phosphorylation, Chkl and Chk2 functions in cell survival and checkpoint regulation differ strikingly. Chk2 function is time- and cell type-dependent, and generally limited to DSB-induced checkpoints (i.e., by IR). Chk1 is involved in checkpoints induced by diverse stimuli (e.g., UV light and numerous DNA-damaging agents), as well as by DNA replication stresses (including in unperturbed cells) (3). Thus, Chk1 is an extremely attractive target for multiple reasons, e.g., (a) Chk1 is associated with all checkpoints, e.g., G2/M, G1/S, S, and most recently, the mitotic spindle checkpoint; (b) Chk1 is essential for maintenance of genomic integrity, whereas Chk2 is conditional; (c) Chk2 function is to some extent replaceable by Chk1 (or other kinases), but the reverse is not true; (d) Chk1 plays a central role in DNA replication checkpoints e.g., by exposure to agents (e.g., nucleoside analogs) that target replication; and (e) Chk1 is involved in other critical functions (e.g., DNA repair and apoptosis inhibition). Therefore, Chk1 has been viewed as the "workhorse," whereas Chk2 is the "amplifier" kinase (16). Consequently, Chk1 currently represents the most important target for anti-cancer therapeutics directed at the DDR network.

#### 1.2.2. Novel Checkpoint Abrogators

The clinical utility of UCN-01, the first Chk1 inhibitor evaluated in humans, is limited by its prolonged plasma half-life due to extensive plasma binding to α1-acidic glycoprotein, and off-target actions (e.g., inhibition of multiple other kinases) resulting in toxicity (e.g., hyperglycemia) (17). These have prompted extensive efforts to develop a new generation of more specific and less toxic inhibitors targeting checkpoint kinases. However, as in the case of UCN-01, the major goal in developing these new agents continues to involve disrupting DNA damage checkpoint responses to genotoxic agents or radiation. Whether strategies combining newer checkpoint abrogators and cytotoxic agents will result in improved therapeutic activity or selectivity remains the subject of intense interest. Currently, numerous clinical trials involving checkpoint abrogators based on this rationale are underway. In such studies, phosphorylation of Chk1 (e.g., serine 345 or 296 [an autophosphorylation site]), histone H3 (e.g., serine 10), Cdc25C (e.g., serine 216), and histone H2A.X (serine 139, designated γH2A.X), as well as levels of Cdc25A, currently serve as potential biomarker for Chk1 inhibition (18). A brief summary of newer checkpoint abrogators follows below.

*AZD7762 (AstraZeneca):* A potent, selective Chk1 inhibitor binds to the ATP-binding site of Chk1 and in vitro inhibits Chk1-mediated phosphorylation of Cdc25C peptide (IC50, 5 nM) (19). AZD7762 is equally potent against Chk2 in vitro. AZD7762 abrogates the S-phase checkpoint (via the Cdc25A/Cdk2 pathway) by gemcitabine or the G2/M-phase checkpoint (via the Cdc25C/Cdk1 pathway) by irinotecan (SN38), resulting in enhanced activity in solid tumor cell lines (particularly p53-mutant cells) and murine xenografts.

*LY2603618 (Lilly):* This inhibitor binds to and blocks Chk1 activity, thereby potentiating the efficacy of various chemotherapeutic agents, possibly by interfering with DNA repair. Preclinical data involving LY2603618 has not been published.

*CBP501 (CanBas):* A peptide corresponding to aa 211-221 of Cdc25C inhibits Chk1 (IC50, 3.4 μM) and Chk2 (IC50, 6.5 μM) in vitro (20). CBP501 diminishes Cdc25C serine 216 phosphorylation, accompanied by Cdk1/cdc2 tyrosine 15 dephosphorylation and increased histone H3 serine 10 phosphorylation, leading to G2/M checkpoint abrogation and enhanced cytotoxicity of bleomycin or cisplatin (CDDP) in vitro and in murine xenografts.

*PF-00477736 (Pfizer):* A selective, potent ATP-competitive Chk1 inhibitor, derived from PF-00394691, inhibits Chk1 (Ki, 0.49 nM) and Chk2 (Ki, 47 nM) in vitro. PF-00477736 abrogates both G2/M-phase (e.g., by camptothecin) and S-phase checkpoints (e.g., by gemcitabine) (21). The latter enhances gemcitabine cytotoxicity in p53-defective tumor cells and in murine xenografts. PF-00477736 also significantly enhances docetaxel efficacy in vitro and in vivo, in association with decreased Cdc25C cytoplasmic phosphorylation (serine 216) and histone H3 phosphorylation (serine 10) (22).

*SCH-900776 (Schering-Plough):* This compound specifically binds to and inhibits Chk1, abrogating the S-phase or G2/M-phase checkpoints, thereby sensitizing tumor cell to IR and alkylating agents. These preclinical data have not yet been published.

*XL844 (Exelixis):* A potent ATP-competitive inhibitor of Chk1 (Ki, 2.2 nM) and Chk2 (Ki, 0.07 nM) (23). XL844 blocks Cdc25A degradation, abrogates the S-phase checkpoints, increases DNA damage in response to gemcitabine, and potentiates gemcitabine activity in vitro and in xenografts.

*CEP-3891 (Cephalon):* This specific Chk1 inhibitor, currently at the preclinical development stage, potently inhibits Chk1 (IC50, 4 nM) as well as other kinases, including TrkA (IC50, 9 nM), MLK1 (IC50, 42 nM), and VEGFR2 (IC50, 164 nM) in vitro. CEP-3891 abrogates S-phase and G2/M-phase checkpoints induced by IR (24). The former event is likely related to delayed IR-induced Cdc25A phosphorylation (serine 123, a residue critical for protein stability). CEP-3891 also accelerates IR-induced mitotic nuclear fragmentation stemming from defective chromosome segregation, accompanied by enhanced lethality (25).

*CHIR-124 (Chiron):* This potent, selective Chk1 inhibitor, which occupies the ATP-binding site, inhibits Chk1 (IC50, 0.3 nM) 2000-fold more potently than Chk2 (IC50, 0.7 μM). In vitro, CHIR-124 also potently targets other kinases such as PDGFR (IC50, 6.6 nM) and FLT3 (IC50, 5.8 nM). CHIR-124 interacts synergistically with topoisomerase I poisons (e.g., camptothecin) in p53-mutant tumor cells and in an orthotopic breast cancer xenograft (26). CHIR-124 also abrogates SN38-induced S-phase (by restoring Cdc25A) and G2/M-phase (via Cdc25C hyperphosphorylation) checkpoints, triggering apoptosis. In addition, CHIR-124 also sensitizes p53–/– HCT116 cells to IR. CHIR-124 is currently in the preclinical development stage.

*PD-321852 (Pfizer):* This compound catalytically inhibits Chk1, leading to Cdc25A stabilization and premature mitotic entry in response to gemcitabine. Inhibition of Chk1-mediated Rad51 responses to gemcitabine-induced replication stress also contributes to chemosensitization by PD-321852 (27). PD321852 is currently in preclinical development.

*MK-1775 (Merck):* This Wee1 inhibitor (IC50, 5.2 nM) potentiates the activity of DNA-damaging agents (e.g., gemcitabine, cisplatin, carboplatin) in vitro and in vivo, particularly in p53-negative cancers (28, 29).

*PD0166285 (Pfizer):* This potent, preclinical inhibitor of Wee1 (IC50, 24 nM) and Myt1 (IC50, 72 nM) inhibits Cdk1/cdc2 phosphorylation at inhibitory sites (i.e., tyrosine 15/threonine14), independently of p53 status. PD0166285 abrogates IR-induced G2/M-phase checkpoints and enhances p53-dependent cell killing (30). In addition, PD0166285 also stabilizes microtubules and downregulates cyclin D (31).

*17-AAG (Tanespimycin or KOS-953, Kosan):* Chk1, but not Chk2, is one of many client proteins of the molecular chaperone Hsp90. Exposure to the Hsp90 inhibitor 17-AAG downregulates Chk1 (32), leading to Cdc25A stabilization and sensitization to gemcitabine, etoposide, and SN38 particularly in p53–/– cells (33). In addition to multiple trials involving 17-AAG that focus on other client proteins, one ongoing clinical trial is based on Chk1 downregulation (34).

## 2. Materials

### 2.1. Cell Cycle Analysis

#### 2.1.1. Flow Cytometry

1. PBS
2. 100% ethanol
3. Propidium iodide (PI) solution: $3.8 \times 10^{-3}$ M Na citrate containing 0.5 mg/mL RNAse A (stock =10 mg/mL in $dH_2O$, store at −20°C) and 0.01 mg/mL propidium iodide (stock = 25 mg/mL in DMSO, store at 4°C). Store at 4°C.

#### 2.1.2. [$^3$H]Thymidine Incorporation

1. [*methyl*-$^3$H]thymidine
2. 0.5 × PBS
3. 0.05% trypsin-EDTA (for adherent cells)

4. 5% and 20% TCA (trichloroacetic acid)
5. 96% ethanol

#### 2.1.3. BrdU Incorporation

1. 1 mM BrdU stock solution in sterile PBS. Store at 4°C.
2. 1% BSA (bovine serum albumin) in PBS.
3. 70% ethanol.
4. 2 M HCl containing 0.5% Triton X-100.
5. 0.1 M $NaB_4O_7$ (pH 8.5).
6. 1% BSA/0.5% Tween-20 in PBS.
7. Anti-BrdU antibody conjugated by fluorescein (e.g., FITC, fluorescein isothiocyanate).
8. 5 μg/ml propidium iodide in PBS. Store at 4°C.

#### 2.1.4. Mitotic Index

##### 2.1.4.1. Microscopy

1. PBS
2. 0.5 × PBS or 0.75 mM KCl
3. 0.1–0.2 mg/L Giemsa dye

##### 2.1.4.2. Flow Cytometry

1. PBS
2. 70% ethanol
3. 0.25% (v/v) Triton X-100 in PBS
4. 5% BSA in PBS. Store at 4°C.
5. Primary antibody specifically recognizing the selected mitotic biomarker (e.g., phosphorylated histone H3 at serine 10 or 28)
6. Secondary antibody conjugated with a fluorescein (e.g., FITC)
7. 1 μg/mL propidium iodide/10 μg/mL RNase A in PBS. Store at 4°C.

#### 2.1.5. Cell Proliferation Analysis

##### 2.1.5.1. Cell Counting

Trypan blue solution (optional)

##### 2.1.5.2. MTS

1. MTS solution (e.g., CellTiter 96R Aqueous One Solution Cell Proliferation Assay, Promega). Store at −20°C.
2. 10% SDS (optional)

##### 2.1.5.3. Clonogenic Assay

1. 2× culture medium (e.g., RPMI 1640): made by doubling amount of medium powder. Store at 4°C.
2. 1.0% agar: freshly dissolve agar in sterile $dH_2O$ by microwaving, place at 45°C, and allow to cool.
3. 1× culture medium containing 10%FBS.
4. FBS.

### 2.2. Analysis of Checkpoint Signaling Components

#### 2.2.1. Western Blot

1. PBS.
2. SDS sample buffer: 62.5 mM Tris–HCl/pH 6.8, 2% SDS, 10% glycerol, 0.1% (w/v) bromophenol blue, 50 mM DTT, 1:1000 CLAPS. Note: the sample buffer without DTT and CLAPS can be formulated and stored at −20°C as aliquots. Before use, add DTT (stock = 1 M, −20°C) and CLAPS (proteinase inhibitors, stock [store at −20°C] containing 1 mg/mL *c*hymostatin, 1 mg/mL *l*eupeptin, 1 mg/mL *a*protinin, 1 mg/mL *p*epstatin, and 5 mg/mL trypsin inhibitor/*s*erpins; for formulation of stock solution, 5 mg of chymostatin and pepstatin should be dissolved in 250 μL DMSO or 500 μL methanol, respectively, after which deionized $H_2O$ [$dH_2O$] is added to a total of 1 mL). Importantly, for blotting phosphorylated proteins, also add phosphatase inhibitors including Na vanadate (stock = 0.5 M, −20°C) and Na pyrophosphate (stock = 0.1 M, freshly made) at final concentrations of 1 mM each.
3. 1× running buffer: diluted from 5× stock (1 L containing 94 g glycine, 15.2 g Tris, 5 g SDS, pH 8.3. Store at room temperature.).
4. 1× transfer buffer (store at room temperature): dissolve 2.4 g Tris and 15 g glycine, adjust to pH 8.5, then add 200 mL methanol, and add $dH_2O$ to 1 L.
5. TBS-T: 1× TBS containing 0.1% (v/v) Tween-20. Store at room temperature.
6. 5% nonfat dry milk in TBS-T. Store at 4°C.
7. 5% BSA in TBS-T. Store at 4°C or −20°C.
8. Specific primary antibody.
9. HRP-conjugated secondary antibody.
10. ECL (enzyme chemiluminescent) reagent.
11. Antibody against house-keeping proteins (e.g., β-tubulin, α-actin).

#### 2.2.2. Immunofluorescent Staining and Confocal Microscopy

1. PBS ($Ca^{2+}$ and $Mg^{2+}$ free).
2. 4% paraformaldehyde buffered in 0.1 M PBS (store at 4°C): dissolve 4 g paraformaldehyde powder in 50 mL 0.2 M PBS by continuously stirring, heating at 60–65°C, and adjusting pH to 7.4 by adding 1 N NaOH dropwise until the solution becomes clear. Store at 4°C and use within a few days. Before use, dilute 1:1 with $dH_2O$.
3. 0.5% Triton X-100/2% BSA solution (store at 4°C): 0.5% (v/v) Triton X-100 and 2% (w/v) BSA in PBS.
4. 0.05% Triton-X-100/1% BSA solution (store at 4°C): 0.05% (v/v) Triton X-100 and 1% (w/v) BSA in PBS.

5. Specific primary antibody: unlabeled or fluorescence-conjugated specific primary (for direct staining). If using unlabeled primary antibody (indirect staining), then fluorescence-conjugated secondary antibody is required.
6. Mounting medium: e.g., 9 parts of glycerol, 1 part of PBS, and 1–3% *n*-propyl gallate. Optional: many types of mounting medium are commercially available. If possible, choose one containing anti- or slow-fade reagents to preserve the durability of fluorescent signals.

*2.2.3. In Vitro Kinase Assays*

1. PBS.
2. 1× cell lysis buffer (store at 4°C): 20 mM Tris, pH 7.4, 150 mM NaCl, 1 mM EDTA, 1 mM EGTA, 1% Triton X-100, 2.5 mM Na pyrophosphate, 1 mM β-glycerophosphate, 1 mM Na vanadate, 1 μg/mL leupeptin. Add 1 mM PMSF before use.
3. Specific primary antibody.
4. Dynabeads or agarose beads conjugated with protein A, G, A/G, or L.
5. 1× kinase buffer (store at 4°C): 25 mM Tris, pH 7.5, 5 mM β-glycerolphosphate, 2 mM DTT, 0.1 mM Na vanadate, and 10 mM $MgCl_2$.
6. Specific substrate.
7. ATP.
8. [$\gamma^{32}$P]-ATP.
9. 2× SDS-sample buffer (store at −20°C).

*2.2.4. (Co-) Immunoprecipitation*

1. PBS.
2. RIPA buffer (store at 4°C): 1% NP40, 0.5% Na deoxycholate, 0.1% SDS in PBS. Add 1 mM PMSF, 1 mM Na vanadate, and 1:1000 CLAPS before use.
3. Specific primary antibodies for immunoprecipitation and Western blot, respectively.
4. Dynabeads or agarose beads conjugated with protein A, G, A/G, or L.
5. SDS sample buffer containing 1% (v/v) β-ME. Store at −20°C.

## 2.3. Analysis of DNA Damage

*2.3.1. Comet Assay*

1. PBS ($Ca^{2+}$ and $Mg^{2+}$ free).
2. Lysis solution (store at 4°C): 10 mM Tris–HCl, 2.5 M NaCl, 0.1 M EDTA, adjust to pH 10 with NaOH pellets, filter through a 0.5-μm filter.
3. Low melting point (LMP) agarose.

4. Alkaline solution (store at 4°C): 1.2% w/v NaOH containing 1 mM EDTA, pH > 13. Cool to room temperature before use.
5. 1× TBE.
6. 70% ethanol.
7. SYBR Green I in TE buffer (pH 7.5).

### 2.4. Analysis of Apoptosis

#### 2.4.1. DNA Ladder

1. Lysis buffer (store at 4°C): 5 mM Tris, 20 mM EDTA, 0.5% Triton X-100, pH 8.0, add 200 μg/mL proteinase K (stock = 20 mg/mL, store at −20°C) immediately before use.
2. RNase A (stock =10 mg/mL, store at −20°C).
3. Low melting point (LMP) agarose.
4. 1× TBE.
5. Ethidium bromide.
6. DNA loading buffer.
7. 100-bp DNA ladder.

#### 2.4.2. Microscopy

##### 2.4.2.1. TUNEL

1. PBS.
2. 4% formaldehyde in PBS (store at 4°C).
3. Acetic acid/ethanol (1:2, store at −20°C).
4. 1 mg/mL BSA in PBS (store at 4°C).
5. Terminal transferase (store at −20°C).
6. Staining mixture: 20 μL 5× reaction buffer, 1 μL terminal transferase, 10 μL $CoCl_2$ (stock = 25 mM), 68.8 μL $dH_2O$, and 0.2 μL fluorescein-10-dUTP (add immediately before use). Prepare the mixture in an amber tube. Other components of mixture are usually ordered together with terminal transferase from the same manufacturer.
7. Mounting medium containing propidium iodide or DAPI.
8. Nail polish.

##### 2.4.2.2. Hoechst 33342 Staining

1. PBS.
2. Methanol/acetic acid (3:1, store at 4°C).
3. 1 μg/mL Hoechst 33342 in $dH_2O$ or PBS (store at 4°C).

#### 2.4.3. Flow Cytometry

##### 2.4.3.1. $DiOC_6$/7-AAD

1. $10^{-6}$ M $DiOC_6$ (stock = $10^{-2}$ M in DMSO, store at −20°C) in PBS, store at 4°C.
2. 0.5 μg/mL 7-AAD (stock = 10 μg/mL in DMSO, store at −20°C) in PBS, store at 4°C.

2.4.3.2. Annexin V

1. PBS.
2. 1× binding buffer (store at 4°C): 10 mM HEPES/NaOH, pH 7.4, 140 mM NaCl, and 2.5 mM $CaCl_2$.
3. Propidium iodide: 50 μg/mL in PBS, store at 4°C.

*2.4.4. Western Blot*

2.4.4.1. Cytosolic Fraction

1. PBS.
2. Digitonin lysis buffer: 75 mM NaCl, 8 mM $Na_2HPO_4$, 1 mM $NaH_2PO_4$, 1 mM EDTA, 350 μg/mL digitonin, 250 mM sucrose). Dissolve 35 mg digitonin in 200 μL DMSO by vortexing before adding to the solution. Formulate the solution without sucrose and store at 4°C, which may become slightly cloudy in cold. Bring up room temperature until the solution becomes clear, and add sucrose before use.
3. 2× SDS sample buffer (store at 4°C).

## 3. Methods

### *3.1. Cell Cycle Analysis*

*3.1.1. Flow Cytometry*

Propidium iodide is a dye that stains DNA and RNA. After removing RNA by RNAse, propidium iodide specifically stains DNA. Cells in different phases of cell cycle contain distinct contents of DNA (i.e., 1n, G0/G1; 2n, G2/M; between 1n and 2n, S; and less then 1n, sub-G1, indicating apoptotic cells). Therefore, staining of DNA by propidium iodide followed by flow cytometry is classically employed to monitor cell populations in these phases of the cell cycle.

1. Collect 1–2 × $10^6$ cells for each condition or time point.
2. Pellet the cells at 500×*g* for 5 min (4°C), aspirate the supernatant, and mildly vortex to loosen the cell pellet for complete resuspension of the cells in fixative solution as follows (*see* **Note 1**).
3. Resuspend the cells in 1.5 mL PBS followed by the addition of 3 mL 100% ethanol, in which the cells are fixed at a final concentration of 67% ethanol (*see* **Note 1**).
4. Incubate the cells on ice for at least 1 h. Optional: incubate overnight at 4°C on ice.
5. Pellet the cells at 500×*g* for 5 min (room temperature), and carefully aspirate the supernatant to avoid loss of cells from the relatively loose pellet.
6. Gently resuspend the cells in 1.0 ml of propidium iodide solution (*see* **Note 2**).
7. Incubate the cells in the dark on ice for 3 h (recommended) or at room temperature for 15–30 min.

8. Pellet cells at 500×$g$ and carefully aspirate the supernatant.
9. Gently resuspend the cells in 1.0 mL PBS as noted for Step 6, and analyze by flow cytometry.
10. To quantify contents of cells in each phase (i.e., $G_0/G_1$, S, and $G_2/M$, as well as sub-$G_1$ apoptotic cells), use a commercial software such as ModFit as per the manufacturer's instructions (**Fig. 1**, Ref. (35)).

#### 3.1.2. [$^3$H]Thymidine Incorporation

Pulse-labeling of cells with [$^3$H]thymidine has traditionally been used to monitor DNA synthesis and the maximum number of cells in S phase by comparing values with baseline incorporation.

1. Pellet 1 × $10^5$ cells by centrifugation for 5 min at 500×$g$ (room temperature), and aspirate the supernatant. Note: the number of cells can be reduced by maximum of tenfolds, if necessary.
2. Resuspend in an equivalent volume of pre-warmed complete medium containing 3–5 μCi/mL [*methyl*-$^3$H]thymidine (*see* **Note 3**), and incubate for 30 min at 37°C.
3. For suspension cells, centrifuge the cells for 5 min at 500×$g$ (room temperature), aspirate the supernatant, wash twice with 5 mL of ice-cold PBS, resuspend in 0.5 mL of 0.5× PBS. For adherent cells, carefully rinse the cells three times with 1–5 mL ice-cold PBS. After aspirating PBS, trypsinize the cells by incubation at 37°C for 15 min in 0.5 mL of 0.05% trypsin-EDTA.
4. Add an equal volume (e.g., 0.5 mL) of ice-cold 20% TCA (trichloroacetic acid), and incubate for 20 min at 4°C (*see* **Note 4**).
5. Transfer entire precipitate to a pre-wetted (5% TCA or PBS) filter paper under vacuum or an automatic plate harvester according to the manufacturer's directions. Note: for

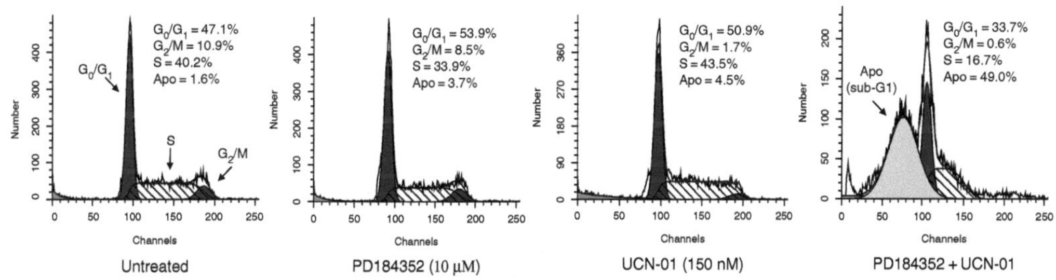


Fig. 1. Cell cycle analysis. Human leukemia U937 cells were exposed (24 h) to the Chk1 inhibitor UCN-01 (150 nM) in the absence or presence of the MEK1/2 inhibitor PD184352 (10 μM), after which cell cycle analysis was performed to monitor distribution of cell population in diverse phases. Whereas these agents administration individually induced a modest increase in population of $G_0/G_1$ and S phase (by UCN-01) or $G_0/G_1$ (by PD184352), co-treatment resulted in a dramatic induction of apoptosis, manifested by a pronounced increase in the population of sub-$G_1$ cells.

complete transfer, rinse culture wells with PBS and aspirate PBS to the same paper filter, and repeat three times.

6. Wash the filter paper with 96% ethanol to remove PBS.
7. Carefully place the filter in a scintillation flask containing 8 ml scintillation fluid, and use a scintillation counter to measure c.p.m., reflecting [*methyl*-$^3$H]thymidine incorporation.

#### 3.1.3. BrdU Incorporation

Bromodeoxyurdine (BrdU) is a thymidine analogue which is capable of being incorporated into DNA in place of thymidine during DNA synthesis. Therefore, pulse labeling with BrdU can monitor the capacity of DNA synthesis and determine precisely the percentage of cells specifically in S phase by flow cytometry. Alternatively, cytospin slides (for suspension cells) or coverslip (for adherent cells) can be prepared after labeling with BrdU, stained as described below, and observed by fluorescence or confocal microscopy. In addition, BrdU incorporation analysis can also be used in vivo by intravenous infusion of 1–5 mg BrdU per kg body weight, after which tissue samples are removed 30–60 min later and stained as follows.

1. Add a final concentration of BrdU (1 mM stock in sterile PBS) to the cell suspension in culture medium to a final concentration of 10 μM, and incubate for 30 min in a $CO_2$ incubator at 37°C.
2. Wash the cells twice with PBS containing 1% BSA (bovine serum albumin) by centrifugation for 15 min at 500×*g*, and resuspend in PBS on ice.
3. Add the cell suspension dropwise into 5 mL of 70% ethanol (–20°C) while gently vortexing, and incubate on ice for 30 min.
4. Centrifuge for 10 min at 500×*g*, aspirate the supernatant, and gently vortex to loosen the cell pellet.
5. Slowly add 1 mL of 2 M HCl containing 0.5% Triton X-100 to the cells while vortexing, and incubate for 30 min at room temperature.
6. Centrifuge for 10 min at 500×*g*, aspirate the supernatant, and resuspend in 0.1 M $NaB_4O_7$ (pH 8.5).
7. Centrifuge for 10 min as 500×*g*, and aspirate the supernatant. Optional: the cells can be resuspended in 70% ethanol and stored at –20°C. Before proceeding to the next step, pellet the cells by centrifugation for 10 min at 500×*g*, and aspirate the supernatant.
8. Resuspend the cells in PBS containing 1% BSA and 0.5% Tween-20, and adjust cell concentration to 1 × $10^6$ per 100 μL.

9. Mix 100 μL cell suspension with a final concentration of 10 μg/mL anti-BrdU antibody conjugated by fluorescein (e.g., FITC, fluorescein isothiocyanate) (*see* **Note 5**).
10. Wash the cells once with PBS containing 1% BSA and 0.5% Tween-20 and centrifuge for 5 min at 500×*g*.
11. Resuspend the cells in PBS containing 5 μg/mL propidium iodide.
12. Analyze by flow cytometry as above (**Fig. 2**, Ref. (36)). This method can also be used to monitor DNA synthesis by determining the percentage of $BrdU^+$ cells (**Fig. 3**).

#### *3.1.4. Mitotic Index*

The mitotic index reflects the number of cells progressing through M phase, during which chromatin condensation occurs. The mitotic index can be obtained by counting cells with condensed chromatin (usually rounded in monolayer cells) or labeled by immunocytochemical staining for biomarkers (e.g.,

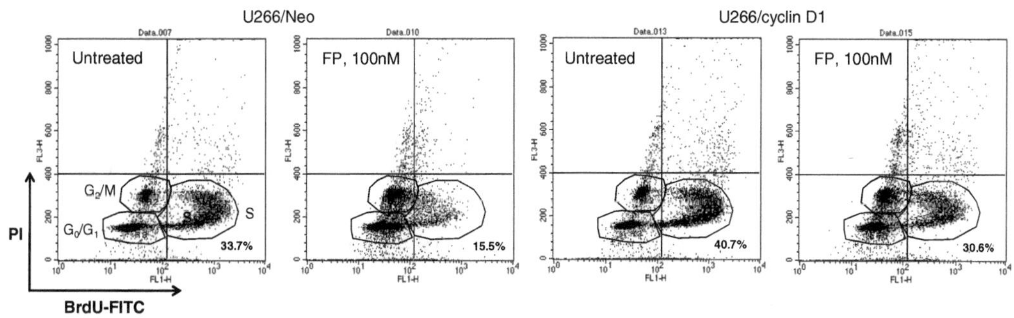


Fig. 2. Determination of S phase content by BrdU incorporation. Human multiple myeloma U266 cells stably transfected with a construct encoding full-length human cyclin D1 or its empty vector (Neo) were exposed (24 h) to the Cdk inhibitor flavopiridol (FP, 100 nM), after which cells were pulse-labeled with BrdU and analyzed by flow cytometry. Administration of FP resulted in a marked decline in the S phase fraction, an event reversed by overexpression of cyclin D1.

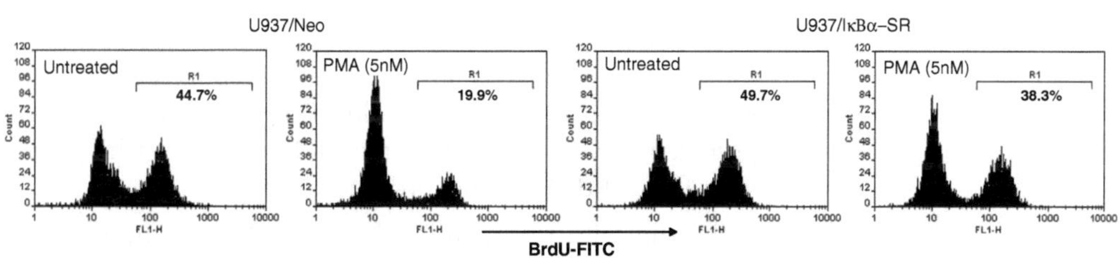


Fig. 3. Analysis of DNA synthesis by BrdU incorporation. U937 cells stably transfected with a construct encoding IκBα "super repressor" (IκBα-SR, mutation of Ser32, 36 to Ala) or its empty vector (Neo) were exposed (24 h) to PMA (5 nM), after which cells were pulse labeled by BrdU and analyzed by flow cytometry. The S phase content was assessed by determining the percentage of BrdU-positive cells. Ectopic expression of IκBα-SR prevented the PMA-mediated decline in percentage of BrdU-positive cells. Because cell proliferation requires DNA synthesis, these data indicate that blockade of the NF-κB pathway by IκBα-SR blocks PMA-induced U937 differentiation characterized by exit of cell cycle but undergoing maturation.

phosphorylated histone H3 at serine 28 or 10) specific for mitotic cells under microscopy or by flow cytometry. These methods are adaptable to suspension cells or trypsinized adherent cells. For adherent cells, complete trypsinization to collect all cells from culture is important to avoid erroneous results. Comparing with the classical microscopic method, immunocytochemical techniques using flow cytometry or fluorescent imaging of multi-well plate represent faster, more reliable, and precise methods to determine the mitotic index.

#### 3.1.4.1. Microscopy

1. Centrifuge the cells for 5 min at 500×*g*, aspirate the supernatant, and wash with 2 volumes of ice-cold PBS.
2. Pellet the cells, resuspend, and incubate on ice in 0.5 mL of ice-cold 0.5× PBS for 10 min or 0.75 mM KCl for at least 20 min (*see* **Note 6**). Optional: resuspend the cells in 6 mL 2% (v/v) of ice-cold 3:1 ethanol/glacial acetic acid, and store overnight at 4°C.
3. Centrifuge for 5 min at 500×*g* (room temperature), discard supernatant, and resuspend at a final density of 1–5 × $10^3$ cells/mL in 0.5 mL of ice-cold 3:1 ethanol/glacial acetic acid while vortexing gently to avoid cell clumps.
4. Add 100 μL of the cells onto glass slides (2–3 slides per condition to obtain mean values for the mitotic index, *see* **Note 7**), and air dry the slides for 30 min to 2 h. Optional: use cytospin apparatus to prepare slides.
5. Stain the slides with 0.1–0.2 mg/L Giemsa dye for 10 min, gently rinse with PBS, and air dry.
6. Count the cells under light microscope to determine the percentage of cells with mitotic figures (i.e., condensed nuclear material without nuclear membrane) relative to the total number of cells counted (generally, 200–500 per condition).

#### 3.1.4.2. Flow Cytometry

1. Centrifuge 2 × $10^6$ cells for 5 min at 500×*g*, resuspend in 200 μl PBS, then add 2 mL of 70% ethanol (−20°C) while vortex to avoid cell clumps, and incubate at −20°C for at least 3 h or overnight (recommended). Optional: after removing medium, add ice-cold 100% methanol to fix and permeabilize the cells for 10 min, followed by washing once with PBS. If this is done, proceed to Step 3.
2. Pellet the cells, resuspend in 1 ml of PBS containing 0.25% Triton X-100, and incubate on ice for 15 min to permeabilize the cells.
3. Pellet the cells, resuspend in PBS/5% BSA, and incubate for 1 h to block the cells.
4. Pellet the cells, resuspend in 100 μL PBS/5% BSA containing a primary antibody specifically recognizing the selected

mitotic biomarker (e.g., phosphorylated histone H3 at serine 10 or 28) at an optimal dilution, and incubate for 1 h at room temperature or overnight at 4°C (recommended).

5. Wash the cells twice with 1 mL PBS (room temperature).
6. Resuspend the cells in 100 μL PBS/5% BSA containing a secondary antibody that is conjugated with a fluorescein (e.g., FITC), and incubate for 30 min to 1 h at room temperature in the dark.
7. Wash the cells twice with 1 mL PBS, resuspend in 1 mL PBS containing 1 μg/mL propidium iodide and 10 μg/mL RNase A, and incubate for 20 min at 37°C in the dark. There is no need to wash the cells prior to immediate analysis.
8. Analyze the cells by flow cytometry to determine the percentage of fluorescence-positive mitotic cells in the total number of cells. Optional: if a fluorescent imaging system is available, this protocol can be used for multi-well plate as well, without trypsinization for adherent cells.

#### *3.1.5. Cell Proliferation Analysis*

##### 3.1.5.1. Cell Counting

Daily counting of cell number is the simplest method to monitor cell proliferation/growth in culture. Cell counting can be done under a light microscope using a hemocytometer or by an automatic cell counter (e.g., Coulter Counter) according to manufacturer's instructions. If necessary, mix cells with the equal volume of trypan blue solution to monitor cell viability. While living cells exclude trypan blue dye, only dead cells will be stained. For this purpose, some new models of automatic cell counters contain a system for trypan blue staining and measurement.

##### 3.1.5.2. MTS

The MTS assay can be used to determine precisely proliferation and viability of cells in culture. For example, the CellTiter 96 Aqueous One Solution Cell Proliferation Assay (Promega) is a colorimetric method used to monitor cell proliferation or cytotoxicity by determining the number of viable cells. A tetrazolium compound MTS [3-(4,5-dimethylthiszol-2-yl)-5-(3-carboxymethoxyphenyl)-2-(4-sulfophenyl)-2H-tetrazolium, inner salt) is an analog of MTT [3-(4,5-dimethylthiszol-2-yl)-5-diphenyltetrazolium bromide), which is bioreduced by cells to a colored formazan product. This conversion is mediated by NADPH or NADH generated by dehydrogenase enzymes in metabolically active cells. The formazan product of MTT reduction forms a crystalline precipitate, and thus requires a step to dissolve the crystals before recording absorbance readings at 570 nm with a plate reader. In contrast, the MTS formazan product is soluble in cell culture medium, which thus simplifies the assay procedure. Moreover, MTS is combined with an electron coupling reagent PES (phenazine ethosulfate), which has enhanced chemical stability, to form a stable solution "One

Solution." This assay can be substituted for a [$^3$H]thymidine incorporation assay.

1. Plate the cells into a 96-well plate in a volume of 100 μL culture medium containing the required amount of FBS and allow to reach a density of $2 \times 10^5$/mL before treatment.
2. Completely thaw the MTS reagent (e.g., 20 mL size) at room temperature for approximately 90 min, or in a water bath at 37°C for 10 min.
3. Add 20 μL of the MTS reagent (317 μg/mL final concentration) into each well, and incubate the plate for 1–4 h at 37°C in dark in a $CO_2$ incubator.
4. Record the absorbance at 490 nm using a 96-well plate reader (see **Note 8**). The net amount of 490 nm absorbance directly reflects the number of living cells in culture.

#### 3.1.5.3. Clonogenic Assay

1. Pellet cells for 5 min at 500×*g* at 37°C, wash three times with complete medium (containing 10% FBS), and resuspend in original volume of complete medium.
2. Count total number of cells, and dilute to 5000 cells/mL in complete medium.
3. Formulate cloning solution by mixing 1.05 mL 2× medium (the final concentration will be 1× after adding equal amount of agar), 0.70 mL FBS (a final concentration of 20%), 0.35 mL 1× complete medium, 1.05 mL 1.0% agar (a final concentration of 0.3%), and 0.35 mL cell suspension (a final density of 500 cells/mL) (*see* **Note 9**)
4. Immediately plate 1 ml/well of the mixture onto a 12-well plate. Plate evenly to cover the whole bottom of wells with the cell suspension mixture.
5. Add sterile $dH_2O$ between wells and in any empty wells immediately after plating (*see* **Note 10**).
6. Allow the plates to sit for approximately 15 min at room temperature until agar has hardened, and then incubate at 37°C, 5% $CO_2$.
7. Score colonies (>50 cells, **Fig. 4**, Ref. (37)) using an inverted microscope 10–12 days (*see* **Note 11**) after plating.

### *3.1.6. Expression/Activation of Cyclins and Cdks*

Expression of cyclins and activation status of cyclin-dependent kinases (Cdk) are characteristic for each phase of the cell cycle. Consequently, Western blot analysis or in vitro kinase assays can be used to monitor levels of cyclins or activity of Cdks, respectively, as indicators of specific cell cycle phases. In general, early G1 phase is characterized by expression of D1–D3 cyclins (D1 in most cases) and activation of Cdk4/6; mid-late G1 by cyclin E and Cdk2; S by cyclin A and Cdk2; and G2/M by cyclin A/B and

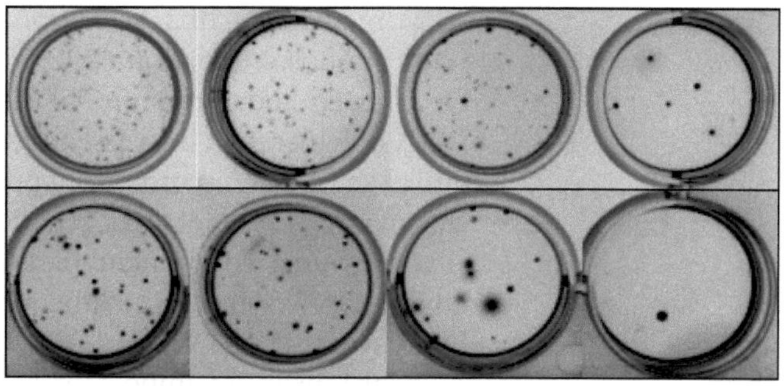


Fig. 4. Clonogenic assay. Human leukemia cells were exposed to UCN-01 (HL-60, 100 nM; Raji, 150 nM) with or without the farnesyltransferase inhibitor L744832 (10 μM), after which cells were washed free of drug and plated in soft agar. After 12 days of incubation, colonies (more than 50 cells) were scored. Combined treatment resulted in the virtual abrogation of colony formation in each of cell lines.

Cdkl/cdc2. However, abnormalities of cyclin expression and/or Cdk activity is (are) features that occur frequently in cancer cells. Therefore, caution should be used with this method of assessing the cell cycle. The procedures for Western blot analysis and kinase assays are described below in the **Section 3.2.1**.

### 3.2. Analysis of Checkpoint Signaling Components

Depending upon the cell cycle phase in which DNA damage occurs, and the specific types of DNA lesions that arise (e.g., single-strand breaks/SSBs, double-strand breaks/DSBs, chemical adducts, mismatches) (38), diverse checkpoint signaling pathways, are activated (38), which block entry into S phase (the Gl/S checkpoint), delay progression through S phase (the intra-S phase or S phase checkpoint), or prevent entry into mitosis (the G2/M phase checkpoint) (39).

*Intra-S (or S) phase checkpoint:* At least two pathways are involved in the S-phase checkpoint, i.e., the ATM/ATR-Chkl/Chk2-Cdc25A-Cdk2 pathway and the Nbsl-dependent pathway that includes the ATM/Nbsl/Smcl and the ATM/Nbsl/FANCD2 pathways (40). The ATM/ATR-Chkl/Chk2-Cdc25A-Cdk2 pathway represents the primary mechanism responsible for the intra-S checkpoint. Following activation by ATR and ATM, Chkl and Chk2 subsequently phosphorylate Cdc25A at multiple serine sites. C-terminal phosphorylation by Chkl, together with N-terminal phosphorylation by an unknown kinase(s), targets Cdc25A to an SCF (Skpl-Cullin-F-Box) complex containing β-TrCP, leading to its ubiqutination and proteasomal degradation. In general, ATR/Chkl plays a dominant role in the S-phase checkpoint in response to replication stresses (also known as the replication checkpoint) and genotoxic insults, whereas Chk2 only plays a minor role. For example, loss of Chkl causes failure to degrade Cdc25A in the

S-phase checkpoint in response to DSBs that primarily involve ATM-mediated Chk2 activation. Degradation of Cdc25A prevents dephosphorylation (activation) of Cdk2 at inhibitory sites tyrosine 15 and threonine 14, thereby blocking recruitment of DNA polymerase α into assembled pre-replication complexes and initiation of DNA synthesis by preventing loading of Cdc45 onto chromatin (41). Notably, genetic or chemical interference with any individual component of this pathway results in failure to inhibit DNA synthesis, a phenomenon known as radioresistant DNA synthesis (RDS).

Chk1 plays an important role in the process of translesion DNA synthesis (TLS), an event mediated by the ubiquitinated form of PCNA (42). Whereas the ATM/ATR-Chk1/Chk2-Cdc25A-Cdk2 pathways are essential for stabilizing stressed replication forks, TLS allows replication forks to progress through certain types of DNA lesions. Therefore, both mechanisms are critical for continuous replication of damaged DNA and avoidance of fork collapse. PCNA is ubiquitinated on chromatin after DNA damage, which is reciprocally regulated by the ubiquitin-ligase E2-E3 complex of Rad6/Rad18 and the de-ubiquitinase Usp1. RPA-ssDNA directly binds and recruits Rad18 to ssDNA lesions, whereas neither ATR nor ATM is required for PCNA ubiquitination in TLS. PCNA directly interacts with several "mediator" proteins (e.g., Claspin, Timeless) as well as Chk1 itself, events required for efficient PCNA ubiquitination. Interestingly, kinase activity of Chk1 appears not to be essential for this function.

In addition, Chk1 activation is required for inhibition of transcriptional elongation when DNA replication is blocked. This may explain the observation that in the S phase checkpoint, a subset of p53 target genes (e.g., $p21^{CIP1/WAF1}$) fails to be induced despite the presence of high levels of p53 (43).

*G2/M phase checkpoint* (44): Cdk1/cdc2 is a kinase that governs entry and exit of cells through mitosis. Activation of Cdk1/cdc2 requires Cdc25-mediated dephosphorylation on the inhibitory sites tyrosine 15 and threonine 14, whereas their phosphorylations are mediated by Myt1- and Wee1. Although Cdc25A and Cdc25B can dephosphorylate Cdk1 at these sites, Cdc25C is a major phosphatase responsible for this event. However, while Cdc25A is known to be involved in the G2/M checkpoint, the role of Cdc25B is less clear. Whereas Chk2 may also involve, Chk1 is a major G2/M checkpoint kinase that phosphorylates Cdc25C (serine 216), and Cdc25A (serine 76 and 123, and very likely other serine residues as well), as well as Cdc25B. Phosphorylation of Cdc25C provides a binding site(s) for the 14-3-3 proteins (Rad24 and Rad25), leading to exclusion of Cdc25C from nuclei and subsequent sequestration in the cytoplasm. As a consequence, signaling via the ATM/ATR-Chk1/Chk2-Cdc25s pathway leads to a rapid, reversible G2/M arrest.

Chk1/Chk2, as well as ATM/ATR, phosphorylate p53 at sites critical for its binding to Mdm2 and subsequent turnover. Accumulation/activation of p53 leads to transcriptional expression of endogenous Cdk inhibitors, including Gadd45 and 14-3-3σ, that bind to and inhibit Cdk1. This p53-dependent mechanism results in a slower, irreversible G2/M arrest.

*G1/S phase checkpoint* (45): Cdk2/cyclin E and /cyclin A govern the G1/S transition. To prevent entry of cells with damaged DNA into S phase, the G1/S checkpoint is activated by disabling Cdk2 via Cdc25A- and p53-dependent pathways. The ATM/ATR-Chk1/Chk2-Cdc25A-Cdk2 pathway is responsible for a faster (acute), transient G1 phase delay. In contrast, activation of the p53-dependent pathway, as described above for the G2/M phase checkpoint, results in transcriptional expression of the endogenous Cdk inhibitor $p21^{CIP1/WAF1}$ that binds to and inhibits Cdk2/cycling E, leading to a slower and sustained (including permanent) cell cycle arrest in G1 phase or at the G1/S border.

*Mitotic spindle checkpoint:* This checkpoint delays anaphase onset in cells with mitotic spindle defects and requires the kinase activity of Chk1, but not Chk2. In response to spindle toxins (e.g., taxol and nocodazole), Chk1 is phosphorylated at residues other than the canonical sites (e.g., serine 317 and 345), and associates with kinetochores in prometaphase, leading to sustained anaphase delay. Moreover, Chk1 phosphorylates Aurora-B and enhances its catalytic activity in vitro, whereas failure of the spindle checkpoint in Chk1-deficient cells correlates with decreased Aurora-B kinase activity and impaired phosphorylation and kinetochore localization of BubR1 (46). Abrogation of Chk1 in mitotic cells results in multiple mitotic defects, in association with mislocalized Aurora B (47). Chk1 is also required for protection of cells from spontaneous chromosome missegregation, as well as optimal tumor cell killing by taxol. In addition, Chk1 also negatively regulates another mitotic substrate, Plk1 (polo-like kinase 1) in the absence or presence of DNA damage (48).

*DNA damage/repair:* Chk1 is involved in DNA repair by targeting repair kinases (e.g., DNA-PK), which, together with Ku70-KU80 (designed the DNA-PK complex), are important for DSB repair (49). Moreover, Chk1-dependent phosphorylation of Rad51 is required for DNA damage-induced homologous repair/HRR (50). Lastly, Chk1-mediateded FANCE phosphorylation is critical for the FA (Fanconi Anemia)/BRCA-mediated DNA repair pathway (51, 52). Conversely, abrogation of Chk1 by either inhibitors or siRNA causes ssDNA formation and DNA strand breaks (53).

*Apoptosis:* p53 is a central downstream checkpoint signaling protein responsible for apoptotic responses. However, ATR/Chk1 signaling is essential for suppression of a caspase-3-dependent apoptotic response following replication stress (54). Moreover, Chk1, but not Chk2, also blocks a caspase-2-dependent apoptotic response independently of p53, Bcl-2, and caspase-3 (55). Interestingly, caspase-mediated Chk1 cleavage (Asp299/Asp351) promotes its activation (56), raising the possibility of unexplored, direct links between Chk1 and apoptotic signals.

*Transcription:* Chk1 phosphorylates histone H3 (Thr11), responsible for DNA damage-induced transcriptional repression of cell cycle-regulatory genes (e.g., cyclin B1 and Cdk1) by loss of histone acetylation (57).

In general, cell cycle checkpoints are activated through a precisely orchestrated signal transduction process involving recruitment to DNA lesions, followed by subsequent phosphorylation and/or activation of signaling components. As a consequence, many signaling components (e.g., "sensor," "proximal transducer," and "mediator") interact with each other to form multiprotein complexes microscopically visible as nuclear foci at the sites of DNA damage. Based on these events, several approaches have been developed to evaluate activation status and functional roles of checkpoint signaling proteins. In general, Western blot analysis is used to monitor phosphorylation at specific sites (primarily in studies involving kinases and their substrates), total levels of proteins that vary specifically in different phases of cells cycle, as well as subcellular distribution of proteins (e.g., Cdc25C, nuclear versus cytosolic fractions). Kinase assays (e.g., in vitro using cell lysates) can be applied to determine the activity of kinases (e.g., Chk1, cdc2/Cdk1) if their substrates are available. Immunofluorescent staining and confocal microscopy have been widely used to assess subcellular localization and colocalization employing multiple labeling for evaluation of protein–protein relationship/interactions of these signaling components. In the latter case, co-immunoprecipitation can be performed to determine further association/interactions between proteins of interest. In addition, many antibodies recognizing only site-specific phosphorylated forms of proteins are also available for immunofluorescent staining. It is now possible to examine phosphorylation status and functional contributions of individual signaling components at sites of DNA damage. For further determination of the functions of individual signaling components at the sites of DNA lesions, confocal microscopy can also be used to monitor localization of target proteins following transfection of cells with constructs encoding fluorescent protein (e.g., GFP)-labeled proteins.

*3.2.1. Western Blot*

Most of the procedures for Western blot analysis are identical, but differ in the use of appropriate antibodies that specifically recognize targeting proteins. In general, Western blot analysis of proteins involving checkpoint signaling pathways include examination of site-specific phosphorylation (e.g., kinases and their substrates) and/or total levels of proteins (e.g., cyclins, Cdc25A, and endogenous Cdk inhibitors such as $p21^{CIP1/WAF1}$). The proteins that are phosphorylated during checkpoint signaling include the "proximal transducer" ATM (e.g., serine 1981), the distal "transducer" Chkl (e.g., serine 317 and 345 by ATR/ATM, and serine 296 via autophosphorylation) and Chk2 (e.g., threonine 68 by ATM, as well as threonine 383 and 387 by intramolecular trans-autophosphorylation), the "effector" Cdc25C (e.g., serine 216), Cdc25A (e.g., serine 123), and p53 (e.g., serine 15), cdc2, and Cdk2 (tyrosine 15 and threonine 14), and a variety of "mediator" proteins such as histone H2A.X (serine 139, designated γH2A.X) and histone H3 (e.g., serine 10). The antibodies for site-specific phosphorylation of these proteins are available through numerous sources (e.g., Cell Signaling, Santa Cruz, Millipore/former Upstate, Abcam, etc.). It is important in the detection of phosphorylated proteins to use TBS rather than PBS throughout, and to add phosphatase inhibitors in the cell lysis buffer or sample buffer to prevent dephosphorylation of substrates.

*3.2.2. Adherent Cells*

1. Aspirate culture medium, rinse the cells with PBS (*see* **Note 12**).
2. Add 100 μL/plate of SDS sample buffer, immediately scrape the cells off the plate/flask, and transfer to a centrifuge tube. Keep on ice, and then proceed to Step 3.

*3.2.3. Suspension Cells*

1. Pellet the cells in a centrifuge tube for 5 min at 500×*g*, aspirate culture medium, and wash once with PBS.
2. Add 100 μL (for $5 \times 10^6$ cells) of SDS sample buffer.
3. Sonicate for 10–15 s to shear DNA and reduce viscosity of the cell lysates. Keep on ice.
4. Boil the cell lysate for 5 min, and then cool on ice.
5. Microcentrifuge for 5 min at top speed, transfer the supernatant to a new microfuge tube (*see* **Note 13**).
6. Quantify the concentration of proteins, and load 20–30 μg protein per sample onto SDS-PAGE gels. Load pre-stained protein marker to determine molecular weight.
7. Run the gel in 1× running buffer at 100 V.
8. Electrotransfer the proteins to nitrocellulose membrane in 1× transfer buffer for 1.5–2 h at 400 mA.
9. Rinse the membrane with TBS-T (*see* **Note 14**).
   For the following steps, gentle agitation is required throughout to prevent drying of membranes.

10. Block the membrane with 5% milk/TBS-T for 1 h at room temperature.
11. Incubate the membrane with the specific primary antibody at the appropriate dilution with 5% BSA/TBS-T overnight at 4°C (*see* **Note 15**).
12. Wash the membrane three times for 10–15 min each with TBS-T.
13. Incubate the membrane with HRP-conjugated secondary antibody at the appropriate dilution with 5% milk/TBS-T for 1 h at room temperature.
14. Wash the membrane three times for 10–15 min each with TBS-T.
15. Drain the membrane, and incubate with ECL (enzyme chemiluminescent) reagent for 1 min at room temperature.
16. Drain the membrane, wrap with a plastic sheet, expose to X-ray film (**Fig. 5**, this research was originally published in *Blood* Ref. (58). © the American Society of Hematology). Optimize the exposure time based on signals.
17. Re-probe with antibody against house-keeping proteins (e.g., β-tubulin, α-actin) by repeating Steps 11–16, which serve as controls for equal loading and transfer. Optional: the membrane can also be stripped and reprobed for other proteins including these house-keeping proteins. Alternatively, the membrane can be stored in wet at 4°C for future use.

#### 3.2.4. Immunofluorescent Staining and Confocal Microscopy

The most important feature of the confocal microscope is its capacity for isolating and collecting a plane of focus within the sample, thus eliminating the out of focus "haze" that often obscures fine detail in a non-confocal, fluorescent microscope. Moreover, the confocal microscope has a stepper motor to control fine focus, enabling the optical section of a three dimensional

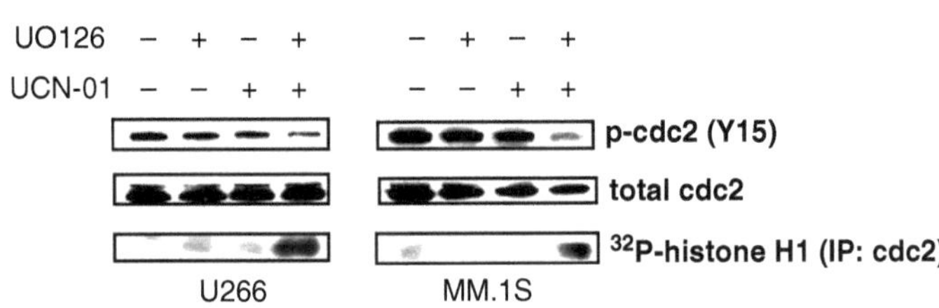


Fig. 5. Western blot and in vitro kinase assay to monitor inhibition of Chk1. Human myeloma U266 and MM.1S exposed to the MEK1/2 inhibitor U0126 (20 μM) ± UCN-01 (150 nM) for 24 h, after which Western blot analysis was performed to monitor total and phosphorylated $p34^{cdc2}$ at its inhibitory site tyrosine 15 (Y15). Alternatively, kinase assays were performed after immunoprecipitation with the $p34^{cdc2}$ antibody. Co-administration of U0126 and UCN-01 resulted in a clear activation of $p34^{cdc2}$, an event occurring downstream of Chk1 inhibition.

object and the collection of a series of images that can then be used for a two or three dimensional reconstruction. The thickness of optical sections (or the z-resolution) depends upon several factors, including the wavelength of the excitation/emission light, pinhole size, numerical aperture of the objective lens, refractive index of components in the light path, and the alignment of the instrument. Moreover, multi-color fluorescent signals (e.g., double or triple labels with different fluorophores) can be captured with a confocal microscope, which allows precise colocalization of proteins that indicate their interactions at DNA lesions.

The antibodies for immunofluorescent staining are as discussed above in Section **3.2.1**. However, some antibodies suitable for Western blot cannot be used or have not been evaluated for immunofluorescent staining. Therefore, it is important to investigate the manufacturer's product description before purchasing antibodies for this purpose.

1. Prepare cytospin slides (for suspension cells) or growing cultured cells on coverslips or slides (for adherent cells) and wash briefly with PBS ($Ca^{2+}$ and $Mg^{2+}$ free).
2. Fix the cells by submerging the slides for at least 20 min at room temperature in 4% paraformaldehyde buffered in 0.1 M PBS (*see* **Note 16**).
3. Wash the slides three times with PBS for 5 min each. After washing, encircle the area of cells/tissues using a PAP pen to avoid excessive spreading of the solution in the following procedures. Also, add 80–100 μL of solution per slide to cover the cell/tissue area but not spread outside of the circles. Importantly, do not allow the slides to dry, and thus use a humidified chamber.
4. Permeabilize and block the cells with 0.5% Triton X-100/2% BSA solution (*see* **Note 17**) for 1 h at room temperature.
5. Incubate the cells with primary antibody appropriately diluted in 0.05% Triton X-100/1% BSA solution, overnight at 4°C. If fluorescence-conjugated primary antibody (direct immunofluorescent staining) are used, incubation should be performed in the dark, after which proceed to Step 8.
6. Wash the slides three times with 1× PBS for 5 min each.
7. Incubate the slides with fluorescence-conjugated secondary antibody (*see* **Note 18**) appropriately diluted in the buffer, as described above for the primary antibody, for 1 h at room temperature in the dark.
8. Wash the slide three times with PBS for 5 min each. Note: if available, wash once with PBS, and twice with commercially available anti-fade buffer.

9. Mount the slides using a mounting medium.
10. View the samples as soon as possible, although they can be stored in the dark at 4°C for as long as several months.

*3.2.5. In Vitro Kinase Assays*

This assay is based on in vitro phosphorylation of a specific substrate by the intracellular kinase. Whereas the assay can theoretically be used to evaluate activity of all kinases, it is primarily dependent upon availability of the substrates. In addition to the use of whole protein as a substrate, peptides containing the site phosphorylated by the kinase are more commonly employed. For example, a peptide of Cdc25C containing the Chk1-phosphorylation site (e.g., serine 216) and histone H1 protein have been used for kinase assays of Chkl and cdc2/Cdk1, respectively. The following procedure is designed for the use of peptide substrates (available from Santa Cruz, etc.). Alternatively, if protein substrates (available from Millipore, Cell Signaling, etc.) are used, Western blot can also be employed to examine assay kinase activity by monitoring substrate phosphorylation using the corresponding antibodies (e.g., phospho-histone H1 for cdc2/Cdk1). Importantly, although crude cell lysates may also be used with the addition of multiple inhibitors of other potential or unknown kinases that may mediate substrate phosphorylation, it is strongly recommended to pull down the kinase by immunoprecipitation and subsequently assay the immunoprecipitates to optimize the specificity of the kinase assay.

1. Harvest cells under non-denaturing conditions; remove culture medium and wash cells twice with ice-cold PBS.
2. Remove PBS, lyse cells by incubating on ice for 5 min in 1× cell lysis buffer (*see* **Note 19**). For adherent cells, scrape cells off the plate, and transfer to a centrifuge tube.
3. Sonicate four times for 5 s each on ice, and transfer cell lysates to a new microfuge tube on ice. Optional: instead of sonication, cells can also be disrupted by repeated passage through a 21-gauge needle, followed by incubation on ice for 30 min.
4. Microcentrifuge at 16,000×*g* for 10 min at 4°C, and transfer the supernatant to a new tube. If necessary, these cell lysates can be aliquoted and stored at –80°C.
5. Add appropriate amount of primary antibody to the cell lysates (1 μg/μL of protein concentration, 200–300 μg total protein per condition), and incubate 4 h or overnight at 4°C with gentle rocking. If available, use beads or agarose conjugated with primary antibody. If this is done, proceed to Step 7.

6. Apply Dynabeads (washed before use, *see* **Note 20**) according to the manufacturer's instructions, and incubate for an additional 4 h or overnight at 4°C with gentle rocking.
7. Wash the beads twice with 1× cell lysis buffer (or RIPA buffer), and twice with 1× kinase buffer on ice.
8. Resuspend the beads in 1× kinase buffer, add substrate mix (1:3) containing appropriate amount of the specific substrate with 300 μM ATP and 150 μCi/mL [$\gamma^{32}$P]-ATP (*see* **Note 3**) in 1× kinase buffer, and incubate for 20–30 min at 30°C. Then place on ice. If protein substrates and Western blot are used, perform the reaction without [$\gamma^{32}$P]-ATP.
9. Terminate the reaction by adding an equal volume of 2× SDS-sample buffer and boiling for 5 min.
10. Pellet the beads by microcentrifugation for 2 min and apply the supernatant to SDS-PAGE gels.
11. After electrophoresis, dry the gel for autoradiography (**Fig. 6**, Ref. (35)) (*see* **Note 21**).

*3.2.6. (Co-) Immunoprecipitation*

1. Harvest 20 × $10^6$ cells per condition, and wash twice with ice-cold PBS. Note: the cell pellet may be frozen and stored at −20°C for later use.
2. Lyse cells with 150 μL RIPA buffer by repeated passage through a 21-gauge needle (*see* **Note 22**), followed by incubation on ice for 30 min.
3. Spin cell lysates at 15,000×*g* for 30 min at 4°C and transfer the supernatant to a new microfuge tube.
4. Quantify protein concentrations, take 200 μg protein per condition, and dilute with RIPA buffer to 1 μg/μL (200 μL total, *see* **Note 23**).

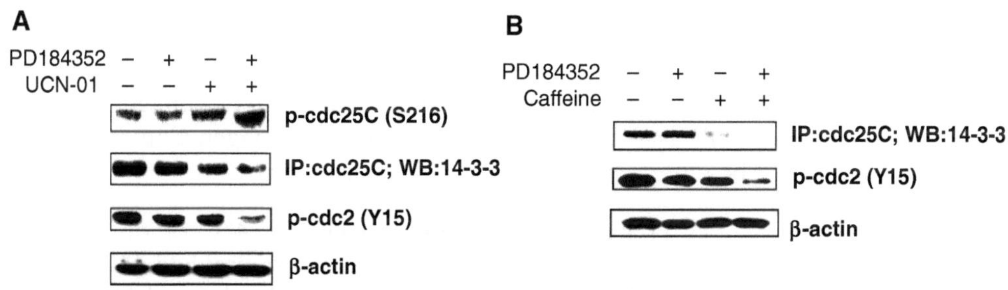


Fig. 6. Co-immunoprecipitation to assess protein–protein interactions. U937 cells were exposed to UCN-01 (A, 150 nM) or caffeine (B, 2 mM) ± PD184352 (10 μM) for 18 h, after which cells lysates were immunoprecipitated with cdc25C antibody and subsequently subjected to Western blot analysis with 14-3-3 antibody. Co-administration of PD184352 markedly promoted dissociation between the cdc25C and 14-3-3 proteins due to increased phosphorylation of cdc25C at Ser216 following exposure to UCN-01 or caffeine. This event leads to nuclear export of cdc25C and resulting dephosphorylation (activation) of $p34^{cdc2}$.

5. Add 1 μg primary antibody, and rock for 4 h or overnight (recommended) at 4°C.
6. Add 20 μL Dynabeads (washed with RIPA before use) per condition in RIPA, and rock for 4 h at 4°C. Optional: other beads or agarose for immunoprecipitation can also be used as described above (**Section 3.2.5**).
7. Wash beads 3–5 times with RIPA buffer using a magnetic rack.
8. After aspirating RIPA buffer, add 20 μL 1× SDS sample buffer containing 1% β-ME, vortex, boil for 5 min, and vortex again.
9. Pull off the beads using the magnetic rack and load the supernatant (~20 μL) on a SDS-PAGE gel.
10. Follow the procedure for Western blot analysis (**Section 3.2.1.**), probing the blots with antibodies recognizing the co-immunoprecipitated proteins (**Fig. 6**, Ref. (35)). If necessary, load Western blot samples of cell lysates without immunoprecipitation for comparison, which is helpful in identification of the co-immunoprecipitated proteins on the blots.

### 3.3. Analysis of DNA Damage

#### 3.3.1. Comet Assay

The comet assay is a simple and effective method for evaluating DNA damage in cells via single cell gel electrophoresis. In an electric field, damaged DNA (e.g., due to denaturation, breaks, or cleavage) migrates out of the cell and forms a DNA "comet" tail, while undamaged DNA remains within the nucleus. Therefore, evaluation of "comet" tail shape, length, and migration pattern allows assessment of DNA damage. The electrophoresis can be performed using TBE buffer or alkaline solution. The former permits use of the tail length rather than the tail moment for data analysis, which is preferred for evaluation of massive DNA damage such as that which occurs during apoptosis. In contrast, the latter is more sensitive and used to detect smaller amounts of DNA damage (e.g., single-strand breaks). However, it employs non-buffered alkali for electrophoresis, which may result in poor DNA migration and difficulty in controlling voltage. Commercially available reagents or kits such as Trevigen CometAssay may be more reliable and less time consuming.

##### 3.3.1.1. TBE Electrophoresis

1. Prepare lysis solution and chill at 4°C or on ice for at least 20 min (*see* **Note 24**). Before use, add 1% Triton X-100 (solution may turn slightly cloudy) and 10% DMSO (optional) that prevents oxidation during lysis incubation.
2. Melt LMP agarose in boiling water bath for 5 min, and then maintain in a 37°C water bath for at least 20 min to cool (*see* **Note 25**).

3. Harvest cells (*see* **Note 26**) by centrifugation for 5 min at 500×*g*, and resuspend at 1 × $10^5$ cells/mL in ice-cold PBS ($Ca^{2+}$ and $Mg^{2+}$ free). For adherent cells, gently scrape the cells using a rubber policeman, pellet cells by centrifugation, wash once with ice-cold PBS, and resuspend at 1 × $10^5$ cells/mL. Thoroughly remove culture medium which can reduce the adhesion of the agarose on glass slide.
4. Mix the cells with molten agarose (37°C) at a ratio of 1:10 (v/v) by gently pipetting once or twice, immediately pipette the mixture onto slide, and spread using the side of the pipette tip. If agarose/cell mixture does not spreading evenly on the slide, warm the slide at 37°C before use.
5. Place the slide flat at 4°C in the dark for at least 10 min (*see* **Note 27**).
6. Immerse the slide in pre-chilled lysis solution on ice or 4°C for 30–60 min.
7. Drain the slide to remove excess lysis solution, and immerse in freshly made alkaline solution (*see* **Note 28**) for 20–60 min at room temperature in the dark.
8. Drain the slide to remove excess alkaline solution, wash twice by immersing in 1× TBE buffer for 5 min each.
9. Place the slide flat onto a gel tray, transfer the slide to a horizontal electrophoresis apparatus, and align the slide equidistant from the electrodes.
10. Carefully pour 1× TBE buffer until it covers samples, run the gel under 1 V/cm (distance between two electrodes) for 10 min.
11. Very gently drain the slide to remove excess TBE, immerse in 70% ethanol for 5 min.
12. Air dry the slide to bring all the cells into a single plane. At this stage, samples can be stored at room temperature with desiccant.
13. Stain the slide with SYBR Green I in TE buffer (pH 7.5) at an appropriate dilution according to manufacturer's instructions, in a refrigerator for 5 min in the dark. Note: after drying, the sample can also be stained by a silver staining solution, which can be then be observed under regular light microscope. This procedure allows for long term storage.
14. Gently drain the slide to remove excess SYBR solution, and completely dry the slide at room temperature in the dark.
15. View the slide by a fluorescence microscope, and acquire the image via a digital camera (**Fig. 7**, this research was originally published in *Blood* Ref. (59) © the American

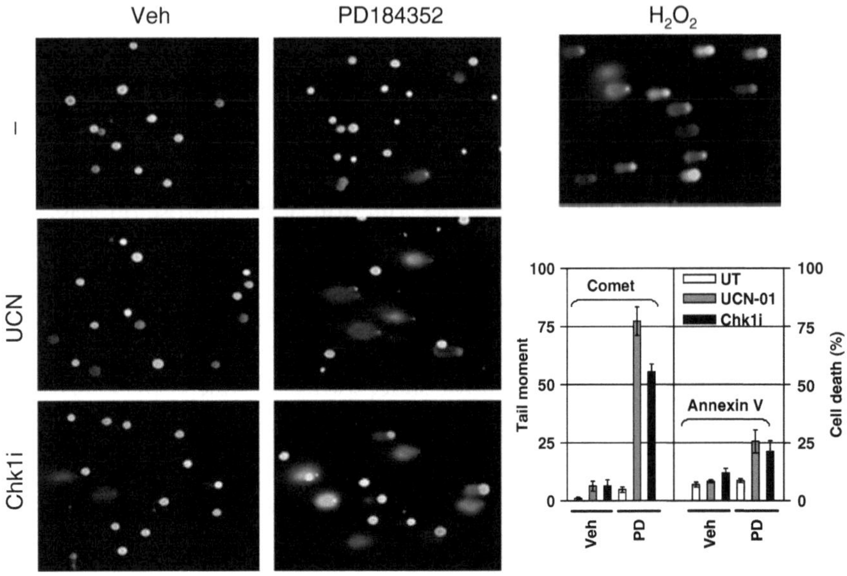


Fig. 7. Comet assay. U266 cells were treated with 150 nM UCN-01 or a specific Wee1/Chk1 inhibitor (Chk1i, 2 μM) with or without PD184352 (5 μM) for 24 h, after which a comet assay was performed to assess DNA breaks. As control, U266 cells were treated with 100 μM hydrogen peroxide ($H_2O_2$) for 20 min. Co-administration of the MEK1/2 inhibitor markedly promoted DNA damage induced Chk1 inhibitors, an event occurring prior to induction of apoptosis.

Society of Hematology). The comet tail can be scored according to DNA intensity (e.g., nominal, medium, and high) using untreated cells as a control. Using a commercially available image analysis software, comet assay data can be quantified by calculating tail length and tail moment calculated based on both tail length and distribution of DNA in the tail (*see* **Note 29**).

3.3.1.2. Alkaline Electrophoresis

1. Process the sample through Steps 1–7 as described for TBE electrophoresis, directly transfer the slide to a horizontal electrophoresis apparatus, and carefully pour the alkaline solution until it covers the sample.
2. Set the voltage to approximately 1 V/cm (distance between two electrodes), add or remove buffer until the current is approximately 300 mA, and run the gel for 20–40 min (*see* **Note 30**).
3. Gently drain the slide to remove the alkaline solution, rinse several times in $dH_2O$, and then immerse in 70% ethanol for 5 min.
4. Process the sample through Steps 12–15 as described for TBE electrophoresis, and analyze data (*see* **Note 31**).

*3.3.2. Nuclear Foci*

After DNA damage, numerous proteins are phosphorylated and recruited to sites of DNA breaks, forming multi-protein macrocomplexes that are visible as nuclear foci under a microscope following immunofluorescent staining. Among others, foci of phosphorylated ATM (serine 1981 in particular) or H2A.X (serine 139, designated γH2A.X) are the most common markers for DNA damage. However, H2A.X phosphorylation (including at serine 139) can also be induced by DNA fragmentation associated with apoptosis, a possibility that should be taken account when primary DNA breaks (e.g., double-stranded breaks) rather than DNA damage secondary to apoptosis are investigated (**Fig. 8**, this research was originally published in *Blood* Ref. (59) © the American Society of Hematology). Although a regular fluorescent microscope can be used to view the samples, a confocal microscope is best suited choice to define clearly nuclear foci in cells. The procedure for immunofluorescent staining is described above in **Section 3.2.4**). This method can also be used to evaluate DNA damage in vivo (**Fig. 9**, this research was originally published in *Blood* Ref. (59) © the American Society of Hematology).

### 3.4. Analysis of Apoptosis

#### 3.4.1. DNA Ladder

One of the most characteristic events of the apoptotic process is activation of endonucleases, leading to the internucleosomal fragmentation of genomic DNA. Agarose gel electrophoresis can be used to reveal a typical pattern of oligonucleosomal DNA fragments corresponding to steps of approximately 180 bp, the so-called DNA ladder. This phenomenon represents the point of no return within the apoptotic signaling cascade.

1. Pellet $20 \times 10^6$ cells for each sample by centrifugation at $500\times g$ for 10 min at 4°C.

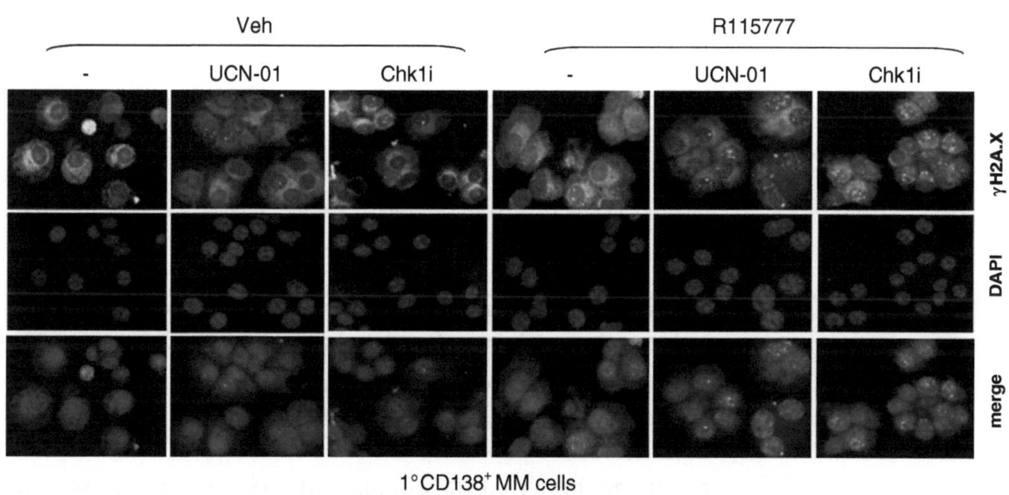


Fig. 8. Immunofluorescence staining of nuclear foci. Primary myeloma CD138$^+$ cells isolated from bone marrow of a multiple myeloma patient were exposed to UCN-01 (150 nM) or Chk1i (2 μM) with or without R115777 (5 μM) or PD184352 (5 μM) for 16 h, after which cells were stained with Alexa Fluor (AF) 488–conjugated phospho-H2A.X (Ser139, γH2A.X) antibody. Combined treatment resulted in a marked increase in formation of nuclear foci, compared to treatment with agents individually.

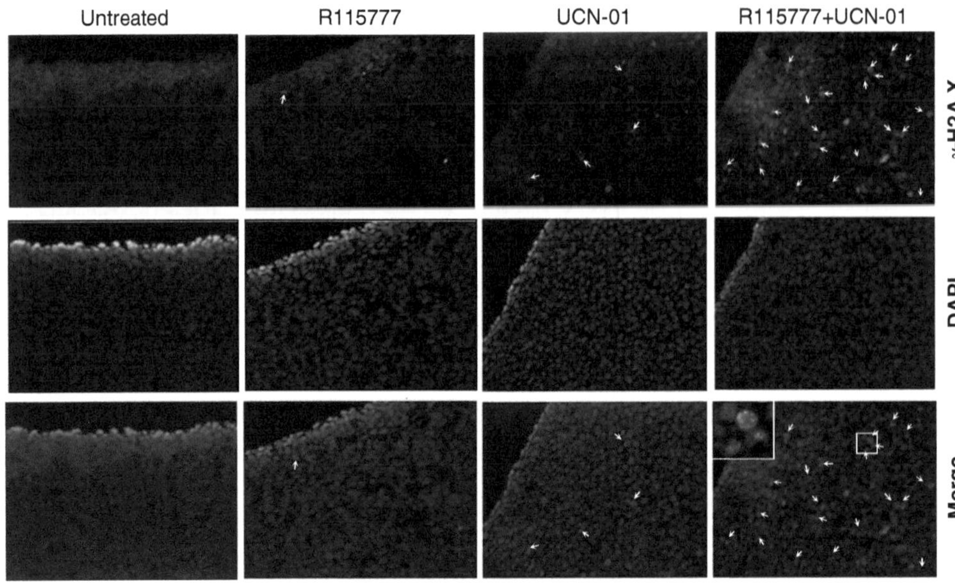


Fig. 9. Formation of nuclear foci in vivo in a murine xenograft model. Nude mice were inoculated subcutaneously with $10^7$ MM.1S cells into the right rear flank. After tumors were measurable, 25 mg/kg R115777 with or without 0.5 mg/kg UCN-01 were administrated intraperitoneally daily for 12 days. After the final dose, tumors were excised and tumor tissue sections were stained with Alexa Fluor 488–conjugated H2A.X antibodies to monitor γH2A.X expression and nuclear foci formation. *Arrows* indicate cells with γH2A.X nuclear foci. As observed in vitro, co-administration of R115777 markedly increased expression and nuclear foci formation of γH2A.X in vivo.

2. Aspirate the supernatant, and gently loosen the cell pellet by tapping the tube.
3. Add 450 μL of lysis buffer, vortex, transfer to a microfuge tube, and incubate in 56°C water bath for 18 h.
4. Centrifuge the sample at 50,000×*g* for 45 min at 4°C (*see* **Note 32**).
5. Carefully transfer the supernatant to a fresh tube, add a final concentration of 200 μg/mL RNase A, and incubate at 37°C for 3 h. During transfer of the supernatant, do not disturb the pellet. Alternatively, remove the viscous pellet first and then transfer the supernatant. After this step is performed, samples can be stored at 4°C overnight.
6. Prepare 1.8% low melting point agarose gel in 1× TBE by adding a final concentration of 0.5 μg/mL ethidium bromide.
7. Mix 30–50 μL of the sample with DNA loading buffer, dry load into each well, and add 1× TBE to overlay the gel. Load a 100-bp DNA ladder as a molecular-weight marker.
8. Run the gel at 100 V for 1 h, and view under UV light (**Fig. 10**, Ref. (37)).

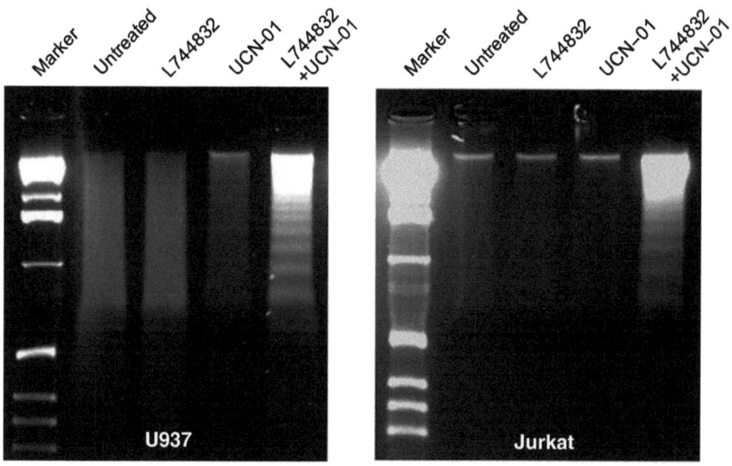


Fig. 10. Analysis of DNA ladder to determine apoptosis. Human leukemia cells were exposed to UCN-01 (U937, 100 nM; Jurkat, 150 nM) with or without the farnesyltransferase inhibitor L744832 (10 μM), after which cells were lysed, and DNA was extracted, separated by agarose gel electrophoresis, and stained with ethidium bromide. Combined treatment induced DNA fragmentation/ladders, classic hallmarks of apoptosis.

#### 3.4.2. Microscopy

##### 3.4.2.1. TUNEL

Terminal deoxynucleotidyl transferase dUTP nick end labeling (TUNEL) is a common method for detecting DNA fragmentation that results from apoptotic signaling cascades by labeling the terminal end of nucleic acids. The assay relies on the presence of nicks in the DNA which can be identified by terminal deoxynucleotidyl transferase, an enzyme that will catalyze the addition of dUTP that are secondarily labeled with a marker. It may also label cells that have suffered severe DNA damage.

1. Fix cytospin slides (for suspension cells) on coverslip/slides (for adherent cells) in 4% formaldehyde/PBS for 10 min at room temperature in the dark. The fixative solution can be stored for approximately 1 week at 4°C. Be sure to bring to room temperature before use.
2. Rinse slides twice in PBS for 5 min each at room temperature. Shake off slides briefly and rinse out the container each time to avoid crossover.
3. Immerse in 1:2 acetic acid/ethanol for 5 min at –20°C.
4. Rinse slides twice in PBS as above to remove all acid.
5. Encircle around the cells with a Pap pen, lay slides in a humidified chamber, and add 1 mg/ml BSA in PBS for 30–60 min at room temperature in the dark.
6. Rinse slides twice in PBS as above.
7. Mark the area of cells on the back of slides. It will be hard to find the area of cells later.

8. Make 100 μL per slide of staining mixture in an amber tube.
9. Add the staining mixture to the area of cells, and incubate for 1 h at 37°C in the humidified chamber.
10. Rinse twice in PBS as above.
11. Mount slides with mounting medium containing propidium iodide or DAPI, and seal with nail polish.
12. When dry, view slides (*see* **Note 33**), or store in the dark (**Fig. 11**, this research was originally published in *Blood* Ref. (58) © the American Society of Hematology).

This methods can also be used to monitor apoptosis in vivo (**Fig. 12**, this research was originally published in *Blood* Ref. (59) © the American Society of Hematology).

#### 3.4.2.2. Immunocytochemistry of Caspase-3 Activation

Caspase-3 (CPP-32) is a critical executioner of apoptosis, as it is either partially or totally responsible for the proteolytic cleavage of many key proteins such as the nuclear enzyme poly (ADP ribose) polymerase (PARP). Activation of caspase-3 requires proteolytic processing of its inactive zymogen at Asp175. Thus, apoptosis can be examined by immunofluorescent staining using a fluorescein-conjugated antibody that specifically recognizes the cleaved form (Asp175) of caspase 3 (**Fig. 13**). *See* **Section 3.2.4** for the procedure of immunofluorescent staining. In addition, flow cytometry can be used to quantify apoptosis by using the same antibody (for the procedure, *see* **Section 3.1.4.2**).

#### 3.4.2.3. Hoechst 33342 Staining

Hoechst 33342 nucleic acid stain is a popular cell-permeant nuclear counterstain that emits blue fluorescence when bound to dsDNA. This dye is often used to distinguish condensed pyknotic nuclei in apoptotic cells and for cell cycle studies in combination with BrdU.

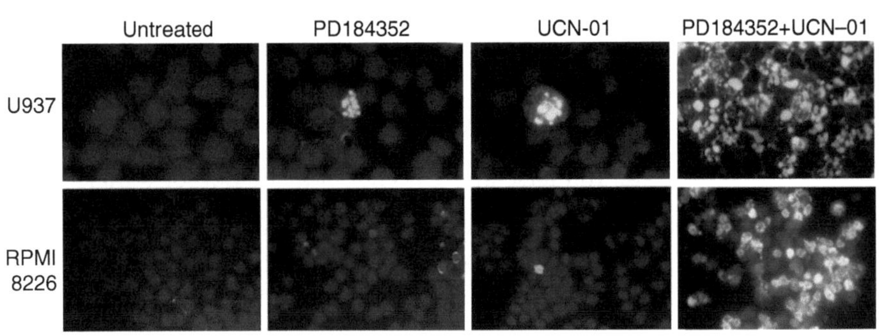


Fig. 11. TUNEL staining for apoptosis. Human leukemia U937 and myeloma RPMI8226 cells were exposed to UCN-01 (150 nM) ± PD184352 (10 μM) for 18 and 24 h, respectively, after which TUNEL staining with propidium iodide counterstaining was performed and viewed under fluorescence microscopy. Co-treatment with PD184352 and UCN-01 strikingly induced apoptosis in both leukemia and multiple myeloma cells.

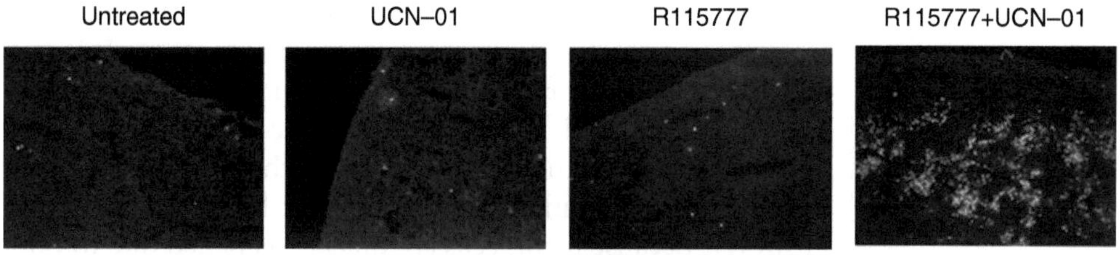


Fig. 12. TUNEL staining to detect apoptosis in vivo in a murine xenograft model. Nude mice were inoculated subcutaneously with $10^7$ MM.1S cells into the right rear flank. After tumors were measurable, 25 mg/kg R115777 with or without 0.5 mg/kg UCN-01 were administrated intraperitoneally daily for 12 days. After the final dose, tumors were excised. Tumor tissue sections were stained by TUNEL and viewed under fluorescence microscopy. Co-administration of R115777 dramatically potentiated apoptosis in vivo.

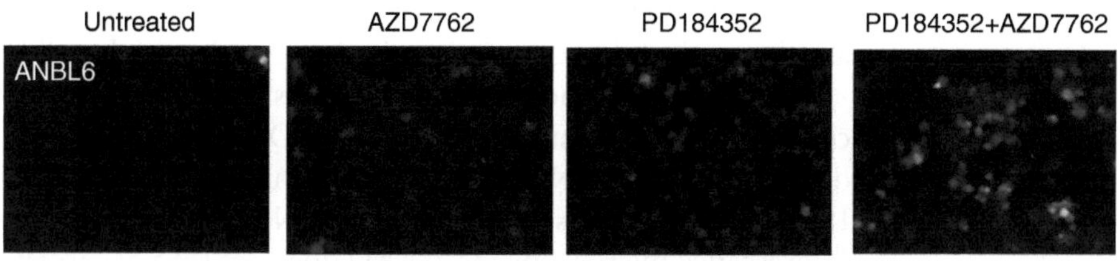


Fig. 13. Immunofluorescence staining to monitor caspase activation. Human myeloma IL-6-dependent ANBL6 cells were exposed to the new generation of Chk1 inhibitor AZD7762 (50 nM) with or without PD184352 (5 μM) for 24 h, after which caspase-3 activation was determined by immunofluorescence staining using an antibody specifically recognizing cleaved/active form of caspase-3 with DAPI counterstaining. Co-exposure to AZD7762 and PD184352 resulted in the marked activation of casapse-3.

1. After air drying, fix cytospin slides (for suspension cells) or coverslips/slides in 3:1 methanol/acetic acid for 30 min.
2. Wash slides three times with PBS for 10 min each.
3. Stain slides with 1 μg/mL Hoechst 33342 for 30 min at room temperature.
4. Wash slides three times with PBS to remove excessive dye, air dry, and mount with mounting medium for fluorescent staining.
5. View (*see* **Note 34**) or store slides at 4°C in the dark.

*3.4.3. Flow Cytometry*

3.4.3.1. $DiOC_6$/7-AAD

Energy released during oxidation reactions in the mitochondrial respiratory chain is stored as an electrochemical gradient consisting of a transmembrane electrical potential (Δψ), which is then able to drive the synthesis of ATP to fuel the cell's energy-dependent processes. Many membrane permeable lipophilic cations can be accumulated by living cells exhibiting a negative interior membrane potential. $DiOC_6$ (3,3′-diehexiloxadicarbocyanine iodide) is a carbocyanine derivative with short alkyl tails (< 7 carbon atoms), which accumulates on hyperpolarized membranes by translocation into the lipid bilayer and aggregation that decreases its green fluorescence. It is widely used for cell membrane potential measurements, including by flow cytometry and fluorescence microscopy. In both live and fixed cells, it stains more selectively mitochondria at low concentrations, while staining the ER (Golgi) at higher concentrations. However, because living cells are quickly damaged due to photodynamic toxicity, cells stained with this dye can only be exposed to light for short periods of time. A critical early event in apoptosis is mitochondrial damage, which causes a reduction in mitochondrial transmembrane potential (Δψm). Therefore, flow cytometry to detect reduction or loss of DiOC6 uptake has been used to monitor apoptosis upon the collapse of Δψm. According to the same principle, other fluorescent dyes such as 5,5′,6,6′tetrachloro-1,1′,3,3′-tetraethylbenzimidazol-carbocyanine iodide (JC-1) can also be used to detect variations in Δψm at the single cell level.

7-AAD (7-aminoactinomycin D) is a fluorescent chemical compound with a strong affinity for DNA via intercalation in double-stranded DNA with a high affinity for GC-rich regions. It is used as a fluorescent marker for DNA in flow cytometry and fluorescence microscopy. As 7-AAD does not readily pass through intact cell membranes, it is used as a cell viability stain. Cells with compromised membranes will stain with 7-AAD, while live cells with intact cell membranes will remain dark.

1. For $DiOC_6$ staining (**Fig. 14**, Ref. (35)), transfer 0.5 mL cells into to a Falcon tube, add 20 μL of $10^{-6}$ M $DiOC_6$, and incubate for 15 min at 37°C in the dark.

2. For 7-AAD staining (**Fig. 15**, Ref. (58)), add 5 μl of 0.5 μg/ml 7-AAD into 0.5 ml cells in a Falcon tube, and incubate for 15 min at 37°C in the dark.

3. For $DiOC_6$/7-AAD double staining (**Fig. 16**, Ref. (60)), add 20 μL of $10^{-6}$ M $DiOC_6$ and 5 μL of 0.5 μg/mL 7-

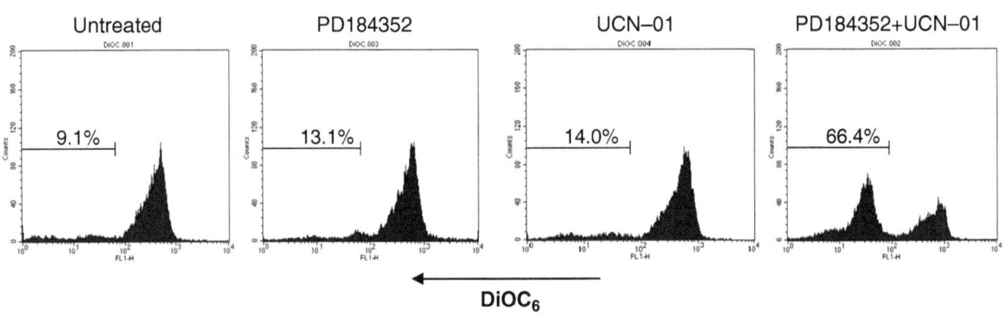


Fig. 14. $DiOC_6$ staining and flow cytometry. U937 cells were exposed to UCN-01 (150 nM) ± PD184352 (10 μM) for 18 h, after which the percentage of cells exhibiting loss of mitochondrial membrane potential (Δψm) was determined by monitoring $DiOC_6$ uptake. Co-treatment with PD184352 and UCN-01 led to a dramatic increase in mitochondrial injury, reflected by loss of Δψm.

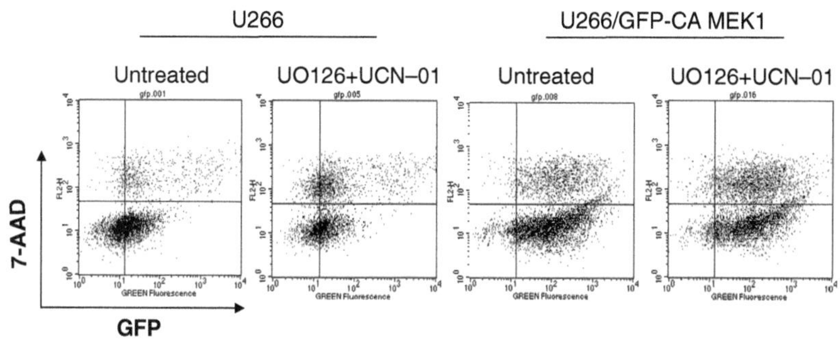


Fig. 15. 7-AAD staining and flow cytometry. U266 cells transiently transfected with GFP-tagged constitutively active form (CA, mutation of serine to aspartic acid at sites 218 and 222) of MEK1 were exposed to UCN-01 (200 nM) ± U0126 (20 μM) for 24 h, after which 7-AAD staining was performed to monitor cell death by flow cytometry. Ectopic expression of CA-MEK1 partially prevented cell killing by MEK/Chk1 inhibitor regimen.

AAD into 0.5 mL cells and incubate for 15 min at 37°C in the dark.

For suspension cells, it is not necessary to remove the culture medium. For adherent cells, trypsinize and pellet the cells, and resuspend in 0.5 ml PBS before staining.

3.4.3.2. Annexin V

In living cells, phospholipids are asymmetrically distributed between the inner and outer leaflets of the plasma membrane with phosphatidylcholine and sphingomyelin exposed on the external leaflet of the lipid bilayer, and phosphatidylserine predominantly observed on the inner surface facing the cytosol. A critical stage of apoptosis involves the acquisition of surface changes by dying cells that eventually culminates in the recognition and uptake of these cells by phagocytes. Cells undergoing apoptosis display disruption of the phospholipid asymmetry of their plasma membrane and expose phospholipid-like phosphatidylserine (PS) on the outer layer of the membrane. This event occurs in the early phases of apoptosis, while the cell membrane remains intact. Annexin V, a member of annexin family, has proven to be a useful tool in detecting apoptotic cells as it preferentially binds to negatively charged phospholipids like PS in the presence of $Ca^{2+}$, while displaying minimal binding to phosphatidylcholine and sphingomyeline. Changes in PS asymmetry, which is analyzed by measuring Annexin V binding to the cell membrane, are detected before morphological changes associated with apoptosis have occurred and before membrane integrity has been lost. Flow cytometry is utilized to identify and quantify apoptotic cells on a single-cell basis by using fluorescein (e.g., FITC)-conjugated Annexin V. Staining cells simultaneously with FITC-Annexin V (green fluorescence) and the non-vital dye propidium iodide (red fluorescence) allows, via bivariate analysis, discrimination between living cells ($FITC^+/PI^+$), early apoptotic ($FITC^+/PI^-$), and late apoptotic or necrotic cells ($FITC^-/PI^+$).

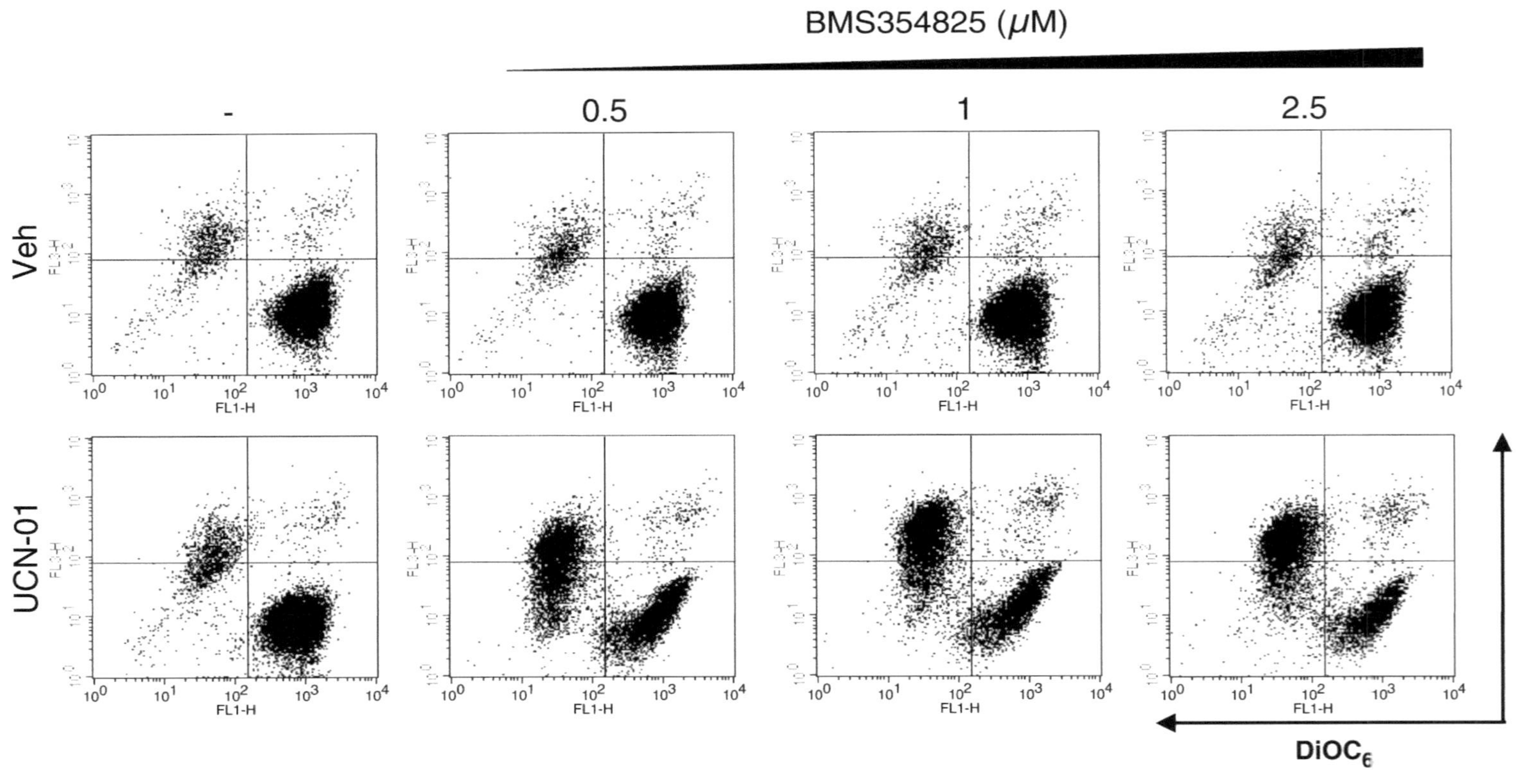


Fig. 16. $DiOC_6$/7-AAD double staining and flow cytometry. Human myeloma H929 cells were exposed to UCN-01 (100 nM) in the presence or absence of the Src inhibitor BMS354825 at the indicated concentrations for 24 h, after which cells were stained with both $DiOC_6$ and 7-AAD to simultaneously monitor loss of $\Delta\psi m$ and cell death. Co-administration of BMS354825 dramatically potentiated UCN-01 lethality.

1. Pellet 5–10 × $10^5$ cells at 500 × *g* for 5 min, wash twice with cold PBS, and then resuspend in 1× binding buffer at a concentration of 1 × $10^6$/mL.
2. Transfer 100 μL of the cells to a Falcon tube, and add 5 μL Annexin V-FITC and 10 μL propidium iodide.
3. Gently mix the cells and incubate for 15 min at room temperature in the dark.
4. Add 400 μL of 1× binding buffer, and analysis by flow cytometry as soon as possible within 1 h (**Figs. 17** and **18**, Refs. (37, 58)).

#### 3.4.4. Western Blot

Proteolytic processing of many proteins including caspases and poly (ADP ribose) polymerase (PARP) is a primary mechanism responsible for transduction of death signals. Consequently, cleavage of these proteins represents an important feature of apoptosis. In general, apoptotic stimuli result in mitochondrial damage, releasing cytochrome c from mitochondria to cytosol where it associates with procaspase-9 (ICE-LAP6, Mch6) and Apaf 1 to form a complex designated the apoptosome. This complex processes the 47 kDa procaspase-9 into a large active fragment (35 kDa or 17 kDa) and a small fragment (10 kDa) by self-cleavage at Asp315. Active caspase-9 then cleaves full length caspase-3 (35 kDa) at Asp175 into activated p17 and p12 fragments. Activated casapse-3, a critical executioner, mediates proteolytic degradation of many key proteins. PARP, a 116-kDa nuclear protein, is one of the main cleavage targets of caspase-3. Cleavage of human PARP between Asp214 and Gly215 separates the N-terminal DNA binding domain (24 kDa) from the C-terminal catalytic domain (89 kDa), and thereby disables its function. Western blot analysis is frequently employed to evaluate the apoptotic signaling cascade by using antibodies that specifically recognize cleaved fragments of caspases and PARP, such as cleaved caspase-3 (Asp175), cleaved caspase-9 (Asp315), and cleaved PARP (Asp214), etc. (**Fig. 19**, Ref. (35)). Moreover, Western blot analysis can also be used to monitor release of mitochondrial apoptotic proteins (e.g., cytochrome C, Smac/Diable, and AIF) to cytosol (**Fig. 19**, Ref. (35)). For the procedure of Western blot analysis, *see* **Section 3.2.1**.

##### 3.4.4.1. Cytosolic Fraction

1. Pellet 4 × $10^6$ cells, and wash in PBS.
2. Resuspend in 100 μL digitonin lysis buffer.
3. Centrifuge at high speed for 1 min, transfer supernatant to a new microfuge tube, and add 2× sample buffer.
4. Load 10–30 μg protein for Western blot using antibodies against cytochrome c (Santa Cruz, **Fig. 19**, Ref. (35)), Smac/DIABLO (Millipore), or AIF (Santa Cruz).

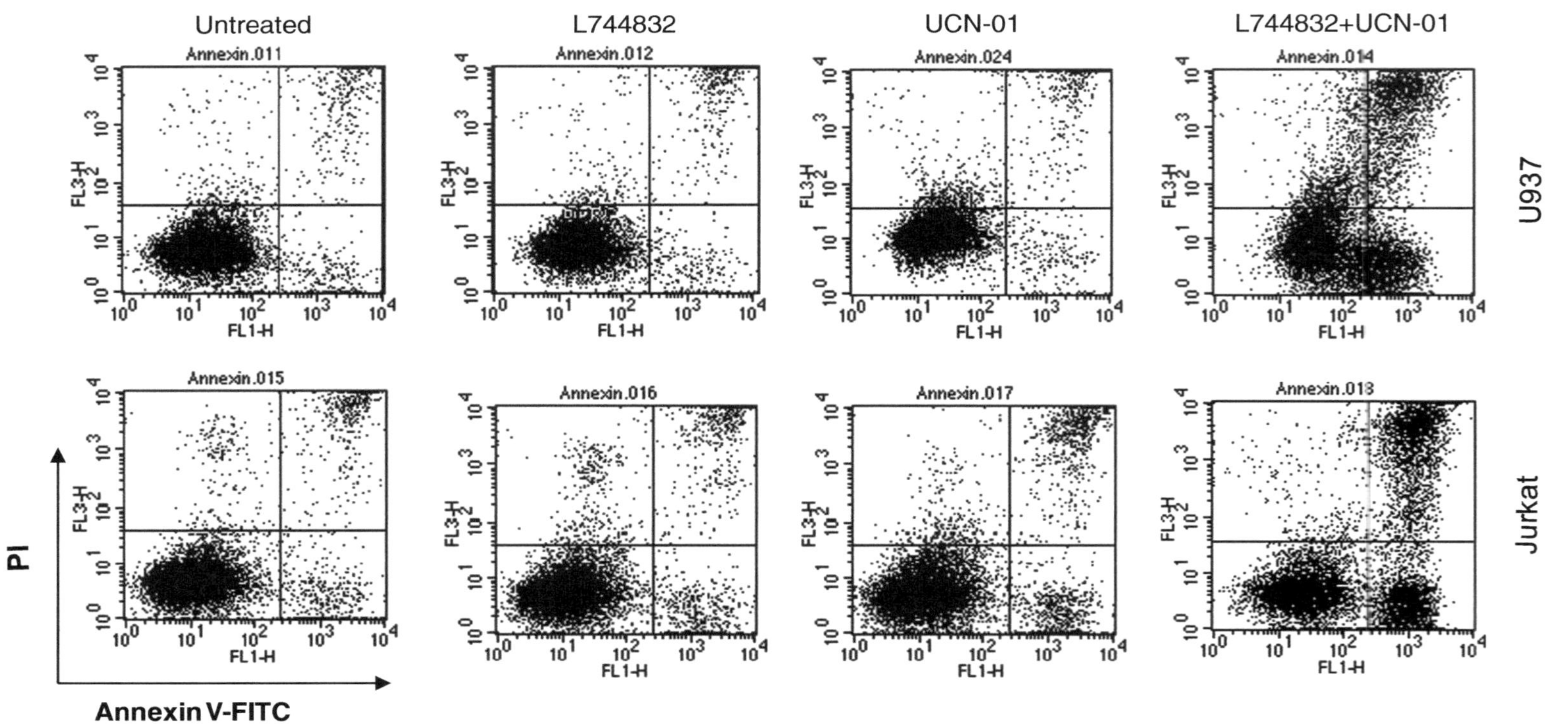


Fig. 17. Annexin V-FITC staining to distinguish early and late apoptotic cells by flow cytometry . Human leukemia cells were exposed for 18 h to 10 μM L744832 (farnesyltransferase inhibitor) with or without UCN-01 (U937, 100 nM; Jurkat, 150 nM), respectively, after which cells were stained with Annexin V-FITC and propidium iodide (PI) to determine the percentage of Annexin $V^+/PI^-$ (early apoptotic) and Annexin $V^+/PI^+$ (late apoptotic) cells by flow cytometry. Combined treatment resulted in a striking increase in both early and late apoptotic cells.

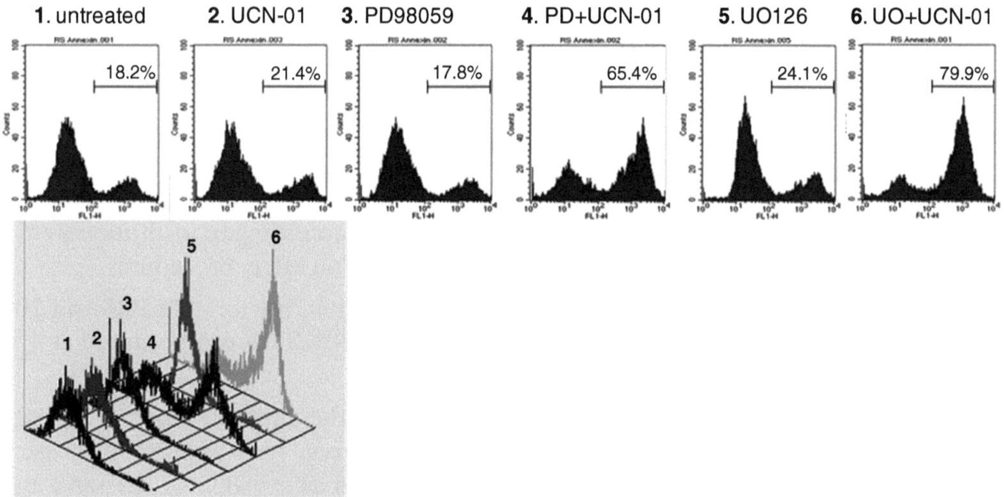


Fig. 18. Determination of total Annexin V-positive cells. Human myeloma MM.1S cells were exposed to 150 nM UCN-01 with or without the MEK1/2 inhibitor PD98059 (50 μM) or U0126 (20 μM) for 24 h, after which the percentage of cells exhibiting Annexin V positivity was determined by flow cytometry. Co-administration of either MEK1/2 inhibitor significantly increased apoptosis induced by UCN-01.

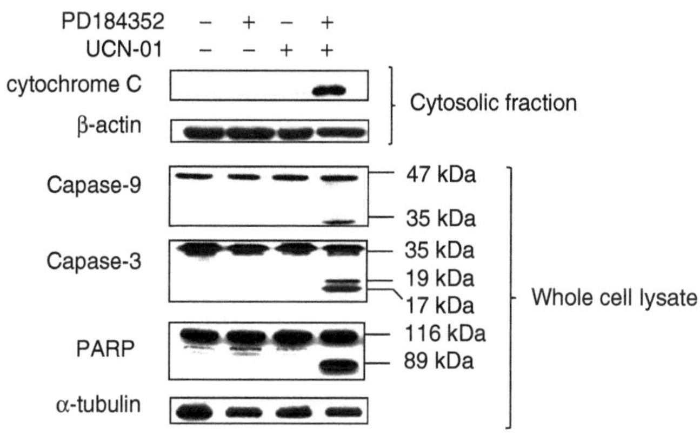


Fig. 19. Western blot to analyze activation of apoptotic signaling cascade. U937 cells were exposed for 18 h 150 nM UCN-01 ± 10 μM PD184352, after which cells were lysed and subjected to Western blot to monitor activation of caspases. Alternatively, cytosolic fractions were prepared and expression of cytochrome *c* in cytosolic fraction was examined.

## 4. Notes

1. In this protocol, the cells may form solid clumps, making it difficult to resuspend the cells as single cell suspension as required for flow cytometric analysis. Therefore, it is critical to loose cell pellet before adding ethanol, as well as add

ethanol dropwise while mildly vortexing to avoid formation of cell clumps.

2. Gently pipette up and down, and then mildly vortex to resuspend the cells completely. This will prevent cells from clumping due to propidium iodide.
3. [*methyl*-$^{3}$H]thymidine and [$\gamma^{32}$P] ATP are radioactive, and thus should be handled according to institution's environmental health and radiation safety procedures.
4. For optimal DNA precipitation, use ice-cold PBS and 20% TCA. DNA precipitates in 20% TCA can be stored at 4°C for several days.
5. The final concentration of the antibody may be adjusted according to different sources. It is important to use a fluorescein-conjugated non-reactive IgG of the same isotype as a negative control.
6. Incubation in hypotonic solution causes cell swelling without rupture, which makes it easier to distinguish mitotic figures.
7. Adjust concentrations of cells to avoid overlap of cells on the slides.
8. Monitor color changes (light to dark brown) every 30 min to ensure precise comparison between untreated and drug-treated conditions before the culture medium turns to deep dark brown or black. Alternatively, add 25 μL of 10% SDS to each well to stop the reaction, and the plate the can be stored in the dark in a humidified chamber at room temperature for up to 18 h. Importantly, because the culture medium itself or with the MTS reagent might generate a slight amount of spontaneous 490 nm absorbance, use equal volume (e.g., 100 μL) of the same culture medium as a blank control.
9. First, mix medium and FBS, then add agar, and lastly, add cells. For the latter two steps, mix gently with pipette without forming bubbles.
10. This step is to keep moisture for agar, which is critical to maintain the humidity required for cell growth (colony formation).
11. The time for cells to form colonies will vary depending upon cell type. Therefore, observe the plates at least twice a week after first week.
12. Under certain circumstance, such as in cells undergoing cell death (e.g., apoptosis) or which are in mitosis, adherent cells may lose attachment and become suspended in culture medium. To harvest all cells, collect culture medium and

washing PBS, centrifuge at 1000 × *g*, discard the supernatant, and pool the pellet together with the cell lysate from the following step.

13. At this step, the pellet should not be easily visible at bottom of the original tube. Otherwise it indicates incomplete cell lysis. If this is the case, optimize time and settings for sonication.
14. Although PBS can also be used, TBS should be used throughout for blotting phosphorylated proteins.
15. Incubation for 1 h at room temperature can produce faster results, but may result in weaker signals and stronger background.
16. 4% paraformaldehyde is the most common fixative for immunofluorescent staining. Although paraformaldehyde penetrates rapidly, it fixes cells/tissues slowly. Thus, use the longest fixation time that still results in good antibody labeling. Commercial formalin contains methanol (6–15%) and other impurities that may affect fixation. If necessary, use methanol-free formalin. Methanol and acetone precipitate proteins, causing cell/tissue shrinkage and distortion, and thus are not good fixatives for preserving fine structures, particularly for confocal microscopy. Glutaraldehyde produces autofluorescence. If necessary, subsequently treat with sodium borohydride to eliminate autofluorescence.
17. Triton X-100 may affect membrane extraction or destroy fixed cells. If the cells are fragile, use 0.2% saponin to replace Triton X-100, and also handle the samples gently after fixation.
18. It is critical to select the fluorophore(s) for which excitation and emission spectra coincide with the filters on the available confocal microscope.
19. RIPA buffer can also be used. However, RIPA buffer may not be optimal for some kinases, which may require adjustment of its composition to maintain kinase activity.
20. Use of Dynabeads (brown color) requires a magnetic rack. If available, select one with magnets on the side rather than the bottom, which make it easier to remove washing and antibody solutions. Optional: agarose beads conjugated with protein A, G, A/G, or L can also be used to replace Dynabeads. Use of the agarose beads requires centrifugation for each washing step. Therefore, carefully aspirate the supernatant to avoid loss of the beads, particularly when the beads are colorless.
21. After Step 8, $^{32}$P-labeled substrates can be separated from free [$\gamma^{32}$P]-ATP by acid precipitation, collected on a phos-

phocellulose paper for scintillation counting to determine radioactivity, quantifying activity of the kinase. If protein substrates are used, follow the procedure for Western blot to detect the phosphorylated protein substrates using appropriate specific antibodies.

22. For this step, sonication can also be used to lyse cells, while energy and time of sonication should be optimized to avoid disruption of protein–protein associations when used for co-immunoprecipitation.
23. For proteins with low abundance, the amount of input protein may be increased to 500–800 μg per condition after optimization. However, the protein concentration should be kept at 1 μg/μL, while increasing the amount of primary antibody and Dynabeads with increases in volume of samples. Importantly, the same volume of blank RIPA buffer should be used as a control.
24. Because this assay uses unfixed cells, pre-chilling buffer is important to inhibit endogenous DNA damage and repair. For the same reason, other notes throughout the procedure are also important as discussed below (*see* **Notes 25** and **26**).
25. Cooling the agarose to 37°C is critical to avoid heat shock of cells. Microwaving and heat blocks are not recommended to melt agarose.
26. Importantly, cell samples should be prepared immediately before initiating the assay and handled under dimmed or yellow light to avoid DNA damage caused by UV light. Always use a sample of untreated cells as control for endogenous DNA damage. If necessary, use cells treated (20 min at 4°C) with 100 μM hydrogen peroxide or 25 μM $KMnO_4$ as a positive control.
27. In high humidity environments, increasing gelling time to 30 min will improve adherence of the gel on the slide.
28. The alkali treatment unwinds the supercoiled DNA, allowing damaged DNA or DNA fragments to migrate from the cells under an electric field. However, the comet assay can be performed under neutral conditions (i.e., without treatment with alkaline solution), which will detect primarily double-stranded breaks and DNA fragments induced by apoptosis.
29. For TBE electrophoresis, the tail length may be correlated with the degree of DNA damage. Although tail length increases steadily for low degrees of damage, by itself it is not the best indicator because it reaches a plateau beyond a certain point in the case of higher degree of damage. Under

the latter conditions, the amount of DNA migrating into the tail continues to increase, while values for tail length eventually reach a limit.

30. Because the alkaline solution is a not a buffered system, temperature control is highly recommended. It is also recommended to use a large electrophoresis apparatus (25–30 cm between electrodes along with recirculation system of electrophoresis buffer). Alternatively, run the gel at lower temperature (e.g., 16°C or 4°C) to avoid background DNA damage, increase sample adherence at high pH, and improve reproducibility.
31. For alkaline electrophoresis, the distribution of DNA between the tail and the head of the comet, including length, width, and DNA content/intensity, should be used to evaluate the degree of DNA damage.
32. This step is intended to separate low-molecular-weight DNA fragments from intact, high-molecular-weight DNA species.
33. For immunofluorescent staining, it is important to view as soon as possible, as fluorescence will fade over time.
34. Because Hoechst 33342 only stains DNA, apoptotic cells should be determined by morphologic features of nuclei (e.g., condensed chromatin, and nuclear fragments).

## Acknowledgments

This work was supported by awards CA63753, CA 93738, and CA100866, and 1 P50 CA130805-01, 1 P50CA142509-01 and its DRP award (29859/98018093) from the National Cancer Institute, Senior Research Award from Multiple Myeloma Research Foundation, award 6045-03 from the Leukemia and Lymphoma Society of America, and an award from the V Foundation.

### References

1. Zhou, B.B., Anderson, H.J. and Roberge, M. (2003) Targeting DNA checkpoint kinases in cancer therapy. *Cancer Biol. Ther.*, **2**, S16–S22.
2. Bartek, J. and Lukas, J. (2007) DNA damage checkpoints: from initiation to recovery or adaptation. *Curr. Opin. Cell Biol.*, **19**, 238–245.
3. Gottifredi, V. and Prives, C. (2005) The S phase checkpoint: when the crowd meets at the fork. *Semin. Cell Dev. Biol.*, **16**, 355–368.
4. Jazayeri, A., Falck, J., Lukas, C., Bartek, J., Smith, G.C., Lukas, J. and Jackson, S.P. (2006) ATM- and cell cycle-dependent regulation of ATR in response to DNA double-strand breaks. *Nat. Cell Biol.*, **8**, 37–45.

5. Tapia-Alveal, C., Calonge, T.M. and O'Connell, M.J. (2009) Regulation of chk1. *Cell Div.*, **4**, 8.
6. Niida, H., Katsuno, Y., Banerjee, B., Hande, M.P. and Nakanishi, M. (2007) Specific role of Chk1 phosphorylations in cell survival and checkpoint activation. *Mol. Cell Biol.*, **27**, 2572–2581.
7. Walker, M., Black, E.J., Oehler, V., Gillespie, D.A. and Scott, M.T. (2009) Chk1 C-terminal regulatory phosphorylation mediates checkpoint activation by de-repression of Chk1 catalytic activity. *Oncogene*, **28**, 2314–2323.
8. Wilsker, D., Petermann, E., Helleday, T. and Bunz, F. (2008) Essential function of Chk1 can be uncoupled from DNA damage checkpoint and replication control. *Proc. Natl. Acad. Sci. USA*, **105**, 20752–20757.
9. Choi, J.H., Lindsey-Boltz, L.A. and Sancar, A. (2007) Reconstitution of a human ATR-mediated checkpoint response to damaged DNA. *Proc. Natl. Acad. Sci. USA*, **104**, 13301–13306.
10. Chou, D.M. and Elledge, S.J. (2006) Tipin and Timeless form a mutually protective complex required for genotoxic stress resistance and checkpoint function. *Proc. Natl. Acad. Sci. USA*, **103**, 18143–18147.
11. Yoshizawa-Sugata, N. and Masai, H. (2007) Human Tim/Timeless-interacting protein, Tipin, is required for efficient progression of S phase and DNA replication checkpoint. *J. Biol. Chem.*, **282**, 2729–2740.
12. Tse, A.N., Carvajal, R. and Schwartz, G.K. (2007) Targeting checkpoint kinase 1 in cancer therapeutics. *Clin. Cancer Res.*, **13**, 1955–1960.
13. Ashwell, S. and Zabludoff, S. (2008) DNA damage detection and repair pathways--recent advances with inhibitors of checkpoint kinases in cancer therapy. *Clin. Cancer Res.*, **14**, 4032–4037.
14. Kastan, M.B. and Bartek, J. (2004) Cell-cycle checkpoints and cancer. *Nature*, **432**, 316–323.
15. Golding, S.E., Rosenberg, E., Valerie, N., Hussaini, I., Frigerio, M., Cockcroft, X.F., Chong, W.Y., Hummersone, M., Rigoreau, L., Menear, K.A. et al. (2009) Improved ATM kinase inhibitor KU-60019 radiosensitizes glioma cells, compromises insulin, AKT and ERK prosurvival signaling, and inhibits migration and invasion. *Mol. Cancer Ther.*, **10**, 2894–2902.
16. Bartek, J. and Lukas, J. (2003) Chk1 and Chk2 kinases in checkpoint control and cancer. *Cancer Cell*, **3**, 421–429.
17. Kortmansky, J., Shah, M.A., Kaubisch, A., Weyerbacher, A., Yi, S., Tong, W., Sowers, R., Gonen, M., O'Reilly, E., Kemeny, N. et al. (2005) Phase I trial of the cyclin-dependent kinase inhibitor and protein kinase C inhibitor 7-hydroxystaurosporine in combination with Fluorouracil in patients with advanced solid tumors. *J. Clin. Oncol.*, **23**, 1875–1884.
18. Bucher, N. and Britten, C.D. (2008) G2 checkpoint abrogation and checkpoint kinase-1 targeting in the treatment of cancer. *Br. J. Cancer*, **98**, 523–528.
19. Zabludoff, S.D., Deng, C., Grondine, M.R., Sheehy, A.M., Ashwell, S., Caleb, B.L., Green, S., Haye, H.R., Horn, C.L., Janetka, J.W. et al. (2008) AZD7762, a novel checkpoint kinase inhibitor, drives checkpoint abrogation and potentiates DNA-targeted therapies. *Mol. Cancer Ther.*, **7**, 2955–2966.
20. Sha, S.K., Sato, T., Kobayashi, H., Ishigaki, M., Yamamoto, S., Sato, H., Takada, A., Nakajyo, S., Mochizuki, Y., Friedman, J.M. et al. (2007) Cell cycle phenotype-based optimization of G2-abrogating peptides yields CBP501 with a unique mechanism of action at the G2 checkpoint. *Mol. Cancer Ther.*, **6**, 147–153.
21. Blasina, A., Hallin, J., Chen, E., Arango, M.E., Kraynov, E., Register, J., Grant, S., Ninkovic, S., Chen, P., Nichols, T. et al. (2008) Breaching the DNA damage checkpoint via PF-00477736, a novel small-molecule inhibitor of checkpoint kinase 1. *Mol. Cancer Ther.*, **7**, 2394–2404.
22. Zhang, C., Yan, Z., Painter, C.L., Zhang, Q., Chen, E., Arango, M.E., Kuszpit, K., Zasadny, K., Hallin, M., Hallin, J. et al. (2009) PF-00477736 mediates checkpoint kinase 1 signaling pathway and potentiates docetaxel-induced efficacy in xenografts. *Clin. Cancer Res.*, **15**, 4630–4640.
23. Matthews, D.J., Yakes, F.M., Chen, J., Tadano, M., Bornheim, L., Clary, D.O., Tai, A., Wagner, J.M., Miller, N., Kim, Y.D. et al. (2007) Pharmacological abrogation of S-phase checkpoint enhances the anti-tumor activity of gemcitabine in vivo. *Cell Cycle*, **6**, 104–110.
24. Sorensen, C.S., Syljuasen, R.G., Falck, J., Schroeder, T., Ronnstrand, L., Khanna, K.K., Zhou, B.B., Bartek, J. and Lukas, J. (2003) Chk1 regulates the S phase checkpoint by coupling the physiological turnover and ionizing radiation-induced accelerated proteolysis of Cdc25A. *Cancer Cell*, **3**, 247–258.
25. Syljuasen, R.G., Sorensen, C.S., Nylandsted, J., Lukas, C., Lukas, J. and Bartek, J. (2004) Inhibition of Chk1 by CEP-3891 accelerates

mitotic nuclear fragmentation in response to ionizing Radiation. *Cancer Res.*, **64**, 9035–9040.
26. Tse, A.N., Rendahl, K.G., Sheikh, T., Cheema, H., Aardalen, K., Embry, M., Ma, S., Moler, E.J., Ni, Z.J., Lopes de Menezes, D.E. et al. (2007) CHIR-124, a novel potent inhibitor of Chk1, potentiates the cytotoxicity of topoisomerase I poisons in vitro and in vivo. *Clin. Cancer Res.*, **13**, 591–602.
27. Parsels, L.A., Morgan, M.A., Tanska, D.M., Parsels, J.D., Palmer, B.D., Booth, R.J., Denny, W.A., Canman, C.E., Kraker, A.J., Lawrence, T.S. et al. (2009) Gemcitabine sensitization by checkpoint kinase 1 inhibition correlates with inhibition of a Rad51 DNA damage response in pancreatic cancer cells. *Mol. Cancer Ther.*, **8**, 45–54.
28. Schellens, J.H., Leijen, S., Shapiro, G.I., Pavlick, A.C., Tibes, R., O'Day, S.J., Demuth, T., Viscusi, J., Xu, Y. and Oza, A.M. (2009) A phase I and pharmacological study of MK-1775, a Wee1 tyrosine kinase inhibitor, in both monotherapy and in combination with gemcitabine, cisplatin, or carboplatin in patients with advanced solid tumors. *J. Clin. Oncol.*, **27**, 3510.
29. Mizuarai, S., Yamanaka, K., Itadani, H., Arai, T., Nishibata, T., Hirai, H. and Kotani, H. (2009) Discovery of gene expression-based pharmacodynamic biomarker for a p53 context-specific anti-tumor drug Wee1 inhibitor. *Mol. Cancer*, **8**, 34.
30. Wang, Y., Li, J., Booher, R.N., Kraker, A., Lawrence, T., Leopold, W.R. and Sun, Y. (2001) Radiosensitization of p53 mutant cells by PD0166285, a novel G(2) checkpoint abrogator. *Cancer Res*, **61**, 8211–8217.
31. Hashimoto, O., Shinkawa, M., Torimura, T., Nakamura, T., Selvendiran, K., Sakamoto, M., Koga, H., Ueno, T. and Sata, M. (2006) Cell cycle regulation by the Wee1 inhibitor PD0166285, pyrido [2,3-d] pyimidine, in the B16 mouse melanoma cell line. *BMC Cancer*, **6**, 292.
32. Arlander, S.J., Eapen, A.K., Vroman, B.T., McDonald, R.J., Toft, D.O. and Karnitz, L.M. (2003) Hsp90 inhibition depletes Chk1 and sensitizes tumor cells to replication stress. *J. Biol. Chem.*, **278**, 52572–52577.
33. Tse, A.N., Sheikh, T.N., Alan, H., Chou, T.C. and Schwartz, G.K. (2009) 90-kDa heat shock protein inhibition abrogates the topoisomerase I poison-induced G2/M checkpoint in p53-null tumor cells by depleting Chk1 and Wee1. *Mol. Pharmacol.*, **75**, 124–133.
34. Tse, A.N., Klimstra, D.S., Gonen, M., Shah, M., Sheikh, T., Sikorski, R., Carvajal, R., Mui, J., Tipian, C., O'Reilly, E. et al. (2008) A phase 1 dose-escalation study of irinotecan in combination with 17-allylamino-17-demethoxygeldanamycin in patients with solid tumors. *Clin. Cancer Res.*, **14**, 6704–6711.
35 Dai, Y., Yu, C., Singh, V., Tang, L., Wang, Z., McInistry, R., Dent, P. and Grant, S. (2001) Pharmacological inhibitors of the mitogen-activated protein kinase (MAPK) kinase/MAPK cascade interact synergistically with UCN-01 to induce mitochondrial dysfunction and apoptosis in human leukemia cells. *Cancer Res.*, **61**, 5106–5115.
36 Dai, Y., Hamm, T.E., Dent, P. and Grant, S. (2006) Cyclin D1 overexpression increases the susceptibility of human U266 myeloma cells to CDK inhibitors through a process involving p130-, p107- and E2F-dependent S phase entry. *Cell Cycle*, **5**, 437–446.
37 Dai, Y., Rahmani, M., Pei, X.Y., Khanna, P., Han, S.I., Mitchell, C., Dent, P. and Grant, S. (2005) Farnesyltransferase inhibitors interact synergistically with the Chk1 inhibitor UCN-01 to induce apoptosis in human leukemia cells through interruption of both Akt and MEK/ERK pathways and activation of SEK1/JNK. *Blood*, **15**, 1706–1716.
38. Branzei, D. and Foiani, M. (2008) Regulation of DNA repair throughout the cell cycle. *Nat. Rev. Mol. Cell Biol.*, **9**, 297–308.
39. Lobrich, M. and Jeggo, P.A. (2007) The impact of a negligent G2/M checkpoint on genomic instability and cancer induction. *Nat. Rev. Cancer*, 7, 861–869.
40. Falck, J., Petrini, J.H., Williams, B.R., Lukas, J. and Bartek, J. (2002) The DNA damage-dependent intra-S phase checkpoint is regulated by parallel pathways. *Nat. Genet.*, **30**, 290–294.
41. Liu, P., Barkley, L.R., Day, T., Bi, X., Slater, D.M., Alexandrow, M.G., Nasheuer, H.P. and Vaziri, C. (2006) The Chk1-mediated S-phase checkpoint targets initiation factor Cdc45 via a Cdc25A/Cdk2-independent mechanism. *J. Biol. Chem.*, **281**, 30631–30644.
42. Yang, X.H. and Zou, L. (2009) Dual functions of DNA replication forks in checkpoint signaling and PCNA ubiquitination. *Cell Cycle*, **8**, 191–194.
43. Beckerman, R., Donner, A.J., Mattia, M., Peart, M.J., Manley, J.L., Espinosa, J.M. and Prives, C. (2009) A role for Chk1 in blocking transcriptional elongation of p21 RNA during the S-phase checkpoint. *Genes Dev.*, **23**, 1364–1377.

44. Kuntz, K. and O'Connell, M.J. (2009) The G(2) DNA damage checkpoint: Could this ancient regulator be the achilles heel of cancer? *Cancer Biol. Ther.*, **8**.
45. de Bruin, R.A. and Wittenberg, C. (2009) All eukaryotes: before turning off G1-S transcription, please check your DNA. *Cell Cycle*, **8**, 214–217.
46. Zachos, G., Black, E.J., Walker, M., Scott, M.T., Vagnarelli, P., Earnshaw, W.C. and Gillespie, D.A. (2007) Chkl is required for spindle checkpoint function. *Dev. Cell*, **12**, 247–260.
47. Peddibhotla, S., Lam, M.H., Gonzalez-Rimbau, M. and Rosen, J.M. (2009) The DNA-damage effector checkpoint kinase 1 is essential for chromosome segregation and cytokinesis. *Proc. Natl. Acad. Sci. USA*, **106**, 5159–5164.
48. Tang, J., Erikson, R.L. and Liu, X. (2006) Checkpoint kinase 1 (Chk1) is required for mitotic progression through negative regulation of polo-like kinase 1 (Plk1). *Proc. Natl. Acad. Sci. USA*, **103**, 11964–11969.
49. Goudelock, D.M., Jiang, K., Pereira, E., Russell, B. and Sanchez, Y. (2003) Regulatory interactions between the checkpoint kinase Chkl and the proteins of the DNA-dependent protein kinase complex. *J. Biol. Chem.*, **278**, 29940–29947.
50. Sorensen, C.S., Hansen, L.T., Dziegielewski, J., Syljuasen, R.G., Lundin, C., Bartek, J. and Helleday, T. (2005) The cell-cycle checkpoint kinase Chkl is required for mammalian homologous recombination repair. *Nat. Cell Biol.*, 7, 195–201.
51. Chen, C.C., Kennedy, R.D., Sidi, S., Look, A.T. and D'Andrea, A. (2009) CHK1 inhibition as a strategy for targeting Fanconi Anemia (FA) DNA repair pathway deficient tumors. *Mol. Cancer*, **8**, 24.
52. Wang, X., Kennedy, R.D., Ray, K., Stuckert, P., Ellenberger, T. and D'Andrea, A.D. (2007) Chkl-mediated phosphorylation of FANCE is required for the Fanconi anemia/BRCA pathway. *Mol. Cell Biol.*, **27**, 3098–3108.
53. Syljuasen, R.G., Sorensen, C.S., Hansen, L.T., Fugger, K., Lundin, C., Johansson, F., Helleday, T., Sehested, M., Lukas, J. and Bartek, J. (2005) Inhibition of human Chkl causes increased initiation of DNA replication, phosphorylation of ATR targets, and DNA breakage. *Mol. Cell Biol.*, **25**, 3553–3562.
54. Myers, K., Gagou, M.E., Zuazua-Villar, P., Rodriguez, R. and Meuth, M. (2009) ATR and Chkl suppress a caspase-3-dependent apoptotic response following DNA replication stress. *PLoS. Genet.*, **5**, e1000324.
55. Sidi, S., Sanda, T., Kennedy, R.D., Hagen, A.T., Jette, C.A., Hoffmans, R., Pascual, J., Imamura, S., Kishi, S., Amatruda, J.F. et al. (2008) Chkl suppresses a caspase-2 apoptotic response to DNA damage that bypasses p53, Bcl-2, and caspase-3. *Cell*, **133**, 864–877.
56. Matsuura, K., Wakasugi, M., Yamashita, K. and Matsunaga, T. (2008) Cleavage-mediated activation of Chkl during apoptosis. *J. Biol. Chem.*, **283**, 25485–25491.
57. Shimada, M., Niida, H., Zineldeen, D.H., Tagami, H., Tanaka, M., Saito, H. and Nakanishi, M. (2008) Chkl is a histone H3 threonine 11 kinase that regulates DNA damage-induced transcriptional repression. *Cell*, **132**, 221–232.
58 Dai, Y., Landowski, T.H., Rosen, S.T., Dent, P. and Grant, S. (2002) Combined treatment with the checkpoint abrogator UCN-01 and MEK1/2 inhibitors potently induces apoptosis in drug-sensitive and -resistant myeloma cells through an IL-6-independent mechanism. *Blood*, **100**, 3333–3343.
59. Dai, Y., Chen, S., Pei, X.Y., Almenara, J.A., Kramer, L.B., Venditti, C.A., Dent, P. and Grant, S. (2008) Interruption of the Ras/MEK/ERK signaling cascade enhances Chkl inhibitor–induced DNA damage in vitro and in vivo in human multiple myeloma cells. *Blood*, **112**, 2439–2449.
60 Dai, Y., Chen, S., Shah, R., Pei, X.Y., Wang, L., Almenara, J.A., Kramer, L.B., Dent, P. and Grant, S. (2011) Disruption of Src function potentiates Chkl-inhibitor-induced apoptosis in human multiple myeloma cells in vitro and in vivo. *Blood*, **117**, 1947–1957.

# Index



W.X. Li (ed.), *Cell Cycle Checkpoints*, Methods in Molecular Biology 782,
DOI 10.1007/978-1-61779-273-1, © Springer Science+Business Media, LLC 2011

## F



## G



## H



## I



## K



## L



## M



## N



## O



## P

MIX
Papier aus verantwortungsvollen Quellen
Paper from responsible sources
FSC® C105338

If you have any concerns about our products,
you can contact us on
**ProductSafety@springernature.com**

In case Publisher is established outside the EU,
the EU authorized representative is:
**Springer Nature Customer Service Center GmbH**
**Europaplatz 3, 69115 Heidelberg, Germany**

Printed by Libri Plureos GmbH
in Hamburg, Germany